Lecture Notes in Physics

Edited by H. Araki, Kyoto, J. Ehlers, München, K. Hepp, Zürich
R. Kippenhahn, München, H. A. Weidenmüller, Heidelberg
and J. Zittartz, Köln

186

Critical Phenomena

Proceedings of the Summer School
Held at the University of Stellenbosch, South Africa
January 18–29, 1982

Edited by F. J. W. Hahne

Springer-Verlag
Berlin Heidelberg New York Tokyo 1983

Editor

F.J.W. Hahne
University of Stellenbosch
The Merensky Institute for Physics
Stellenbosch, 7600 South Africa

ISBN 3-540-12675-9 Springer-Verlag Berlin Heidelberg New York Tokyo
ISBN 0-387-12675-9 Springer-Verlag New York Heidelberg Berlin Tokyo

Printed in Germany

Printing and binding: Beltz Offsetdruck, Hemsbach/Bergstr.
2153/3140-543210

Table of Contents

LECTURERS

A. Aharony, Tel Aviv University

A.L. Fetter, Stanford University

M.E. Fisher, Cornell University

M.J. Stephen, Rutgers University

H. Thomas, University of Basel

ORGANIZING COMMITTEE

C.A. Engelbrecht, University of Stellenbosch

F.J.W. Hahne (Chairman), University of Stellenbosch

W.D. Heiss, NRIMS, CSIR, Pretoria (now at University of the Witwatersrand)

R.H. Lemmer, University of the Witwatersrand

W.S. Verwoerd, University of South Africa, Pretoria

P. du T. Van der Merwe, AEB, Pelindaba

O.A. van der Westhuysen, CSP, CSIR, Pretoria

Mrs. E. Blum (Secretary), CSIR, Pretoria

PARTICIPANTS

D. Bedford, University of Natal, Durban

M.W.H. Braun, University of Pretoria

J.H. Brink, AEB, Pelindaba

J.D. Comins, University of the Witwatersrand, Johannesburg

E.D. Davis, University of Cape Town

P.R. de Kock, University of Stellenbosch

O.L. de Lange, University of Natal, Pietermaritzburg

S.J. Donovan, University of the Witwatersrand, Johannesburg

C.A. Engelbrecht, University of Stellenbosch

E.A. Evangelidis, AEB, Pelindaba

A.G. Every, University of the Witwatersrand, Johannesburg

D. Eyre, NRIMS, CSIR, Pretoria

G.M. Field, University of Cape Town

G.C.K. Fölscher, University of the Witwatersrand, Johannesburg

P.J. Ford, University of the Witwatersrand, Johannesburg

W.E. Frahn, University of Cape Town

W.L. Gadinabokao, University of Bophuthatswana, Mafikeng

M. Gering, University of the Witwatersrand, Johannesburg

F.J.W. Hahne, University of Stellenbosch

S. Hart, NPRL, CSIR, Pretoria

W.D. Heiss, NRIMS, CSIR, Pretoria

J.J. Henning, AEB, Pelindaba
J.D. Hey, University of Cape Town
M.J.R. Hoch, University of the Witwatersrand, Johannesburg
D.P. Joubert, University of Stellenbosch
S. Klevansky, University of the Witwatersrand, Johannesburg
F.J. Kok, University of Pretoria
R.H. Lemmer, University of the Witwatersrand, Johannesburg
P.E. Lourens, AEB, Pelindaba
L. Matthews, University of Pretoria
R.E. Nettleton, University of the Witwatersrand, Johannesburg
P.E. Ngoepe, University of the Witwatersrand, Johannesburg
G.N. v/d H Robertson, University of Cape Town
F.G. Scholtz, University of Stellenbosch
G.J. Shepherd, Rhodes University, Grahamstown
L.C.A. Stoop, University of South Africa, Pretoria
J.H. van der Merwe, University of Pretoria
P. du T. van der Merwe, AEB, Pelindaba
E. van der Spuy, AEB, Pelindaba
C. van Niekerk, AEB, Pelindaba
W.S. Verwoerd, University of South Africa, Pretoria
J. du P. Viljoen, AEB, Pelindaba
D.H. Wiid, Rand Afrikaans University, Johannesburg

PREFACE

The study of critical phenomena and phase transitions has received considerable attention during the past ten years, and many new achievements have been made. Due to the smallness of our physics community, we in South Africa have not been able to participate in this endeavour in any significant way.

It thus became apparent that in order to acquaint physicists in general with this field an advanced course on these subjects within our programme of summer schools was very opportune. The second school, which had as its topic "critical phenomena," was held at the University of Stellenbosch from January 18 to 29, 1982 and was both well attended and enthusiastically received by students and practising physicists alike.

We consider ourselves very fortunate in having had five outstanding experts in the field present the material in very clear terms. The enthusiasm and clarity of the lectures was surely a unique experience for many of the students, and such exposure to excellent physics is most likely to be the best counter to the waning interest in pure science resulting from competition from financially more rewarding disciplines. On behalf of all the participants, I wish to thank all the lecturers for performing their task admirably.

This venture was only possible as a result of financial and organisational support from the Council for Scientific and Industrial Research (CSIR). The CSIR has sponsored these schools since the initiative was taken by the Organization for Theoretical Physicists (OTP) and the South African Institute of Physics (SAIP).

The venue in the university town of Stellenbosch was well suited for holding the course, and the use of the facilities, as well as other support, is gratefully acknowledged.

These lecture notes consist of manuscripts either supplied by the lecturers themselves or, in two cases, compiled from notes by participants who had further lengthy contact with the lecturer concerned. In all cases these notes provide a very readable account of the courses presented.

We are grateful to the editors that these notes can appear in the Springer series "Lecture Notes in Physics."

Stellenbosch, South Africa F.J.W. Hahne

May 1983

SCALING, UNIVERSALITY AND RENORMALIZATION GROUP THEORY

By

Michael E. Fisher
Baker Laboratory, Cornell University
Ithaca, New York 14853, U.S.A.

Lectures presented at the
"Advanced Course on Critical Phenomena"
held during January 1982 at
The Merensky Institute of Physics
University of Stellenbosch, South Africa

Lecture Notes prepared with the assistance of

Arthur G. Every
Department of Physics, University of the Witwatersrand,
Johannesburg, South Africa
On leave at Department of Physics, University of Illinois
at Urbana-Champaign
Urbana, Illinois 61801, U.S.A.

CONTENTS

1. Introduction

My aim in these lectures will be to describe some of the more interesting and important aspects of critical phenomena. I will, in particular, be discussing the ideas of scaling and critical exponents and emphasizing the idea of universality. Following this I will be dealing with the microscopic formulation of statistical mechanics and certain series expansion methods that have been extensively used in the past. These are not only applicable to critical phenomena, but are useful in other areas of physics and engineering as well. The main emphasis and focus of the lectures will, however, be on the collection of rather subtle ideas which underlie renormalization group theory and its applications to critical phenomena. I will be approaching them in roughly two stages: from the microscopics will come some introductory concepts; then I will be describing the general renormalization group ideas which are essentially topological in nature. I will be aiming, in describing these general concepts, at applications beyond critical phenomena. They have for example been used to handle the Kondo problem and to study various aspects of field theories. Finally, I will discuss some of the first practical successes of the renormalization group, based on the so-called "epsilon expansions". These expansions are generated in terms of the parameter $\varepsilon = 4-d$, where d is the spatial dimensionality of the physical system. They were some of the first sweet fruits of the renormalization group ideas!

What is the task of theory? It is worthwhile, when embarking on theory to have some viewpoint as to what theory is. There are different opinions on this subject. Some people feel the task of theory is to be able to calculate the results of any experiment one can do: they judge a theory successful if it agrees with experiment. That is not the way I look at a theory at all. Rather, I believe the task of theory is to try and understand the universal aspects of the natural world; first of all to identify the universals; then to clarify what they are about, and to unify and inter-relate them; finally, to provide some insights into their origin and nature. Often a major step consists in finding a way of looking at things, a language for thinking about things -- which need not necessarily be a calculational scheme. This aspect of renormalization group theory, which I view as very important, is underplayed in a number of articles and books on the subject. "Shapes" are aspects I often regard as important. To make an illustrative point here, the geometrical properties of the circle have been known for a long time, and we tend to take them for granted. The ratio of the circumference to the diameter is called π, which is only the first of many Greek letters that will be introduced in these lectures! Its value, which today we know as 3.14159265358... was, from very early times, felt to be the same for all circles; i.e., that it was a universal property. This is true if space is Euclidean; and, to a very high degree of accuracy, the space we inhabit is, indeed, Euclidean. The value of this ratio is of course of great interest. The Bible has an unambiguous statement[1] that the value of

π is 3, although the people to whom that is attributed probably knew that it was not exactly equal to 3. This "Biblical" theory is the analogue to the so-called "classical" theory of critical exponents that will be referred to frequently in these lectures. We know that the Ancient Greeks already had very good inequalities for π. Of course, the numerical value of π is now known to very many decimal places indeed, and there are numerous series expansions which converge to the exact value, such as

$$\pi = \sqrt{\frac{6}{1^2} + \frac{6}{3^2} + \frac{6}{5^2} + \cdots} \; . \tag{1.1}$$

Also, there are explicit formulae that relate π to the other transcendental numbers, the most famous being

$$e^{i\pi} = -1. \tag{1.2}$$

In a similar way, in the theory of critical phenomena there is a set of important numbers, the <u>critical exponents</u>, and they are also believed to be universal in character. In these lectures evidence will be presented to show that this is so. In addition some formulae, in the form of series expansions, have been derived for these critical exponents, but, so far, only the first few terms in the expansions are known. Also, while the expansion (1.1) for π is convergent, the ε-expansions for the critical exponents are almost certainly not convergent in general, unless they are treated in a special way. We will also see that there are a number of formulae like (1.2) which relate the various critical exponents to one another, although perhaps with not quite the mathematical rigor and generality of (1.2).

These remarks more or less sum up the attitude I will be taking towards the subject matter of these lectures.

2. Critical Phenomena in magnets and fluids: Universality and Exponents

2.1 The gas-liquid critical point

The first critical point to be discovered was in carbon dioxide. Suppose one examines[2] a sealed tube containing CO_2 at an overall density of about 0.5 gm/cc, and hence a pressure of about 72 atm. At a temperature of about 29° C one sees a sharp meniscus separating liquid (below) from vapor (above). One can follow the behavior of liquid and vapor densities if one has a few spheres of slightly different densities close to 0.48 gm/cc floating in the system. When the tube is heated up to about 30° C one finds a large change in the two densities since the lighter sphere floats up to the very top of the tube, i.e., up into the vapor, while the heaviest

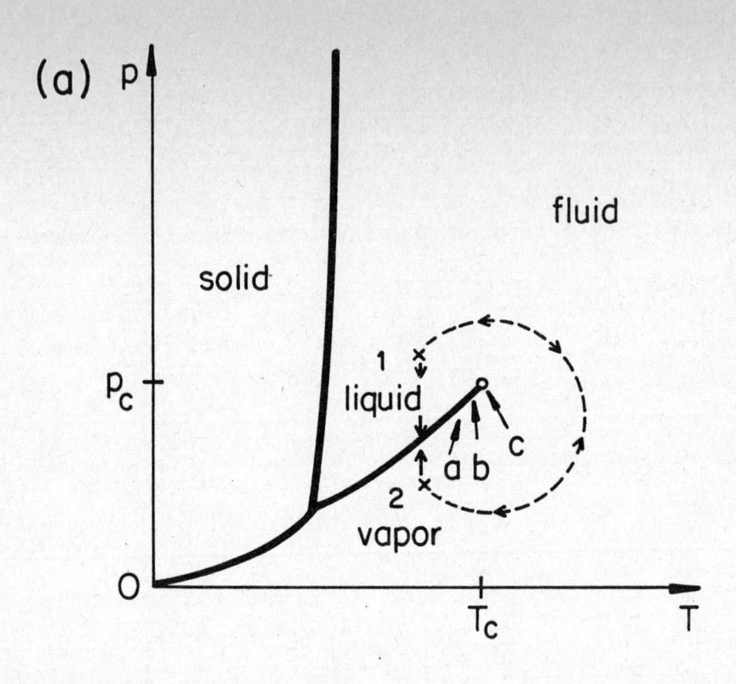

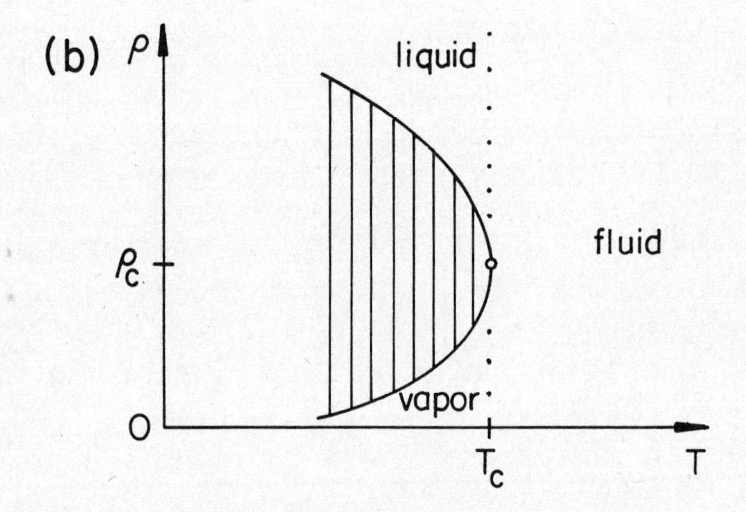

Fig. 2.1 (a) (p,T) diagram for a typical physical system;
(b) corresponding plot of particle number density ρ versus T. The vertical "tie-lines" link coexisting liquid and vapor densities, and span the region of liquid vapor coexistence.

one sinks down to the bottom of the liquid. However, a sphere of about "neutral" density (in fact "critical density") remains floating "on" the meniscus. There is, indeed, still a sharp interface between the two fluids, but they have approached one another closely in density. Further slight heating to about 31° C brings on the striking phenomenon of critical opalescence. If the carbon dioxide, which is quite transparent in the visible region of the spectrum, is illuminated from the side, one observes a strong intensity of scattered light. This has a bluish tinge when viewed normal to the direction of illumination, but has a brownish-orange streaky appearance, like a sunset on a smoggy day, when viewed from the forward direction (i.e., with the opalescent fluid illuminated from behind). Finally, when the temperature is raised a further few tenths of a degree, the opalescence disappears and the fluid becomes completely clear again. Moreover, the meniscus separating "liquid" from "vapor" has completely vanished: no trace of it remains! All differences between the two phases have gone: indeed only one, quite homogeneous, "fluid" phase remains above the critical temperature ($T_c \simeq 31.04^\circ$ C).

These phenomena are best interpreted in the pressure-temperature (p,T) phase diagram shown in Fig. 2.1. The first three stages are represented by the points a, b and c on the vapor pressure curve. Note that T_c and p_c are the critical temperatures and pressures respectively at which critical opalescence is observed. As the temperature is raised further, the system follows a contour of constant overall density (the "critical isochore"). The whole process is completely reversible. Significantly, it is possible to go from liquid (point 1) to vapor (point 2) either smoothly via a route along which the properties of the fluid always change smoothly and continuously, or through the vapor pressure curve, at which a first order transition takes place with a discontinuity in density, internal energy, etc. Any point inside the shaded region of Fig. 2.1(b) corresponds to liquid and vapor coexisting with one another. As the critical point is approached the two densities, $\rho_{liq}(T)$ and $\rho_{vap}(T)$ become closer and closer to each other until they match at $T = T_c$.

2.2 Universal behavior

One finds that the actual variation of $\rho_{liq}(T)$ and $\rho_{vap}(T)$ is close to universal for gases such as argon, krypton, nitrogen, oxygen, etc., in the sense that if the temperature is normalized by the critical temperature, T_c, and the density by the critical density, ρ_c, then the data for the different gases all fit very nearly on the same coexistence curve. The shape of this coexistence curve will be one of the first objects of our investigation. The simplest curve which has the same basic shape as the coexistence curve graphed as T vs ρ is, of course, the parabola $y = Ax^2$. The assertion that the coexistence curve is parabolic (in the critical region) in fact represents the "Biblical" or classical theory of the coexistence curve. At first sight, it seems to be a most natural and unprejudiced starting point. But what really is the shape of this curve near T_c? That is the question one must ask!

To that end we introduce here a variable that will be greatly used, namely, the reduced temperature

$$t = \frac{T-T_c}{T_c}, \tag{2.1}$$

which measures the deviation of the temperature from critical in dimensionless units. Now as T approaches T_c from below, the difference between the liquid and gas densities, ρ_{liq} and ρ_{vap} respectively, is going to vanish as, we might reasonably expect, some power β of $|t|$. Thus we write

$$\rho_{liq}-\rho_{vap} \sim |t|^{\beta} \quad \text{as } T \to T_c-. \tag{2.2}$$

The exponent β is the first of the critical exponents that will be introduced in these lectures. It is the analogue of π because it directly describes the shape of the coesxistence curve. From the way the parabola is oriented in Fig. 2.1(b), we see that the classical or "Biblical" theory prediction is simply $\beta=1/2$. How does this compare with the value of β measured in the real world? The experiments that have been done in this connection are some of the most precise experiments ever performed in Physics. A notable example is provided by the work of Balzarini and Ohrn[3] who measured the coexistence curves for xenon and sulphur hexafluoride using very sensitive optical methods. These two fluids are obviously very different chemically but, nevertheless, their critical behavior is found to be essentially the same. The data on the density jump $\Delta\rho = \rho_{liq} - \rho_{vap}$ span the range from $t \simeq 3 \times 10^{-2}$ down to 3×10^{-6} and on a log-log plot lie very accurately on two straight and parallel lines. This first confirms the power law behavior and then yields a value of β which is quite close to 1/3. The precise value lies somewhere in the interval 0.32 -- 0.34, perhaps closer to 0.32. Despite the experimental accuracy and the great range of the data one cannot, unfortunately, actually determine such critical exponents to much better than ±0.02. We are certain now that it is not a simple fraction, or at least not a very simple fraction like 1/2 or 1/3! Clearly, therefore, the "Biblical" value is quite outrageously wrong. Finally, in line with π being independent of the size of the circle, it is found that β is also quite independent of the type of fluid; the same values are found for water, a highly associated liquid, for liquid metals, and for the 'quantal liquids' helium three and four at their liquid-vapor critical points.

2.3 Binary fluids

Another type of system which has been much investigated is that of a mixture of two chemical compounds, say A and B, that at high enough temperatures are mutually soluble, but at lower temperatures separate out into two phases as oil separates from water, which we will call α and β (see Fig. 2.2). There are a great many combinations that can be used: organic liquids such as aniline and cyclohexane or carbontetrachloride and perfluoroheptane are favourites because the interesting behavior occurs (under atmospheric pressure) at temperatures close to room temperature. The vapor phase is usually present, as shown in Fig. 2.2, but plays no essential role. The denser phase at the bottom could be, for example, A-rich, while the less dense one floating above it would then be B-rich. As the temperature is increased a liquid-liquid critical point or consolute point is reached and critical opalescence is exhibited just as for a one-component fluid. Beyond this point only a single, homogeneous liquid phase exists.

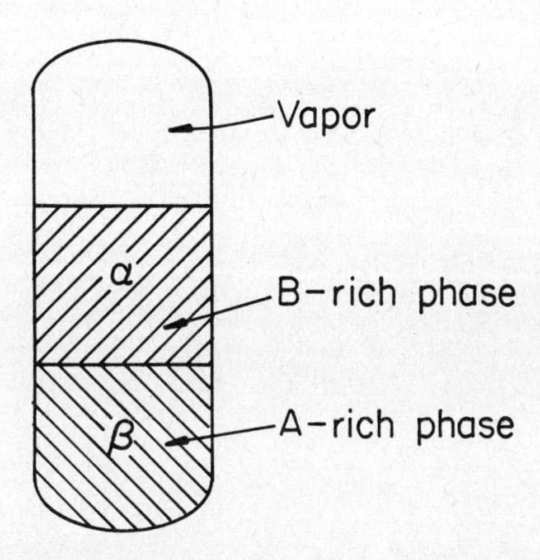

Fig. 2.2 Illustrating phase separation in a binary liquid mixture of two chemical species A and B.

Now what should one focus on instead of the density difference? We will use symbols such as x_A^α to denote the mole fraction of A molecules in the A-rich phase α and so on. As the critical temperature is approached from below one observes that the composition difference between α and β phases varies as

$$x_A^\alpha - x_A^\beta \sim |t|^\beta, \quad (t \to 0-). \tag{2.3}$$

The question is "Does β have the same value as before?" The answer is an unequivocal "yes" as can be seen from experiments on very many binary fluid systems (including molten metal mixtures). An interesting comparison has been published by Sienko[4]. He finds, for example, that on a normalized log-log plot the coexistence curve for CCl_4 and C_7F_{14} shows a form which is almost indistinguishable from that of the liquid-vapor coexistence curve for CO_2, so that β again lies close to 1/3. Sienko and coworkers also studied the metal-ammonia systems in which alkali metals Na, Li and Ca are dissolved in NH_3. At first sight these mixtures appear to provide an exception to the $\beta \simeq 1/3$ rule. For temperatures deviating from (below) T_c by from 1% to 10% (i.e., $t = 0.01$ -- 0.1) the coexistence curve on a log-log plot has a steeper slope than for other systems and, indeed, seems to conform to a $\beta = 1/2$ relation as predicted by the "Biblical" theory. But accurate data that are taken closer into the critical (or consolute) point fall clearly into line with all the other systems: the slope changes quite rapidly around $t \simeq 0.007$ to 0.009 and decreases to yield again $\beta \simeq 1/3$. So we are forced to accept this universality of behavior, but we learn that the universality does not extend indefinitely out of the critical region. Indeed it is really a matter of extrapolating in towards the critical point if one wants to determine the true, universal, asymptotic behavior. So when I discuss critical behavior it is always a matter of approaching close enough to the critical point. It is worthwhile to embody this point in a formal definition of a critical exponent which can then be used for more exact and rigorous theoretical arguments and analyses.

2.4 Critical exponents defined precisely

Generally, when we say a function $f(x)$ behaves like x^λ, or write

$$f(x) \sim x^\lambda \quad \text{as} \quad x \to 0+, \tag{2.4}$$

it will be taken to mean that

$$\lim_{x \to 0+} \frac{\ln[f(x)]}{\ln x} = \lambda. \tag{2.5}$$

In this way we can avoid introducing a constant for the coefficient of x^λ as would be essential if we wrote $f(x) \approx Ax^\lambda$ or $f(x) \propto x^\lambda$. At a more subtle level suppose we have a function such as

$$f(x) = A|\ln x|^\mu x^\lambda. \tag{2.6}$$

This does not vary as a simple or "pure" power law but rather has a "confluent" logarithmic singularity. From a theoretical viewpoint one can still use eqn. (2.4), and in this way one obtains a critical exponent equal to λ. Thus even functions of this type with more complex singularities are covered. One of the important contributions of renormalization group theory is that it reveals the circumstances under which such logarithmic factors should be anticipated. One must _always_ expect, of course, that over any finite range there will be some correction terms: thus even for an _asymptotically_ pure power law one will generally have

$$f(x) = Ax^{\lambda} \{1 + a_{\theta}x^{\theta} + \ldots + a_1 x + a_2 x^2 + \ldots.\}, \qquad (2.7)$$

where the confluent "correction" exponent, θ, may well be less than unity (although it must be positive for the form written to make sense). On a log-log plot corrections such as these can and do actually alter the slope and lead to erroneous values for measured critical exponents. The most serious correction terms are those where $\theta < 1$, the smaller the value of θ the worse the problem. In fact, values of around 1/2 are expected on theoretical grounds in many real situations. This assertion reflects another valuable contribution of the renormalization group since it has enabled us to give a sensible estimate of the exponent θ and to explain why this sort of behavior is what one should expect in most circumstances.

There have been people in the past who have questioned whether nature really is required to conform to power law behavior near a critical point. The evidence, both experimental and theoretical, is now compelling that, apart from logarithmic factors in special cases and certain correction terms, power law behavior is the _rule_. One would have to be a brave scientist indeed to hold out against this conviction and this point. Nevertheless there are still those -- some would call them "cranks" -- who argue that perhaps the "Biblical" theory is still correct if one goes _really close_ to T_c, so that $\beta = 1/2$ after all! However, I am afraid that in science, new and more correct ideas often win out only after their opponents die or retire. Evidently many people are not as open to rational conviction by new thoughts, as might be desirable!

Another problem that arises in the handling of experimental data is that the critical temperature T_c is, of course, not known in advance. Usually one treats T_c in the expression $t = (T-T_c)/T_c$ as a fitting parameter. When the data extends over several decades, the data close in to the transition point will sometimes be used to determine T_c, while that further out then serves to determine the critical exponents. Sometimes T_c will be determined separately from both sides in similar or distinct experiments. All in all, great care has to be exercised when interpreting even the very best data if one is not to assign misleadingly small "error" estimates to parameters such as T_c, β, and the amplitudes A, etc.

2.5 Specific heats

In 1963 Voronel' (then in the Soviet Union) and his coworkers[5] made some historically important measurements of the specific heat at constant volume, $C_V(T)$ of argon in the vicinity of its critical point. More precisely, they observed the specific heat at constant overall density along the critical isochore $\rho = \rho_c$. Below T_c the system will, as seen, then consist of a mixture of vapor and liquid, and the proportions of the two will actually change as the temperature is varied. So this "specific heat" actually contains a latent heat contribution. Nevertheless that is, both experimentally and theoretically, the most appropriate function to measure for the study of critical behavior. Now the "Biblical" or classical theory predicts that (I will not say "this" anymore) the specific heat merely has a jump discontinuity at the critical point, i.e., $C_V(T_c-) \neq C_V(T_c+)$. Actually $C_V(T_c-) > C_V(T_c+)$ is predicted as indicated by the dashed curve in Fig. 2.3. Voronel' was the first one to do sufficiently careful and accurate measurements to show unambiguously that this was not so! On the contrary, $C_V(T)$ rose up smoothly but very steeply on both sides of T_c as sketched in Fig. 2.3. Asymptotically the variation has the form

$$C_V(T) \sim |t|^{-\alpha}, \qquad (t \to 0+), \tag{2.8}$$

where the specific heat exponent α has a value in the region of 1/8 to 1/9 for most fluids. Because of the small value of α, correction terms now assume much greater importance and make α hard to determine precisely. Also one might question whether C_V does, indeed, diverge to infinity, or whether it just has a sharp spike or cusp at T_c.

On this latter question microscopic models are able to provide us with some definite guidance. These models come in various shapes and sizes: but the most famous is undoubtedly the Ising model, which I will be discussing in more detail later in these lectures. Onsager's celebrated solution of the 2-dimensional Ising model in 1944 gave the specific heat as

$$C_V(T) = A \ln|t| + \text{finite "background" terms.} \tag{2.9}$$

The singular behavior is carried mainly by the leading logarithmic term (although terms like $t\ln|t|$ appear in the "background"). As is readily confirmed by application of the formal definition (2.5), a logarithmic divergence corresponds to the limiting case of $\alpha \to 0+$. [Consider the function $\ell_\alpha(t) = (|t|^{-\alpha}-1)/\alpha$.] To draw attention to the fact that the logarithm is present, this case is usually reported as

$$\alpha = 0 \ (\log).$$

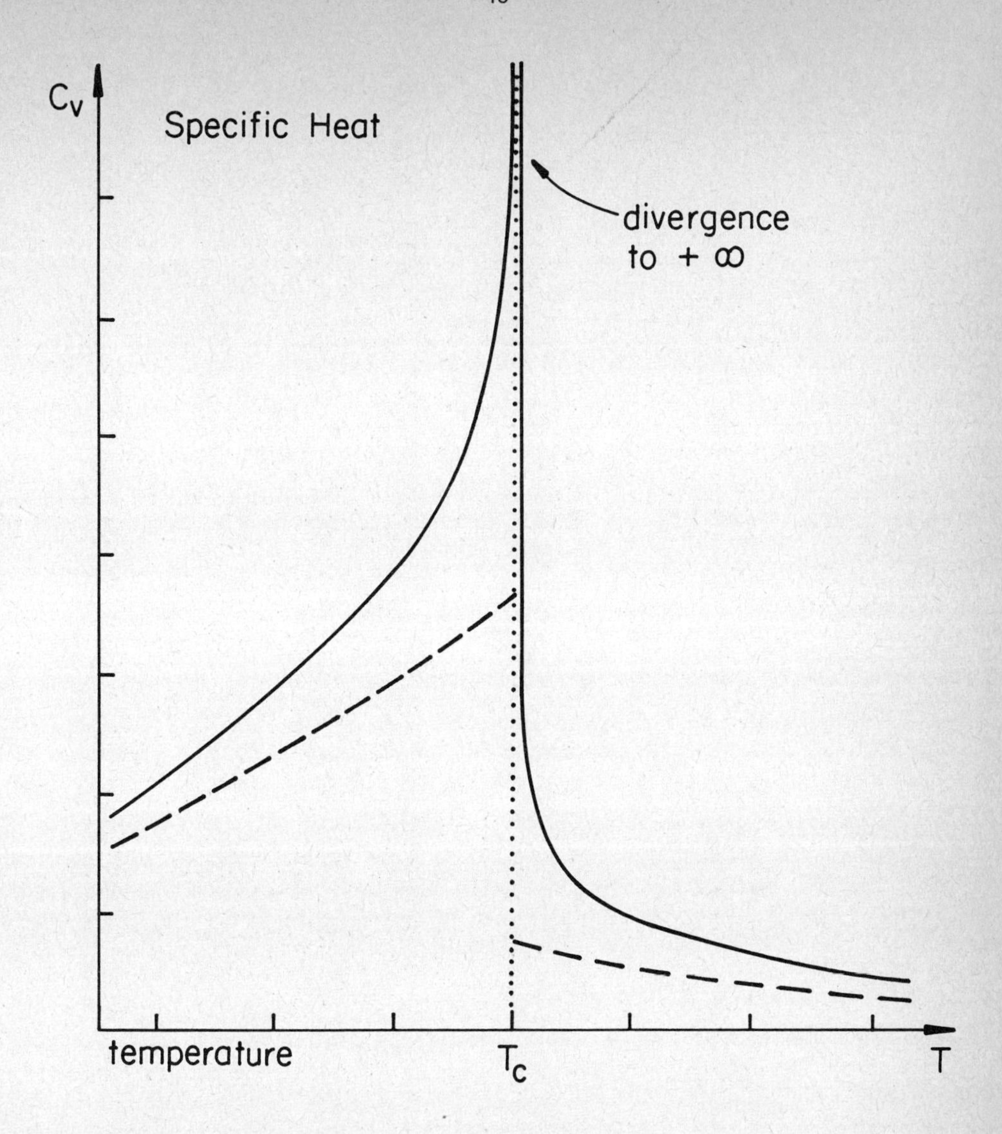

Fig. 2.3 Sketch showing the variation of the specific heat, $C_V(T)$, of argon and other fluids through the gas-liquid critical point. The dashed curve represents the prediction of the classical (or "Biblical") theories.

One small detail that Fig. 2.3 suggests one should take into account is that the specific heat does not mirror itself around the critical point. Thus one should, properly, define two exponents: α' for $T < T_c$ and α for $T > T_c$. The convention is that primed exponents refer to $T < T_c$ and unprimed to $T > T_c$ (except where, like β, the definition makes sense only for $T < T_c$). Nowdays it is rather well established on both experimental and theoretical grounds that $\alpha = \alpha'$, so the distinction is often dropped unless one has reason for being circumspect.

Modern experiments on critical specific heat obtain temperature resolutions of 10^{-6} or 10^{-7} in t. Some of the best experiments are those of Ahlers[6] on liquid helium at its lambda point, $T_\lambda \equiv T_c \cong 2.18$ K, where the normal fluid becomes superfluid. The transition is seen to remain sharp down to a tenth of a microdegree. More recently Lipa[7] has pushed the resolution still further down to only tens of nanodegrees. The specific heat seems to continue rising down to these very small deviations from T_c.

It is worthwhile asking the question at this point if, with continuing experimental refinements, one can expect to observe the specific heat continuing to diverge indefinitely close to T_c. Naturally, precautions must be taken to allow for gravity and other small disturbing factors. However, ultimately the basic theoretical answer is "No, the specific heat cannot increase without bound". The reason is that in the laboratory one would always be dealing with a finite system, with a finite number of atoms confined in a bounded region of space. A perfectly sharp phase transition can take place only in a truly infinite system, i.e., in the thermodynamic limit where the system is infinitely large in extent but its density, pressure, and all other intensive quantities are fixed and finite. However large a system is in practice, it will still be finite and, ultimately then one will reach the point where the specific heat singularity is seen to be rounded off. Experiments deliberately done on small samples certainly show these rounding effects. So in talking about a phase transition one really should always have in mind the thermodynamic limit.

The specific heat anomaly at the lambda transition in He^4 is now believed to be very close to logarithmic. Thus Ahlers quotes $\alpha = \alpha' \simeq -0.02 \pm 2$ (the uncertainty being in the last decimal place) signifying that α is probably very slightly negative. This suggests that the specific heat does not quite diverge to ∞ but rather comes up to form a sharp cusp at which point $C(T_c)$ is finite but the slope $(dC/dT)_c$ is infinite.

Similar behavior is also observed at magnetic phase transitions: a notable case being the specific heat of the ferromagnet nickel near its Curie or critical point, T_c. Magnetic systems are in many ways much simpler to think about theoretically because magnetic field H=0 is a point of symmetry. One finds that the zero-field specific heat of nickel displays a sharp cusp, but it is much less strong than in the case of superfluid helium or some of the other fluid systems. In this case, and that of other magnetically isotropic magnetic systems, one finds that α is definitely negative although still quite small say $\alpha = -0.10$ to -0.15, high precision being again difficult to attain.

2.6 The order parameter

In the case of simple fluids the parameter of apparently central interest is the density, ρ. Following Landau's general conception of phase transitions, we name this special quantity the order parameter and denote it generally as Ψ. So for single-component fluids we write $\Psi = \rho$. For fluid mixtures we saw that what mattered was the difference between the mole fractions, Δx, which measures differences in composition: so here we have $\Psi = \Delta x$. For superfluid He^4, the crucial theoretical concept, which embodies our understanding of superfluidity, is an effective macroscopic wave function, $\psi = \psi' + i\,\psi''$. As a wavefunction this has both real and imaginary parts. While ρ and Δx are both simple scalar quantities, a complex number is best thought of as a two-component vector. Thus the superfluid order parameter, $\Psi = \psi$, is a two-component vector which has the symmetry of a circle, i.e., can point in any direction in the complex plane. It is the phase of ψ which is in fact responsible for the existence and nature of superfluidity. In the case of ferromagnetism, there are various possibilities, but certainly it is the magnetization, $\vec{M}$, which should be the order parameter. In the case of a magnet like nickel, the magnetization can point freely in any direction; i.e., nickel is spatially, highly isotropic; then the magnetization can be thought of as a three-component vector $\vec{M} = (M_x, M_y, M_z)$.

In summary, we see that the order parameter, Ψ, has a tensorial character which may depend on the class of systems considered. Theoretically it is natural to distinguish between these various cases, and the renormalization group has enabled us to make this distinction meaningful and effective. In particular we often refer to n, the number of components of the order parameter. Then we have:

$n = 1$ for simple fluids, binary fluids, uniaxial ferromagnets, binary alloys, etc.

$n = 2$ for superfluid He^4 and $He^3 + He^4$ mixtures, XY-magnets (easy plane of magnetization).

$n = 3$ for isotropic magnets, etc.

As regards values of the critical exponents, none of which conform to "Biblical" or classical theory, there is found to be a subtle dependence on n. Specifically, one has $\alpha(n{=}1) \simeq 0.11$, $\alpha(n{=}2) \simeq 0.0$ and $\alpha(n{=}3) \simeq 0.14 \pm 4$. Similar slight differences are found for the critical exponent β, viz. $\beta(n=1) \simeq 0.32$, $\beta(n=2) \simeq 0.34$ and $\beta(n=3) \simeq 0.35 - 0.37$. The $n = 2$ value applies to XY-magnets but the corresponding superfluid order parameter is essentially inaccessible to experiment. Clearly then, the symmetry or tensorial character of the order parameter is important. The three cases described above are often referred to as Ising-like (n=1), XY-like (n=2), and Heisenberg-like (n=3). Larger values of n are not just of theoretical interest; they are also required for describing real physical systems, in particular various magnetic crystals of more complex structure and symmetry.

2.7 Fluid-magnet analogy

The close analogies that exist between fluids and ferromagnets are worth emphasizing, even though ferromagnets have an intrinsic symmetry that makes them easier to think about. Conjugate to the order parameter, Ψ, in any thermodynamic system, is a "thermodynamic field" variable, h. In the case of fluid the pressure, p, has traditionally been treated as this conjugate variable, but often it is better to regard h as the chemical potential, μ. The pressure p or chemical potential μ is the variable that directly allows one to alter the density (at constant temperature). The analogous variable for a magnet should therefore be the magnetic field H, which is the variable primarily coupled to the magnetization. Fig. 2.4 illustrates clearly how far the analogy can be taken. In the case of the magnet, in the (h,T) plane, there is a line of first order transitions separating the "up" and "down" ferromagnetized states; this line ends at the critical (or Curie) point. The first order transition line is analogous to the vapor pressure curve, but differs from it in one minor respect in that it is entirely confined to the H=0 or T-axis. This, of course, is a consequence of symmetry under $H \to -H$. In the (Ψ,T) plane there is a coexistence curve in both cases. Inside this curve the magnet breaks up into domains; this is analogous to gas-liquid coexistence in fluids. For the magnet the coexistence or "spontaneous magnetization" curve is symmetric about the T-axis while for the fluid this symmetry is apparently absent. Below T_c the order parameter variation for the fluid is given by $\rho_{liq} - \rho_{vap} \sim |t|^\beta$ while for the magnet it is $M_0(T) \sim |t|^\beta$, where the spontaneous magnetization should be defined as

$$M_0(T) = \lim_{H \to 0+} M(H,T). \tag{2.10}$$

This careful definition of $M_0(T)$ is neccessary because M takes different limiting values depending on whether H=0 is approached from positive or negative values. The specific heat exponent is also defined in an analogous way for the two systems, and so on. Thus while most emphasis will be placed on magnetic systems, analogous effects and similar results hold for other types of systems in nearly all cases.

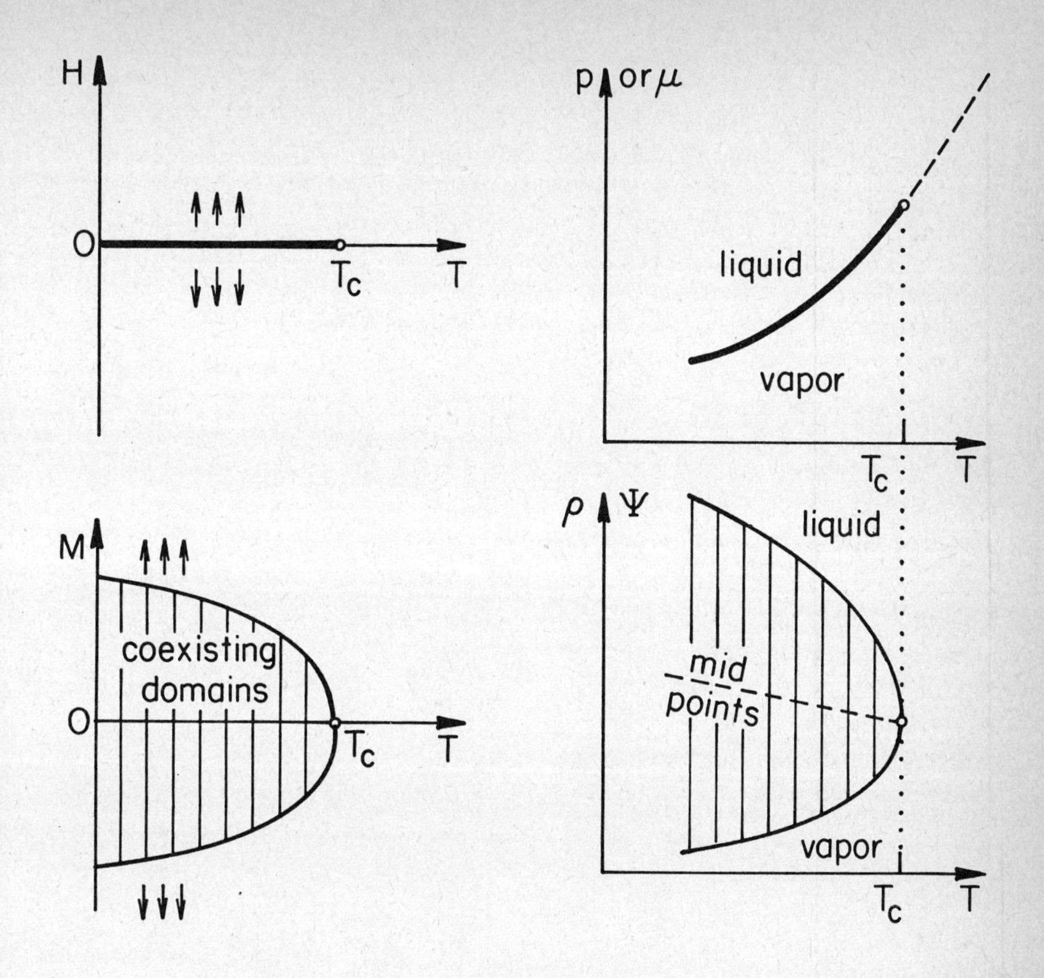

Fig. 2.4 Phase and coexistence diagrams illustrating the magnet-fluid analogy. Note magnetization corresponds to density and magnetic field to pressure or, better, chemical potential.

The question of how the perfect symmetry of the spontaneous magnetization curve is reflected in the less than fully symmetric nature of the fluid coexistence curve is a fairly subtle one. For the magnet the natural field variable to take, because of the symmetry, is H. One suspects that for a fluid the most suitable variable by analogy should be

$$h = p - p_\sigma'(T-T_c), \tag{2.11}$$

where p_σ' is the limiting slope of the vapor pressure line at T_c. In this way h would measure the deviation from the limiting tangent (shown dashed in Fig. 2.4), which one expects might be the analogue of the H=0 symmetry axis of the magnet. This is sometimes called a scaling axis. A remarkable feature of the coexistence curve is that the line of mid points between the liquid and vapor phases is surprisingly straight. Furthermore, one can clearly define two different exponents, β_- and β_+, with respect to deviations below and deviations above critical density, ρ_c, i.e., for the vapor and liquid sides of the coexistence curve. There is no obvious (or known) symmetry between liquid and gas that should tell us *a priori* that these two exponents should be the same; yet to an exceedingly high degree of accuracy they are identical in value! Somehow the system builds itself an asymptotic symmetry from a Hamiltonian which does not, in the first place, possess this symmetry at all. Again, the renormalization group is able to explain how a system is able to build up a symmetry on approach to a critical point, and to decide when a symmetry can be built (or, on the contrary, when a weakly broken near symmetry of the Hamiltonian is amplified).

2.8 Magnetic susceptibility

Above T_c the spontaneous magnetization of a ferromagnetic material is identically zero, but magnetization can be induced by applying a magnetic field, H. Fig. 2.5 illustrates the type of isotherms observed.

The isothermal susceptibility is defined quite generally as a function of H and T by

$$\chi_T(T,H) = \left(\frac{\partial M}{\partial H}\right)_T . \tag{2.12}$$

One usually measures, and is most interested in, the so-called initial susceptibility

$$\chi_T^o(T) = \lim_{H\to 0+} \chi_T(T,H), \tag{2.13}$$

which measures the slope of the magnetization isotherm at zero field (as shown by

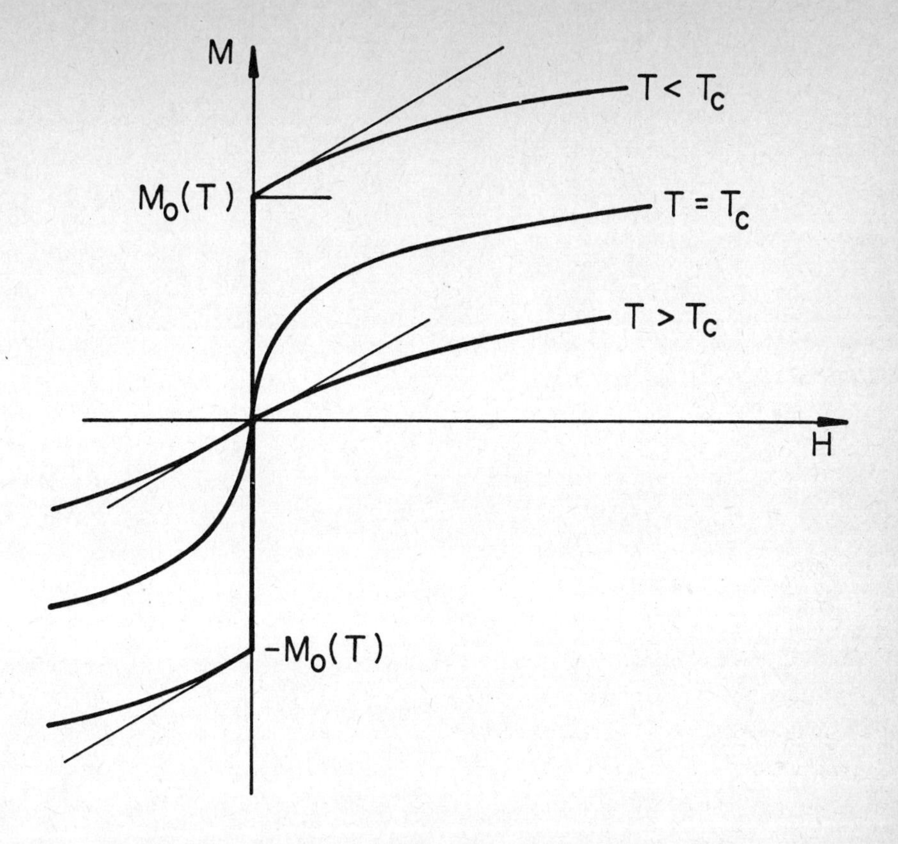

Fig. 2.5 Typical ferromagnetic magnetization curves (isotherms) above T_c, at T_c and below T_c (for a scalar, Ising-like or n=1 system).

the tangents in Fig. 2.5). In practice one often drops both the superscript o and the adjective "initial" and just refers to "the susceptibility". Clearly χ_T measures the ease of magnetizing a ferromagnet and hence is expected to grow large and, indeed, diverge at the Curie point where, after all, a ferromagnet essentially magnetizes itself! This divergence can be seen in Fig. 2.5: the slope of the critical, $T=T_c$ isotherm is actually infinite at zero field. For theoretical purposes it is usually convenient to define the reduced susceptibility $\chi = \chi_T/\chi_T^{ideal}$, where χ_T^{ideal} is the isothermal susceptibility of an ideal paramagnet (with no spin-spin interactions). Evidently, χ, which is dimensionless, measures the enhancement in magnetic responsiveness caused by the interactions, which are, of course, responsible for the ferromagnetic critical behavior. The analogous reduced susceptibility for a fluid is $\chi = K_T/K_T^{ideal}$, where

$$K_T = \frac{1}{\rho} \left(\frac{\partial \rho}{\partial p}\right)_T, \tag{2.14}$$

is the isothermal compressibility of the fluid and $K_T^{ideal} = 1/p$ is the corresponding quantity for an ideal gas. Since, as explained, χ measures the ease with which the order parameter is changed in response to the conjugate field, it is often known as the response function. (See also the lectures by A. L. Fetter).

The divergence of $\chi(T)$ at criticality is very strong and is characterised by an exponent γ defined as expected via

$$\chi \sim 1/t^{\gamma}, \qquad (t \to 0+;\ h = 0). \tag{2.15}$$

Measured values of γ are typically $\gamma(n=1) \simeq 1.23$ -- 1.24, $\gamma(n=2) \simeq 1.31$ --1.32 and $\gamma(n=3) \simeq 1.35$ — 1.38. In the case of superfluid He^4, one does not know how to measure χ: thus only the exponent α can be measured (of the thermodynamic properties we have defined). As can be seen, γ has a small n-dependence, but in all cases deviates markedly from the "Biblical" value which is simply $\gamma=1$.

Below T_c the situation is more complex. Even at $H=0$ there is a nonzero spontaneous magnetization, $M_0(T)$. Nevertheless, (as mentioned), one can still define the initial susceptibility as the limiting slope of the magnetization curve when $H \to 0+$. The temperature dependence of χ, so defined, provides one with the further exponent γ'. These last remarks apply, however, only to the Ising-like case of $n=1$. If $n=2$ or 3, so that a continuous (rotational) symmetry is present it can be shown theoretically, although experimentally it is not so easy to observe, that this limiting slope is infinite, so that $\chi_T(T,H)$ diverges as $H \to 0+$ for $T < T_c$ and the exponent γ' cannot be defined in the usual way.

2.9 Critical isotherm

The order parameter variation on the critical isotherm is generated by fixing the temperature precisely at T_c, varying the order field, h, and observing the change in Ψ, i.e., M or ρ as the case may be. For a magnet one finds that for small H this variation is given by (see Fig. 2.5)

$$M(T = T_c) \sim H^{1/\delta}, \qquad (H > 0,\ T=T_c), \tag{2.16}$$

which defines the critical exponent δ. Values of δ are typically: $\delta(n=1) \simeq 4.8$, $\delta(n=2) \simeq 4.7$, and $\delta(n=3) \simeq 4.6$. These should, perhaps, be regarded as more theoretical than experimental, since δ is extremely difficult to measure accurately owing to the steepness of the critical isotherm. The classical value is $\delta = 3$ which corresponds to a cubic curve for the critical isotherm. Of course, this is just the simplest analytic function which has the correct shape.

Naturally the critical isotherm near a fluid critical point displays completely analogous behavior. The relation is often written in reverse form as

$$|p-p_c| \sim |\Delta\rho|^\delta, \qquad (T=T_c), \tag{2.17}$$

where $\Delta\rho = \rho-\rho_c$, but this clearly corresponds precisely to the expected magnet-fluid analogy. Likewise, "Biblical" theory (in this case the original prophet is van der Waals) predicts δ=3, a cubic relation, but experiment yields $\delta \simeq 4.2$ to 4.8.

3. Scaling

3.1 Introduction: thermodynamic functions

The "Biblical" or classical theories break down completely in the region of a critical point. What then, can replace them? It turns out that the simplest phenomenological theories that come anywhere close to explaining critical behavior embody the concept of <u>scaling</u>. In order to make the discussion reasonably comprehensive one needs to couch it in terms of the full thermodynamics. Let us consider a ferromagnet since its symmetry allows us to make certain convenient (but inessential) simplifications. The Helmholtz free energy, $F(T,H)$, is associated with the basic differential thermodynamic relation

$$dF = -SdT - MdH, \tag{3.1}$$

where S is the total entropy. From this one can, by means of a Legendre transformation, generate the alternative free energy function, $A(T,M) = F+MH$, and it is then a simple matter to show that the basic differential relation becomes

$$dA = -SdT + HdM. \tag{3.2}$$

The magnetic field and susceptibility are obtained from A by differentiation according to

$$H = \left(\frac{\partial A}{\partial M}\right)_T \quad \text{and} \quad \chi^{-1} = \left(\frac{\partial^2 A}{\partial M^2}\right)_T . \tag{3.3}$$

Note that the susceptibility will diverge when $T \rightarrow T_c$, but it is intrinsically non-negative: Indeed a negative static compressibility or magnetic susceptibility is thermodynamically inconceivable. This is equivalent to the statement that the free energy A as a function of M must be a <u>convex function</u>: although the graph of A versus M can have a flat portion, its curvature must, otherwise, be strictly positive (See Fig. 3.1 below).

3.2 The classical phenomenological approach or Landau theory

The simplest type of phenomenological theory in this context derives from mean field theory; it was developed to a fine art by Landau and now frequently goes under his name. It consists, first of all, in identifying the order parameter, Ψ, (physically if this is possible but otherwise just as an abstract quantity), and then expanding the appropriate free energy as a Taylor series in powers of the order parameter. For a magnet the issue is straightforward: we have $\Psi = M$ and the power series expansion reads

$$A(T,M) = A_0(T) + A_2(T)M^2 + A_4(T)M^4 + \dots . \qquad (3.4)$$

By symmetry under $M \leftrightarrow -M$ no odd powers of M can be present. At high temperatures this expansion can be justified for all reasonable models on fully rigorous grounds, but near T_c it turns out to be dangerous! By differentiating twice one obtains the inverse susceptibility, which in zero field above T_c is thus given by

$$\chi^{-1} = 2A_2(T) \text{ for } T > T_c, \quad (H,M = 0). \qquad (3.5)$$

The next assumption is that the coefficients $A_j(T)$ can also be expanded in powers of $t \propto (T-T_c)$ so that, in particular, we may write

$$\chi^{-1} = 2A_{2,0} + 2A_{2,1}t + O(t^2). \qquad (3.6)$$

When $T \to T_c+$ the susceptibility, by definition of T_c, diverges to infinity, so that $\chi^{-1} \to 0$ as $t \to 0+$, and hence $A_{2,0} = 0$. The predicted behavior of χ near T_c is thus

$$\chi \approx C/t \quad \text{as} \quad t \to 0+, \ (H,M = 0). \qquad (3.7)$$

This, of course, corresponds to $\gamma = 1$. The fact that this theory gives an incorrect value for γ can be traced directly to the unjustifiable assumption that A(T,M) can be expanded in a power series near and, indeed, at a critical point. Nevertheless this seems to be a very natural assumption of the sort which is frequently made in physics and engineering. Furthermore, it can also be shown to be the essentially inevitable outcome of any of the wide variety of more microscopically-based mean field theories that have been proposed in this and many other related contexts.

In spite of the evident shortcomings of the classical phenomenological theory, let us continue to explore its consequences by considering the effect of the term of fourth order in M. Its coefficient is

$$A_4(T) = A_{4,0} + O(t) = \tfrac{1}{4}u + O(t), \qquad (3.8)$$

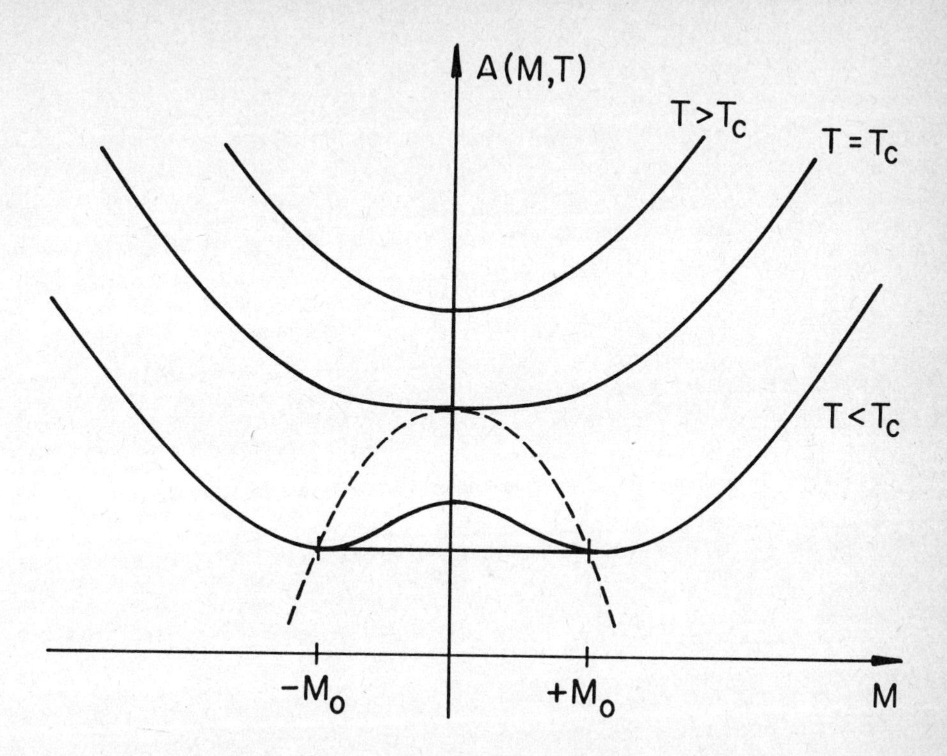

Fig. 3.1 Variation of the free energy A(T,M) according to classical phenomenological theory. The non-convex section of the isotherms for $T < T_c$ must be "corrected" by drawing in the flat, tangential segment, so forming the "convex cover" of the underlying, approximate function.

where the replacement of $A_{4,0}$ by $\frac{1}{4}u$ is purely a matter of convenience. We will assume $u > 0$ to ensure thermodynamic stability (although, in fact, the case $u \leqslant 0$ is required for dealing with tricritical points). Let us now examine the equation of state, which is the relationship connecting T, H and M, near T_c. It is obtained by differentiating A with respect to M and is easily seen to be of the form

$$H \approx M(ct + uM^2), \tag{3.9}$$

where we have put $2A_{2,1} = c$ so that, from (3.5) and (3.6), $A_2(T) \approx \frac{1}{2}ct$. For a fluid the corresponding equation would, for example, follow from van der Waal's equation with M replaced by $\rho - \rho_c$ and H by $p - p_c$.

On setting $t = 0$, we obtain the critical isotherm as $H \sim M^3$ and, thence, the erroneous prediction $\delta = 3$. For $T < T_c$ and $H \rightarrow 0-$ one obtains an equation with three roots, namely,

$$M = 0$$

and $$M = \pm\, M_0(T) \approx B|t|^{\frac{1}{2}}$$

with $$B = (c/u)^{\frac{1}{2}} \quad . \tag{3.10}$$

The first root turns out to have a higher free energy than the other two (see Fig. 3.1) and therefore is of no real physical interest. The other two roots provide two equivalent states of equilibrium spontaneous magnetization. We see clearly that the predicted value of the exponent β is $\frac{1}{2}$, the incorrect classical result.

One of the difficulties of the classical theory is associated with the necessary convexity of the free energy. If one follows through in graphical terms the arguments just presented, one obtains for the variation of A as a function of M for various values of T the results shown in Fig. 3.1. Above T_c the variation predicted by (3.4) is quadratic in M for small M and obviously convex. At T_c the coefficient of the quadratic term vanishes and A has a pure fourth power dependence on M. The graph is extremely flat but still convex as it should be. Below T_c, however, the coefficient of M^2 is negative, so the curve starts off at $M = 0$ like an inverted parabola, although it is ultimately turned around by the positive quartic term. The resulting concave portion of the curve for small M is clearly unphysical, and this should be taken as an indication that the theory has gone wrong! This defect in the theory can, however, be repaired in a more-or-less ad hoc way by means of the "Maxwell construction", which essentially consists of drawing a straight line between the two minima at $-M_0(T)$ and $+M_0(T)$. This process generates the so-called "convex cover" of the original A(M) plot. But what can be done about the totally incorrect values of the critical exponents that come from this theory? Can anything be salvaged? The temptation is to somehow or other graft on the correct values! This desire brings one naturally to the idea of scaling.

3.3 The scaling concept

There are several ways in which the desired modifications of Landau theory can be introduced. One of the earliest and most direct approaches was that of Widom. The gist of the argument goes as follows: Consider the exponent β. The incorrect classical value of $1/2$ arises from the presence of the M^2 term in (3.9). Let us therefore try to patch up the theory by replacing M^2 by $M^{1/\beta}$ where β is now a free parameter that can be fitted to experiment. If this were all, the equation of state would thus become

$$H \approx M(ct + uM^{1/\beta}), \qquad (3.11)$$

and so the spontaneous magnetization below T_c would come out correctly! Likewise, however, one might try to get the susceptibility exponent, γ, right by replacing t in (3.9) by t^γ. For $T > T_c$ it follows that $H \approx cMt^\gamma$ and so $\chi \sim t^{-\gamma}$ as desired. If this modification is to apply also for negative values of t then t should obviously be replaced by $|t|$. This, however, is easily seen to lead one into trouble since it introduces non-analytic behavior into the equation of state everywhere on the critical isotherm $t = 0$ (even for H or M nonzero). This has quite unphysical consequences since, in fact, the equation of state is, both theoretically and experimentally, completely free of singularities on crossing the critical isotherm <u>away</u> from $H = M = 0$. Similar problems arise in (3.11) for small values of M <u>above</u> T_c, where the expansion (3.4) <u>should</u> be valid but is not unless $1/\beta$ is an even integer!

To avoid these problems let us rewrite (3.9) by dividing through by $c|t|^{3/2}$ to obtain the equivalent form

$$D\frac{H}{|t|^\Delta} = \left(\frac{M}{B|t|^\beta}\right)\left\{\pm 1 + \left(\frac{M}{B|t|^\beta}\right)^{1/\beta}\right\}, \qquad (3.12)$$

where we have replaced $3/2$ by Δ while B and D are simply related to the original constants c and u. However, from this point on we may release Δ from its constrained value and treat it as a second free exponent, which, hopefully, can be adjusted to get, say, γ correct. Now the spontaneous magnetization varies as $|t|^\beta$ so the quantity $M/B|t|^\beta$ can be viewed as the magnetization <u>scaled</u> by the spontaneous magnetization, $M_0(T)$. Similarly, on the left hand side of (3.12) we have the magnetic field, H, scaled by a characteristic power of the temperature, namely, $|t|^\Delta$. Next we notice that the full equation of state is a relation connecting M,T and H which we could express as $M = \mathcal{M}(T,H)$. Widom's original suggestion was that, perhaps, when $T \to T_c$, and M and H become small the equation of state in general simplifies if M is replaced by the suitably scaled magnetization, namely $M/B|t|^{\beta}$ and H is replaced by a suitably scaled field, namely $DH/|t|^{\Delta}$. This evidently applies to the special case (3.9) which embodies classical theory but perhaps it also holds asymptotically for the true equation of state in the critical region! More explicitly, the nature of the proposed simplification is that, in the critical region, the equation of state reduces from a function $\mathcal{M}(T,H)$ of two variables to a function of only one variable, but which relates the two scaled variables together. In other words, we make the scaling postulate

$$\frac{M}{|t|^{\beta}} \approx BW\left(D\frac{H}{|t|^{\Delta}}\right), \tag{3.13}$$

where W is some sufficiently general function of a single argument. This assertion, the scaling ansatz, must, at this stage, be regarded purely as a guess, albeit, as we shall see, a remarkably successful guess!

In the classical theory we have $\beta = 1/2$ and $\Delta = 3/2$ and these values are universal for all systems: they simply arise from the integral exponents in the assumed Taylor series expansion. Additionally in classical theory, as one sees from (3.12), the full scaling function, W(y), is also universal. Thus we may expect more generally that β and Δ are universal exponents and W(y) is a universal function, even though the values will differ from their classical counterparts. On the other hand, the parameters B and D, like T_c itself, must reflect the details of the particular ferromagnet: thus they are referred to as non-universal amplitudes. The exponent Δ is often termed the gap exponent.

Let us now examine some of the implications of this simple but, in fact, far-reaching assumption. The susceptibility for $t > 0$ and $H \to 0$ is given by

$$\chi \propto \left(\frac{\partial M}{\partial H}\right)_{T,H=0} \approx |t|^{\beta-\Delta}\, BDW'(0), \tag{3.14}$$

where W'(0) must just be some number. Since, by definition, we have $\chi \sim t^{-\gamma}$, we see that

$$\Delta = \beta + \gamma. \tag{3.15}$$

This shows how Δ should be chosen to give the right value of γ. Otherwise it tells us nothing new.

To find a new result let us look at the critical isotherm, $T = T_c$, for which purpose the limit $t \to 0$ must be studied. In this limit the scaled magnetic field evidently diverges since

$$y = D\frac{H}{|t|^{\Delta}} \to \infty \qquad \text{as} \qquad t \to 0. \tag{3.16}$$

In the spirit of the enterprise let us then assume that $W(y)$ also varies as some power when y becomes large, i.e., suppose

$$W(y) \approx W_\infty y^\lambda \quad \text{as} \quad y \to \infty, \tag{3.17}$$

where W_∞ and λ are constants. It follows that

$$M \approx |t|^\beta D^\lambda B W_\infty H^\lambda / t^{\lambda\Delta}. \tag{3.18}$$

When $t \to 0$, the temperature variable should drop out of this expression since M then becomes a function of H only; consequently we must demand $\lambda\Delta = \beta$ which fixes the exponent λ as

$$\lambda = \beta/\Delta. \tag{3.19}$$

Thus there is, in reality, no free choice of λ! Moreover, from (3.17) we now see that $M \sim H^\lambda$; but, by definition we have $M \sim H^{1/\delta}$ for $T = T_c$. Thus we conclude $\delta = 1/_\lambda$ and hence

$$\delta = \frac{\Delta}{\beta} = 1 + \frac{\gamma}{\beta}. \tag{3.20}$$

This novel equation relating the three exponents β, γ, and δ is known as Widom's relation. It is our first nontrivial scaling law or, simply, exponent relation.

In a similar way, by integrating $M = -(\partial F/\partial H)_T$ to obtain the free energy $F(T,H)$ and then differentiating with respect to T one derives expressions for the entropy and specific heat and hence establishes the so-called Essam-Fisher relation

$$\alpha' + 2\beta + \gamma' = 2. \tag{3.21}$$

A little further investigation using the fact that there must be no singularities as one crosses the critical isotherm, $t = 0$, at nonzero H or M reveals that one must also have

$$\alpha = \alpha', \tag{3.22}$$

for the specific heats above and below T_c and

$$\gamma = \gamma', \tag{3.23}$$

for the susceptibilities. Evidently there are four relations connecting the six exponents α, α', β, γ, γ' and δ, and so only two of them can be independent: this is a striking prediction, by now verified many times experimentally!

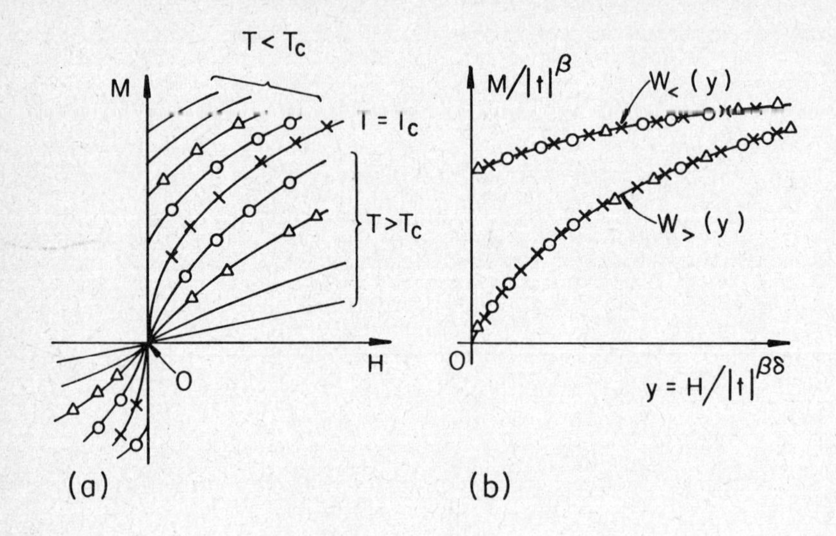

Fig. 3.2 (a) A schematic plot illustrating equation of state data, i.e., M versus H isotherms, for a ferromagnet through the critical region; (b) a scaled plot of the same data illustrating the "collapse" of the data onto a single scaling function W(y), with two branches $W_>(y)$ and $W_<(y)$ corresponding to $t \gtrless 0$.

The success of scaling can be illustrated graphically by replotting equation of state data for magnets, fluids, etc. in scaled form. Thus consider Fig. 3.2(a) where M versus H isotherms are sketched for a ferromagnet. This data may be replotted in scaled form as $M/|t|^{\beta}$ versus $y = H/|t|^{\beta\delta}$ (recalling that $\Delta = \beta\delta$ by the scaling laws) for appropriate choice of the exponents β and δ, which might be determined separately from the spontaneous magnetization curve and critical isotherm. Scaling is confirmed if one observes, as in fact is found,[8] a "collapse" of the data for the different isotherms onto a common locus, which represents the scaling function $W(y)$. Actually in this representation one finds two branches, asymptotically matching as $y \to \infty$, corresponding to $W_{>}(y)$ and $W_{<}(y)$, the scaling function for $T \gtrless T_c$. When the procedure is repeated for different magnets one finds similar results with, indeed, the same scaling function† up to different scaling amplitudes B and D. The scaling function that emerges for fluids is, likewise, the same for all fluids, and furthermore it agrees, as do the exponents, with that found for magnets!†

To sum up then, the scaling postulate proves to be a remarkably successful guess. Our theoretical task from here on is to set scaling theory in a broader context, to explain why it works, and to ask if we can actually calculate the exponents and, also, the scaling functions. The renormalization group approach provides many of the explicit answers and, further, explains the circumstances under which scaling can break down and how it fails.

3.4 Scaling of the free energy

It is useful at this stage to recapitulate by taking a somewhat different approach to scaling, and to be a little more precise. Specifically we will again use the symbol "$\sim$" to mean "behaves like" and take the symbol "$\approx$" to mean "asymptotically equal to" i.e., if $f(x) \approx g(x)$ as $x \to 0$, the ratio, $f(x)/g(x)$ approaches unity when $x \to 0$. As before, the discussion will be couched in magnetic language but, as previously emphasized, the same types of behavior are to be found in many other systems if one merely identifies the analogous quantities properly.

In the critical region the free energy, $F(T,H)$, will have a singular part which embodies the leading critical behavior. Let ΔF be the deviation of the free energy from its value at the critical point with other non-singular contributions (the "background" terms) also subtracted off. We define a normalized or reduced free energy by

---

†We restrict attention here to uniaxial, $n = 1$ or Ising-like magnets. For isotropic, $n = 3$, Heisenberg-like or XY, $n = 2$ magnets of different symmetry the exponents differ slightly, (see later below) and, necessarily, the scaling functions must also differ slightly for these distinct "universality classes."

$$f_{\text{singular}} = \frac{-\Delta F}{k_B TV}. \tag{3.24}$$

Division by $k_B T$ produces a dimensionless quantity which, however, is still proportional to the size of the system; thus, we have also divided by the volume, V, in order to obtain an intensive quantity which contains the bulk thermodynamics. The dimensions of f are thus inverse volume or number density. As we will be working in the thermodynamic limit, f is independent of V (which we suppose becomes infinite through a sequence of domains of reasonable shape).

In the previous section scaling was introduced via the equation of state in (3.13). If we integrate M(T,H) with respect to H, this leads to the free energy which (after background subtraction) will be similarly scaled. Alternatively, we could introduce a scaling postulate directly for the free energy. In this way, the scaling ansatz becomes the assertion

$$f_{\text{sing.}}(T,H) \approx A_0 |t|^{2-\alpha}\, Y\left(D \frac{H}{|t|^{\Delta}}\right) \quad \text{as } t,H \to 0, \tag{3.25}$$

where A_0 and D are non-universal scaling amplitudes which depend on the details of the system. The first, A_0, sets the scale of the free energy while D sets the scale of the magnetic field. As before, there appear two universal exponents α and Δ. A technical point that arises here concerning the scaling function Y(y) was already alluded to before: specifically, the universal function Y(y) should really be considered in two parts: $Y_{>}(y)$ for $t > 0$ and $Y_{<}(y)$ for $t < 0$. These two parts must match analytically as $y \to \infty$, but to pursue that point here would be unecessarily distracting. The reason for writing the power of the temperature prefactor in (3.25) as $2 - \alpha$ is to get the specific heat exponent correct, as is easily seen. Let us set $H = 0$ and normalize the scaling function by setting $Y(0) = 1$, which we may do because of the presence of the factor A_0. Recalling that entropy is given by $S = -(\partial F/\partial T)_{H=0}$ it follows that the singular part of the entropy varies as

$$\Delta S(T) \propto \frac{\partial f}{\partial t} \approx A_1 t^{1-\alpha}, \quad (H=0). \tag{3.26}$$

The internal energy behaves similarly. The specific heat then follows as

$$C(T) = T\left(\frac{\partial S}{\partial T}\right) \propto \frac{\partial^2 f}{\partial t^2} \approx A_2 t^{-\alpha}. \tag{3.27}$$

In these expressions A_1 and A_2 are amplitudes proportional to A_0. (Note that the variation of the prefactor T in the definition of C(T) is smooth and so does not affect the critical behavior of the specific heat. For $t > 0$ the symbol C referred to here could be subscripted either M or H since in zero field (H=0) one has $C_M = C_H$: this is a special feature resulting from the symmetry of a simple ferromagnet. More generally, one should consider $C_M(T)$ or, for a fluid system, $C_V(T)$ and so on.

The reader should sketch the variation of the zero field entropy S(T), noting that S is monotonic, and also continuous through the critical point, $T = T_c$, but exhibits a vertical tangent there, varying in the vicinity as $\pm |t|^{1-\alpha}$ where, as is typically true, we have supposed α has a positive, albeit small value. Moreover, the internal energy and many other quantities 'driven' by the critical behavior, such as the resistance, exhibit precisely the same form near T_c. Thus it is not these quantities themselves but rather their temperature derivatives which diverge at T_c (or, if $-1 < \alpha < 0$ exhibit a sharp cusp there).

The equation of state $M = \mathcal{M}(T,H)$ is obtained by differentiating with respect to H. This yields

$$M = -\left(\frac{\partial F}{\partial H}\right)_T \propto \frac{\partial f}{\partial H} \approx A_0 D|t|^{2-\alpha-\Delta}\, Y'\left(D\frac{H}{|t|^{\Delta}}\right). \tag{3.28}$$

The reason for calling Δ the gap exponent can now be seen. Each successive differentiation with respect to H, to form $\chi = (\partial M/\partial H)$, $\chi_2 = (\partial \chi/\partial H)$, etc. changes the exponent of the $|t|$ prefactor by the constant decrement Δ. For $T < T_c$ and $H \to 0$ the scaling function Y' will approach a nonzero constant value and so, as before, $M_0(T) \approx B\,|t|^{2-\alpha-\Delta}$. But since, by definition, $M_0(T) \sim |t|^{\beta}$, it follows that

$$\beta = 2 - \alpha - \Delta. \tag{3.29}$$

Adopting $M/|t|^{\beta}$ as the scaled magnetization, we see from (3.28) that this is a function only of the scaled magnetic field, $y \propto H/|t|^{\Delta}$, thus recapturing the original scaling postulate (3.13).

To obtain the critical isotherm we let $t \to 0$, and in line with the arguments used in the previous section, assume that $Y(y) \approx y^{\lambda+1}$ when $y \to \infty$. The choice $\lambda = \beta/\Delta$ ensures that $|t|$ cancels out when $t \to 0$. In this way, repeating the details for the sake of completeness, we obtain

$$M \sim t^{\beta}\left(\frac{H}{t^{\Delta}}\right)^{\beta/\Delta} \sim H^{\beta/\Delta}. \tag{3.30}$$

But since $M \sim H^{1/\delta}$ we conclude

$$\Delta = \beta\delta, \tag{3.31}$$

as before [see (3.20)].

Finally, the susceptibility is given by $\chi_T = (\partial M/\partial H)_T$ and for the reduced susceptibility above T_c we find

$$\chi \propto \frac{\partial^2 f}{\partial H^2} \approx A_0 D^2 t^{2-\alpha-2\Delta} \approx Ct^{-\gamma}, \tag{3.32}$$

so that we obtain the scaling relation

$$\gamma = -2 + \alpha + 2\Delta. \tag{3.33}$$

By combining (3.29), (3.31) and (3.33) one readily establishes the various scaling relations

$$\alpha + \beta(1 + \delta) = 2, \tag{3.34}$$

$$\alpha + 2\beta + \gamma = 2, \tag{3.35}$$

and

$$\Delta = \beta + \gamma = \beta\delta. \tag{3.36}$$

Quite clearly, the classical values of the exponents, viz. $\alpha = 0$, $\beta = 1/2$, $\gamma = 1$ and $\delta = 3$ satisfy these relations! Even before the full advent of scaling, Rushbrooke had shown on rigorous thermodynamic grounds that, because of the convexity of the free energy, the exponent inequality

$$\alpha' + 2\beta + \gamma' \geq 2, \tag{3.37}$$

was a thermodynamic necessity. Note that this is a rigorous result that does not depend on any assumption as scaling theory does. Similarly, Griffiths later proved the inequality

$$\alpha' + \beta(1 + \delta) \geq 2, \tag{3.38}$$

corresponding to (3.34). Evidently, then, scaling theory certainly does not conflict with thermodynamics even though it asserts that the rigorous inequalities hold as equalities. Nor, however, can the scaling laws be obtained by pure thermodynamic arguments although quite a few theorists have been tempted to think so and to try to demonstrate it! Occasional reports in the past of measured values of critical exponents violating the above inequalities have all proved to be poorly founded (which is just as well, since otherwise a violation of the Second Law of Thermodynamics would have been observed!)

It seems that, at least as far as systems belonging to the same symmetry class are concerned, the critical exponents are universal quantities satisfying the scaling laws. Similarly, the scaling function Y is a universal function only of the scaled field y for such systems. However, as mentioned before, one does expect some change in Y as n, the number of components of the order parameter and d, the spatial dimensionality, are varied. So we can write $Y = Y(y;n,d)$. To emphasize this point, note, as will be shown, that there are good grounds for believing that classical theory is correct when $d > 4$. To see how the scaling function depends on d, consider the behavior of the (zero-field) susceptibility above and below criticality. We can write

$$\chi \approx C^+/t^\gamma \qquad \text{as } T \to T_c+,$$

$$\approx C^-/|t|^\gamma \qquad \text{as } T \to T_c-, \tag{3.39}$$

where the amplitude ratio, C^+/C^-, should be universal but, clearly, depends on the particular form of the scaling function. Within Landau theory it is an easy exercise to prove $C^+/C^- = 2$ (which is, indeed, universal). We can accept this for $d > 4$, but for the Ising model ($n = 1$) and $d < 4$, we find $C^+/C^- \simeq 5.03$ for $d = 3$ while for $d = 2$ one knows the exact univeral value $C^+/C^- = 37.693562...$.

3.5 Fluctuations, correlations and scattering

What is the 'cause' of the failure of mean field theory and Landau theory? Why do they yield wrong exponents and wrong scaling functions? The short answer is "Because they neglect fluctuations". To understand the significance of this piece of now conventional wisdom and to explore further striking critical phenomena that provide a key to the renormalization group approach, let us study fluctuations in the critical region and introduce the correlation and scattering functions which serve to quantify them and to describe relevant observations.

Much can be learned about criticality by scattering radiation --- light, x-rays, neutrons, etc. --- off the system of interest. In a standard scattering experiment, a well-collimated beam of light, or other radiation, with known wavelength, λ, is directed at the sample, fluid, magnetic crystal, etc., and one measures the intensity, $I(\theta)$, of the light scattered at an angle θ away from the "forward" direction of the main beam. The radiation undergoes a shift in wave vector, $\underset{\sim}{k}$, which is simply related to θ and λ by

$$|\underset{\sim}{k}| = \frac{4\pi}{\lambda} \sin \tfrac{1}{2}\theta. \tag{3.40}$$

The scattered intensity $I(\theta)$ is determined by the fluctuations in the medium. If the medium were perfectly uniform (i.e., spatially homogenous) there would be no scattering at all! If one has in mind light scattering from a fluid, then the relevant fluctuations correspond to regions of different refractive index and, hence, of particle density $\rho(\underset{\sim}{R})$. For neutron scattering from a magnet, fluctuations in the spin or magnetization density are the relevant quantities, and so on. We need to study the normalized scattering intensity $I(\theta,T,H...)/I^{ideal}(\theta)$, where $I(\theta;T,H..)$ is the actual scattering intensity observed at an angle θ, which will normally depend on such factors as temperature, magnetic field, etc. while $I^{ideal}(\theta)$ is the scattering that would take place if the individual particles (spins, etc.) doing the scattering could somehow be taken far apart so that they no longer interacted and thus were quite uncorrelated with one another. Now this normalized scattering intensity is proportional to the fundamental quantity

$$\hat{G}(\underset{\sim}{k}) = \int d\underset{\sim}{R}\, e^{i\underset{\sim}{k}\cdot\underset{\sim}{R}}\, G(\underset{\sim}{R}), \tag{3.41}$$

which represents the Fourier transform of the appropriate real space correlation function $G(\underset{\sim}{R})$ (of density-density, spin-spin, etc.)

As the critical point of a fluid or fluid mixture is approached one observes enormously enhanced values of the scattering, especially at low angles, corresponding via (3.40) and (3.41), to long wavelength density fluctuations in the fluid. In the immediate critical region the scattering is so large as to be visible to the eye, particularly through the phenomenon of *critical opalescence*. This behavior is not, however, limited to fluids. Thus if, for example, one scatters neutrons from iron in the vicinity of the Curie point one likewise sees a dramatic growth in the low-angle neutron scattering intensity as sketched in Fig. 3.3. (With neutrons care must be taken to ensure that the total elastic scattering is observed since the proportionality of $I(\theta)$ to $\hat{G}(\underset{\sim}{k})$ holds only if inelastic scattering processes can be neglected.) As can be seen, for small angle scattering there is a pronounced peak in $I(\theta,T)$ as a function of temperature, and this peak approaches closer and closer to T_c as the angle is decreased. Of course, one could never actually observe zero-angle scattering directly, since this would mean picking up the oncoming main beam, but one can extrapolate to zero angle. When this is done one finds, in fact, that the zero-angle scattering $I(0,T)$, actually *diverges* at T_c. This is the most dramatic manifestation of the phenomenon of critical opalescence and is quite general, being observed whenever the appropriate scattering experiments can actually be performed.

In order to understand these effects we need to examine the correlation function for the relevant quantity, which, in general, is the locally fluctuating order parameter, $\Psi(\underset{\sim}{R})$, for the transition in question. Thus $\Psi(\underset{\sim}{R})$ could, for instance, describe how the spin varies from lattice site to lattice site in a magnetic crystal. The overall spatial average of this quantity is what was previously referred to as the (total) order parameter, Ψ. We will define the correlation function $G_{\Psi\Psi}(\underset{\sim}{R})$, or, for brevity, just $G(\underset{\sim}{R})$ by

$$G_{\Psi\Psi}(\underset{\sim}{R}) = \langle\Psi(\underset{\sim}{0})\Psi(\underset{\sim}{R})\rangle - \langle\Psi(\underset{\sim}{0})\rangle\langle\Psi(\underset{\sim}{R})\rangle. \tag{3.42}$$

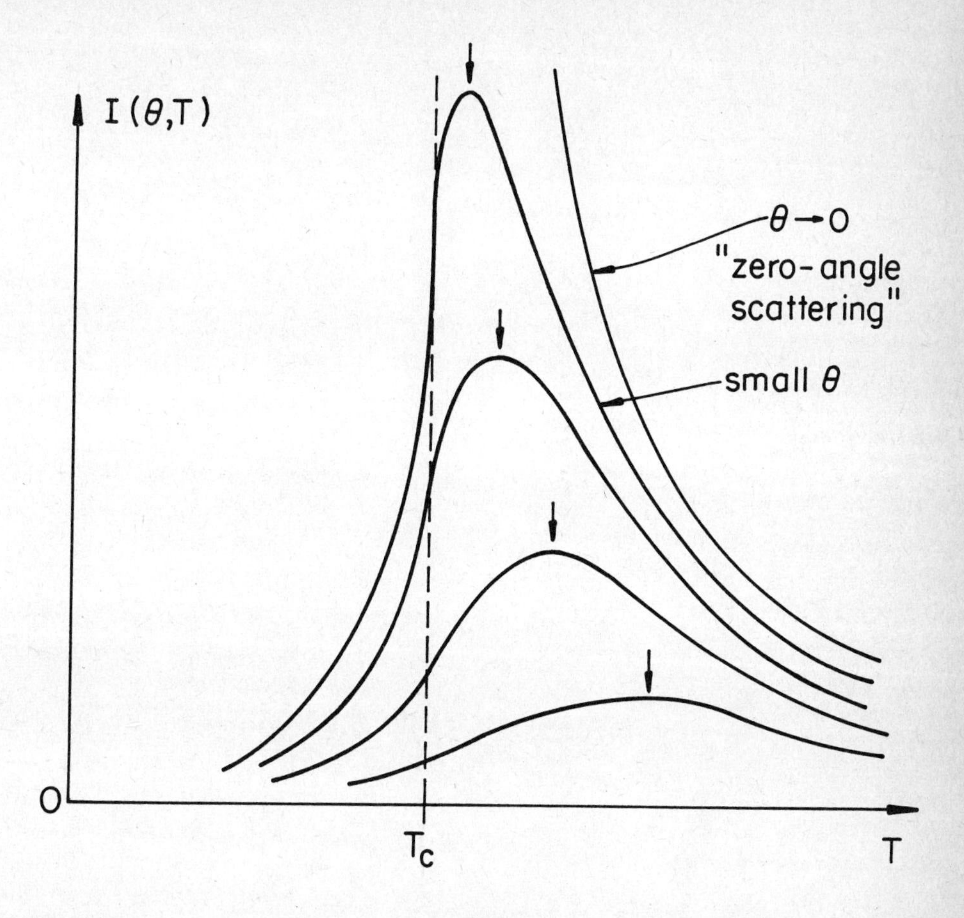

Fig. 3.3 Schematic plot of the elastic scattering intensity of neutrons scattered at fixed angle, θ, from a ferromagnet, such as iron, in the vicinity of the Curie or critical point. The small arrows mark the smoothly rounded maxima (at fixed θ) which actually occur <u>above</u> T_c (in contrast to classical and most mean field theories which yield a nonanalytic maximum <u>at</u> $T = T_c$).

We will always presuppose a macroscopically large system, so that there is translational symmetry. Likewise, we suppose that inhomogeneous effects due to gravity, etc. can be ignored. This means that the two average quantities, $\langle\Psi(\underset{\sim}{0})\rangle$ and $\langle\Psi(\underset{\sim}{R})\rangle$ will be equal to one another and to the overall, bulk, thermodynamic order parameter Ψ. We may thus let

$$\delta\Psi(\underset{\sim}{R}) = \Psi(\underset{\sim}{R}) - \langle\Psi\rangle, \tag{3.43}$$

represent the deviation or fluctuation of Ψ about its uniform mean value; then it is a matter of simple algebra to show that the correlation function directly measures the fluctuations since

$$G(\underset{\sim}{R}) = \langle\delta\Psi(\underset{\sim}{0})\delta\Psi(\underset{\sim}{R})\rangle\,. \tag{3.44}$$

For simplicity we will often assume an isotropic system so that G is a function of R rather than $\underset{\sim}{R}$.

3.6 The correlation length

If one thinks of a lattice of spins above T_c in zero field, $H = 0$, one has $\langle\Psi\rangle = 0$ by symmetry. A ferromagnetic exchange coupling between neighboring spins then tends to align the spins parallel to one another whereas thermal energy works to randomize them. Thus at high temperatures one expects the spin-spin correlation function, $G(\underset{\sim}{R})$, to fall off rather rapidly with the distance, R, separating the spins, whereas at lower temperatures the spins should become correlated with each other over longer and longer distances, the correlation function then decaying more slowly with R.

What should the law of correlation decay be? On fairly general grounds one can show* that away from T_c the correlation function should fall off exponentially with R for large distances, i.e., that the leading behavior is given by

$$G(R) \sim e^{-R/\xi} \qquad \text{as } R \to \infty, \tag{3.45}$$

where ξ is a quantity that has the dimensions of length, and is thus called the correlation length. It evidently tells us the scale on which the correlations decay. At high temperatures ξ will be just a few angströms, but near a critical point it becomes very large. This ties in well with our earlier comments on critical opalescence, since if ξ becomes comparable with the wavelength of the

---

*One must assume that the interactions themselves are of finite range or decay rapidly.

radiation, the medium will then contain fluctuations or inhomogeneities on that scale, and this will give rise to strong low angle scattering i.e., to critical opalescence.

There is another, very general theoretical route that tells us that ξ must become large near a critical point. This utilizes the fluctuation-susceptibility relation which reads

$$\hat{G}(\underset{\sim}{0};T,H) = \int d\underset{\sim}{R}\; G(\underset{\sim}{R};T,H) = \chi(T,H). \qquad (3.46)$$

For simplicity we consider here only the magnetic case. Note that $\hat{G}(\underset{\sim}{0};T,H)$ is the limiting value as $\underset{\sim}{k} \to 0$ of the Fourier transform of $G(\underset{\sim}{R};T,H)$. It thus depends only on T and H and is therefore a thermodynamic function: via statistical mechanics one finds it is just the reduced susceptibility, $\chi(T,H) = k_B T\, \chi_T/m^2$, where m is the magnetic moment per spin. Now when $T \to T_c$ for $H = 0$, we know that χ diverges; somehow this divergence must also come out of the integral in (3.46). Since $G(\underset{\sim}{R})$ is a bounded function it cannot, itself, diverge [In the case of $S = \frac{1}{2}$ spins one has $G(\underset{\sim}{R}) \leqslant \frac{3}{4}$]; thus a divergence of the integral can only mean that $G(\underset{\sim}{R})$ decays very slowly when $T = T_c$, certainly more slowly than an exponential. Consequently we are forced to conclude that $\xi(T)$ diverges to infinity when $T \to T_c$. The variation of ξ near T_c can, naturally, be described by

$$\xi(T) \sim 1/t^{\nu}, \qquad (M = 0), \qquad (3.47)$$

where for three-dimensional systems the new exponent, ν, has values around 2/3. This contrasts with the classical prediction $\nu = 1/2$ (which follows from an extension of phenomenological, Landau theory to inhomogeneous situations). More concretely one has $\nu \simeq 0.63$ for Ising-like ($n = 1$) systems, particularly fluids, increasing to $\nu \simeq 0.70$ for Heisenberg-like systems. For the two-dimensional Ising model the divergence of $\xi(T)$ was established by Onsager along with his original calculation of the zero-field free energy which revealed the logarithmic divergence of the specific heat; his results yield $\nu = 1$. In experiments on fluids such as carbon dioxide, the correlation length has been measured down to $t \simeq 10^{-4}$ or 10^{-5} by when ξ is thousands of angströms in magnitude. The divergence of the correlation length is one of the crucial clues to our general understanding of critical phenomena; the renormalization group approach, in particular, focuses on the behavior of the correlation length.

3.7 Decay of correlations at and below criticality

At T_c the correlation length is infinite. If it were not, then the integral in (3.46) would necessarily converge and be finite: then χ would be bounded at T_c

which is certainly not the case! Thus precisely at $T = T_c$ the correlation function, $G(R)$ cannot be an exponential function of R. Moreover, it must, in general, still decay to zero and one should, in fact, anticipate an "algebraic" or inverse power law form such as

$$G_c(R) \approx \frac{D}{R^{d-2+\eta}} \qquad \text{as } R \to \infty, \tag{3.48}$$

(where the nonuniversal amplitude D should not be confused with our previous use of this symbol). The reason for writing the decay exponent in this rather special way is that $G(R)$ frequently appears, as in (3.46), in volume integrals of the form

$$\int d\underset{\sim}{R}\; G(\underset{\sim}{R})X(\underset{\sim}{R}) \propto \int_0^\infty R^{d-1} G(R)X(R)\, dR,$$

so that the d drops out. Evidently η is a new critical exponent which describes how $G(R)$ behaves at T_c. Its numerical value is always rather small and in classical theory, which, as already mentioned, will be found to apply for spatial dimensionalities $d > 4$, one has $\eta = 0$. (Of course this result provides another good reason for writing (3.47) in the form given.) For real three-dimensional systems one finds $\eta \simeq 0.03$ to 0.06, but it proves to be a very difficult parameter to measure reliably and accurately in experiments. For $d = 2$ Ising-like systems the theoretical value is $\eta = 1/4$; this can even be confirmed by experiments on (effectively) two-dimensional systems. Since η is in all cases small, the integral in (3.46) necessarily diverges and the susceptibility is indeed infinite at T_c.

Beneath T_c there is a subtlety that has to be taken into account. The correlation function in general now exhibits long range order, i.e., $G(\underset{\sim}{R})$ does not decay to zero as $R \to \infty$ but rather approaches a nonzero value, say $G(\infty)$. This appearance of long range order is, in fact, one of the notable characteristics of most phase transitions. In a magnet the zero field spin-spin correlation function $\langle S(\underset{\sim}{0})S(\underset{\sim}{R})\rangle$ when $R \to \infty$ becomes proportional to $[M_0(T)]^2$, the square of the spontaneous magnetization. Via the scattering theory this leads to a magnetic Bragg peak in the scattering of strength proportional to $[M_0(T)]^2$. In systems such as antiferromagnets, where $\Psi(\underset{\sim}{R})$ is a "staggered magnetization", this provides a means of measuring the spontaneous order which would, otherwise, be inaccessible to experimental observation. If one subtracts the limiting value, $\langle S(0)S(\infty)\rangle$, from the correlation function one obtains a net correlation function which again decays to zero. In Ising-like systems there is then also an exponent ν' for the correlation length beneath T_c. Experimentally, one finds $\nu' \simeq \nu$ and theoretically, according to scaling, the two exponents should be exactly the same. Experimentally the scattering intensity $I(\theta) \equiv I(\underset{\sim}{k})$ provides us with the information needed to determine ξ. It is not hard to show that $\hat{G}(\underset{\sim}{k})$, which we recall is essentially proportional to $I(\underset{\sim}{k})$, is an even function of k which, rather generally, can be expressed in the form

$$\frac{1}{\hat{G}(k)} = \frac{1}{\chi}\left[1 + \xi^2 k^2 + O(k^4)\right], \qquad (T > T_c). \tag{3.49}$$

In a so-called Ornstein-Zernike analysis one thus plots $1/\hat{G}(\underset{\sim}{k})$ [or $1/I(\underset{\sim}{k})$] in the critical region versus k^2. The data for small k (such that $ka \lesssim 0.1$ where a is a typical molecular dimension) usually fall close to a straight line whose intercept with the $k^2 = 0$ axis determines the susceptibility $\chi(T)$. As $T \to T_c$ this intercept falls to zero but the successive isotherms remain more or less parallel on the Ornstein-Zernike (or OZ) plot. The reduced slopes evidently serve to determine $\xi(T)$. Close to T_c the plots in the case of very good experiments show a slight downward curvature: this is an indication of a nonzero and positive value of the exponent η. Thus at $T = T_c$ we have, by (3.48), the power law decay $1/R^{d-2+\eta}$, and on Fourier transformation this yields

$$\hat{G}_c(\underset{\sim}{k}) \approx \frac{\hat{D}}{k^{2-\eta}}, \tag{3.50}$$

asymptotically for small k. On an Ornstein-Zernike plot the curvature of the critical isotherm thus measures η. It should be stressed, however, that since η has such a small value, it is difficult to measure this curvature unambiguously: extensive data are needed and careful corrections for multiple scattering and other extraneous effects are called for. Nevertheless, a small positive value is definitely established.

3.8 Scaling of the correlation functions

Our treatment of correlation functions has evidently introduced two new exponents, ν and η. Are these independent of each other? Are they independent of the thermodynamic exponents α, β, γ, and δ? Or are all the exponents somehow linked together? Let us see what the idea of scaling has to say in this context. Accordingly, with no loss of generality we write the correlation function and its Fourier tranform as

$$G(\underset{\sim}{R};T,H) = \frac{\mathcal{D}(\underset{\sim}{R};T,H)}{R^{d-2+\eta}} \quad \text{and} \quad \hat{G}(\underset{\sim}{k};T,H) = \frac{\hat{\mathcal{D}}(\underset{\sim}{k};T,H)}{k^{2-\eta}} \tag{3.51}$$

which serves to pull out the critical point behavior. Now we expect (or hope!) that G and $\hat{G}$ will exhibit some simplified behavior as $T \to T_c$. Scaling means that there should be some reduced description, some compression or collapse of the multivariable data. Thus the dependence of the correlation functions on three variables might, perhaps, reduce to a dependence on only two, properly scaled, variables. The behavior at T_c has been extracted in terms of the functions $\mathcal{D}$ and $\hat{\mathcal{D}}$. In line with our previous application of scaling, it is thus natural to postulate that $\mathcal{D}$ and $\hat{\mathcal{D}}$ are functions not of three variables, but only of two scaled variables. We saw before that the scaled magnetic field had to take the form $H/|t|^{\Delta}$; there is no good reason to expect this to be changed in any way for the correlation functions. However, the length R should now also be replaced by a scaled length and, likewise k by a scaled wave number. We have of course, already identified a characteristic length for the problem: this is the correlation length ξ. Since ξ diverges at criticality it is reasonable to guess that it is the only length that really matters in the critical region. We conclude that the natural scaled length is $R/\xi \approx R|t|^{\nu}$ and that the appropriately scaled wave number is $k\xi \approx k|t|^{-\nu}$. Accepting this we can, asymptotically, replace $\mathcal{D}$ and $\hat{\mathcal{D}}$ by scaling functions D and $\hat{D}$ to obtain the scaling postulates

$$G(\underset{\sim}{R};T,H) \approx \frac{D(R|t|^{\nu};H/|t|^{\Delta})}{R^{d-2+\eta}}, \tag{3.52}$$

and, quite equivalently under Fourier transformation,

$$\hat{G}(\underset{\sim}{k};T,H) \approx \frac{\hat{D}(k/|t|^{\nu};H/|t|^{\Delta})}{k^{2-\eta}}, \tag{3.53}$$

where, for simplicity, we have left out the nonuniversal scaling amplitudes needed for full normalization if we wish to explicitly exhibit the expected universality of D and $\hat{D}$.

Notice that the only other lengths that could conceivably play any role are the interatomic spacings or the atomic and molecular diameters on scales, say, $\underline{a}$. But near the critical point all such lengths become extremely small compared to the range of the correlations, and so, being "overwhelmed", become unimportant to the long wavelength behavior of the fluctuations measured by G and $\hat{G}$. This, indeed, gives us some insight as to why there should be universality. Different fluids are found to have the same critical exponents and scaling functions. The same thing applies to magnets (if they have the same symmetry number n). Where does this universality come from? Clearly the only important differences between different fluids can be traced to the shapes and short range interactions of their constituent molecules, i.e., to differences on a scale of a few angströms. Near the critical point, fluctuations are taking place on the scale of 10^3 Å and beyond, so differences on a scale of a few Å are "washed out" or "averaged over". Thus one can understand, in an intuitive way, universality as a consequence of the fact that the correlation length becomes very large so that the important "effective interactions" no longer take place on an atomic scale but rather on a semi-macroscopic scale set by ξ. On this level the microscopic differences do not matter and one obtains universality. As systems move away from criticality and the correlation length becomes smaller, the differences start to matter. These intuitive ideas, formulated most clearly in the first place mainly by Kadanoff, are capitalized upon and made more concrete in Wilson's development of renormalization group theory.

To see how the correlation function scaling hypothesis implies connections between the critical exponents, let us examine the fluctuation integral in (3.46) that leads to the zero-field susceptibility above T_c, namely,

$$\chi = \int G(\underset{\sim}{R};T,H=0)\, d^dR,$$

$$\approx C_d \int_0^\infty \frac{D(Rt^\nu;0)}{R^{d-2+\eta}} R^{d-1} dR. \qquad (3.54)$$

We have been explicit here in dealing with a volume integral in d dimensions; because of the assumed (asymptotic) isotropy this reduces to an integral over the radius with the factor $C_d R^{d-1}$ representing the surface area of a d-dimensional sphere: for future reference the relevant coefficient is

$$C_d = 2\pi^{d/2}/\Gamma(\tfrac{1}{2}d).$$

A change of variable to the scaled combination $x = Rt^\nu$ transforms the integral to the form

$$\chi \approx C_d\, t^{-(2-\eta)\nu} \int_0^\infty D(x) x^{1-\eta} dx$$

$$\approx (\text{const.})\, t^{-(2-\eta)\nu}. \qquad (3.55)$$

Comparing with the definition $\chi \sim t^{-\gamma}$ we discover the new scaling relation

$$\gamma = (2-\eta)\nu, \qquad (3.56)$$

which relates ν and η. The classical exponent values $\gamma = 1$, $\eta = 0$ and $\nu = \frac{1}{2}$ obviously satisfy this relation. Experimentally, also, this relation checks very well. If it is accepted, it actually provides the best method of measuring the elusive exponent η!

Our theory at this stage is what might be called "three-exponent scaling", since from only three exponents, say α, Δ and ν, one can obtain all the other exponents for both thermodynamic and correlation functions. Notice that all the exponent relations so far encountered have no explicit dependence on the dimensionality, d (even though the actual values of the critical exponents themselves do depend on d). There is, however, an important further exponent relation which does involve d explicitly: this we consider now. The argument we use may, perhaps, be regarded as not very plausible but it does lead to the desired result, and other arguments are not much more convincing! Following Kadanoff we start by noting from (3.24) and (3.25) that the exponent α makes its appearance in the singular part of the free energy in the following way:

$$f_{sing.} = \frac{-\Delta F}{k_B TV} \approx t^{2-\alpha}, \quad (H = 0), \tag{3.57}$$

and that, as remarked, the dimensions of f are those of reciprocal volume or $1/L^d$ (where L is a length). Now $f_{sing.} \to 0$ when $t \to 0$ and so the relevant "critical volume" is evidently diverging, i.e., there is a significant length which is diverging! But, as we have argued, there should be only one important length in the critical region, namely, the correlation length, ξ, which moreover is also diverging as $t \to 0$. This suggests the identification

$$f_{sing.} \sim \frac{1}{\xi^d} \sim \frac{1}{(t^{-\nu})^d}. \tag{3.58}$$

Then comparing with (3.25) yields the new relation

$$d\nu = 2 - \alpha, \tag{3.59}$$

which explicitly involves the dimensionality. This is called a <u>hyperscaling relation</u> to emphasize the fact that it goes beyond and cannot be derived from the ordinary scaling relations for exponents.

Notice now that this extra relation means that just two exponents, say Δ and ν can be used to predict all the others, i.e., we have achieved a "two-exponent scaling theory". On combining the hyperscaling relation (3.59) with various other exponent relations one can easily derive the further hyperscaling relation

$$2-\eta = d\,\frac{(\delta-1)}{(\delta+1)}. \tag{3.60}$$

Again, following Buckingham and Gunton, one can show by rigorous statistical mechanical arguments that this relation must (for most systems) be satisfied rigorously as an inequality, namely,

$$2-\eta \leqslant d\,\frac{(\delta-1)}{(\delta+1)}. \tag{3.61}$$

Once again, then, we see that scaling comes in as the borderline of a general physical inequality. Notice that the experimental observation $\delta \leqslant 4.8$ for $d = 3$ implies, via this inequality, $\eta \geqslant 0.034$.

A peculiar feature of the hyperscaling relations is that the classical exponent values do <u>not</u> satisfy them unless $d = 4$! To check this, substitute $\delta = 3$ and $\eta = 0$ in (3.61) and $\alpha = 0$ and $\nu = \frac{1}{2}$ in (3.59). This fact also serves to demonstrate that the hyperscaling relations have a rather different status than the other scaling relations. However, they hold exactly in the two-dimensional Ising model, and renormalization group theory is able to show why the hyperscaling relations are to be expected fairly generally, why they hold for $d < 4$ but break down for $d > 4$, and by what mechanisms they can be expected to fail when they do. [See Appendix D.]

As discussed, the critical exponents themselves depend not only on d but also on n, the symmetry of the order parameter; so far, however, no general exponent relations have been discovered in which n appears explicitly. On the other hand, for certain classes of problem there are special relations between exponents in dimension d for one type of system and exponents in dimensions $d + 1$ and $d + 2$ for different types of problem! Many of these relations also owe their genesis to renormalization group ideas.

One might mention, in closing our phenomenological discussion of scaling, that the scaling relations can also be obtained by making certain assumptions concerning the asymptotic homogeneity character of the functional relationship between thermodynamic variables. The formalism is elegant and the end results are the same, but this approach tends to obscure the physics of the situation, which is that near a critical point each important quantity has a natural scale or size. When these natural scales are used, a reduced, universal description emerges. At the critical point itself all the temperature and field scales vanish (or diverge) so that one is left with spatial self-similarity. The fluctuations of the order parameter, for example, look statistically the same on all length scales if the magnitude of $\Psi(\underset{\sim}{R})$ is rescaled appropriately. Likewise for the energy-energy fluctuations, etc., which, in the interests of simplicity, we have not discussed. These and general aspects of scaling theory are summarized briefly in the following subsection (which need not, however, be studied in order to follow the balance of these lectures).

3.9 Anomalous or critical dimensions: general definitions and relations

We present here, in note form, a summary[10] of exponent definitions and scaling relations which emphasizes the correlation functions or, more properly, the cumulants or "connected correlation functions", for a general set of critical operators (or local densities) $A(\underset{\sim}{R})$, $B(\underset{\sim}{R})$,... .

For operators or local variables $A(\underset{\sim}{r})$, $B(\underset{\sim}{r})$, ...$N(\underset{\sim}{r})$, conjugate to fields h_A, h_B,.... including:

the order parameter $\Psi(\underset{\sim}{r})$ conjugate to field $h \equiv h_\Psi$,
the energy density $\mathcal{E}(\underset{\sim}{r})$ conjugate to field $t = h_{\mathcal{E}} \approx \Delta T/T_c$,
anisotropy energy $\mathcal{Q}(\underset{\sim}{r})$ conjugate to field $g = h_{\mathcal{Q}}$, etc.,

the general Cumulant is

$$\mathcal{K}_{AB\ldots N}(\underset{\sim}{r}_1, \ldots \underset{\sim}{r}_{n-1}) = \langle A(\underset{\sim}{0})B(\underset{\sim}{r}_1)C(\underset{\sim}{r}_2)\ldots N(\underset{\sim}{r}_{n-1})\rangle_{\mathcal{E}} \tag{3.62}$$

with

$$\mathcal{K}_A = \langle A\rangle, \ \mathcal{K}_{AB}(\underset{\sim}{r}) = \langle A(\underset{\sim}{0})B(\underset{\sim}{r})\rangle - \langle A(\underset{\sim}{0})\rangle \ \langle B(\underset{\sim}{r})\rangle, \text{ etc.} \tag{3.63}$$

The general scaling hypothesis for critical operators A, B, ... is then

$$\mathcal{K}_{AB\ldots}(\underset{\sim}{r}_j;t) \approx \frac{D(\underset{\sim}{r}_1/R,\ldots,\underset{\sim}{r}_{n-1}/R;Rt^{\nu})}{R^{\omega_A + \omega_B + \ldots + \omega_N - \omega^*}} \quad \begin{array}{l}\text{for } r_1,\ldots r_{n-1} >> a,\\ \text{and } 0 \leqslant t << 1, \text{ other fields being}\\ \text{at their critical values.}\end{array} \tag{3.64}$$

Here the ω_A, ω_B,... are the critical <u>(or anomalous) dimension</u> of A, B,..., while ω^* may be called the <u>anomalous dimension of the vacuum</u>. Note that we write t in place of $|t|$, which is generally needed, merely to reduce the complexity of the formulae.)

Hyperscaling means generally

$$\omega^* = 0, \tag{3.65}$$

as predicted by formal renormalization group analysis (see later); as found in the Ising model (n=1) for d=2, and in the spherical model (n=∞) for d⩽4; but as violated at critical points for d > 4: see Appendix D. (Note that Kadanoff, in a notation adopted by many authors, writes x_A, x_B, etc. in place of ω_A, ω_B, etc. but also assumes hyperscaling and so sets $\omega^* = 0$.

<u>Thermodynamic scaling</u> i.e., the scaling of the free energy $f \equiv f_{sing.} = -\Delta F/k_BTV$, obtained by integration of the cumulants and is expressed by

$$f(t,h,\ldots,h_A,\ldots) \approx t^{2-\alpha}\, Y\left(\frac{h}{t^{\Delta}}, \ldots, \frac{h_A}{t^{\phi_A}}, \ldots\right) \tag{3.66}$$

where (see further below)

$$2 - \alpha = d^*/\lambda_{\mathcal{E}} \equiv (d-\omega^*)/(d-\omega_{\mathcal{E}}) \tag{3.67}$$

and

$$\Delta = \phi_{\Psi} = \frac{\lambda_{\Psi}}{\lambda_{\mathcal{E}}} \equiv \frac{d-\omega_{\Psi}}{d-\omega_{\mathcal{E}}}, \tag{3.68}$$

while the "crossover exponent" is defined by

$$\phi_A = \lambda_A/\lambda_{\mathcal{E}}, \tag{3.69}$$

where we have introduced

$$d^* = d-\omega^*, \tag{3.70}$$

and the general <u>scaling eigenvalues</u> or <u>complementary exponents</u>, λ_A, through

$$\lambda_A + \omega_A = \lambda_B + \omega_B = \ldots = d. \tag{3.71}$$

(Kadanoff uses y_A, y_B, etc. in place of λ_A, λ_B, ...). Hyperscaling implies, of course, $d^* = d$ and other relations such as

$$d\nu = 2\beta + \gamma', \qquad \beta = \tfrac{1}{2}(d-2+\eta)\nu, \tag{3.72}$$

etc, which can be found from the following general

<u>Exponent relations:</u>

<u>Correlation exponents</u>

$$\xi \sim t^{-\nu}, \quad \xi \sim h_A^{-\nu_A}, \quad \nu = \nu_{\mathcal{E}}; \tag{3.73}$$

(unindicated fields set to their critical values)

$$\nu_A = 1/\lambda_A \equiv 1/(d-\omega_A); \tag{3.74}$$

$$G_c(\underset{\sim}{r}) = \langle\Psi(\underset{\sim}{0})\Psi(\underset{\sim}{r})\rangle_c \sim 1/r^{d-2+\eta}: \quad \eta = 2\omega_\Psi + 2-d-\omega^*; \tag{3.75}$$

$$\lambda_{\mathcal{E}} = d-\omega_{\mathcal{E}} = 1/\nu, \quad \omega_\Psi = d-1+\tfrac{1}{2}\eta - \tfrac{1}{2}(2-\alpha)/\nu, \quad \omega^* = d-(2-\alpha)/\nu. \tag{3.76}$$

<u>Thermodynamics</u>

$$\chi_{AA} = \frac{\partial^2 f}{\partial h_A^2} = \int d\underset{\sim}{r}\ \langle A(\underset{\sim}{0})A(\underset{\sim}{r})\rangle_c \sim h_A^{-\alpha_A}; \tag{3.77}$$

where

$$2-\alpha_A = (d-\omega^*)/(d-\omega_A) = d^*/\lambda_A, \quad (\alpha = \alpha_{\mathcal{E}}), \tag{3.78}$$

while for first order cumulants

$$\langle A\rangle \sim t^{\beta_A}, \quad (\beta = \beta_\Psi); \quad \langle A\rangle \sim h_A^{1/\delta_A}, \quad (\delta = \delta_\Psi), \tag{3.79}$$

with

$$\beta_A = (\omega_A - \omega^*)/\lambda_{\varepsilon}, \quad \delta_A = \lambda_A/(\omega_A - \omega^*) = 1/(1-\alpha_A). \tag{3.80}$$

For the cross-susceptibility one has

$$\chi_{AB} = \frac{\partial^2 f}{\partial h_A \partial h_B} = \int d\underset{\sim}{r} \, \langle A(\underset{\sim}{0})B(\underset{\sim}{r})\rangle_{\mathcal{C}} \sim t^{-\gamma_{AB}}, \quad (\gamma = \gamma_{\psi\psi}) \tag{3.81}$$

$$\gamma_{AB} = (d-\omega_A-\omega_B+\omega^*)/(d-\omega_{\varepsilon}). \tag{3.82}$$

<u>Crossover</u> (at multicritical points):

h_A scales as t^{ϕ_A} with

$$\Delta = \phi_{\psi} = \beta + \gamma = \beta\delta, \tag{3.83}$$

and $$\phi_A = \lambda_A/\lambda = \nu/\nu_A = (d-\omega_A)/(d-\omega_{\varepsilon}). \tag{3.84}$$

<u>General relations</u>

$$\alpha + \beta_A + \beta_B + \gamma_{AB} = 2, \quad \phi_A = \beta_A + \gamma_{AA} = \beta_A\delta_A. \tag{3.85}$$

4. Microscopic Models

4.1 The need for models

The traditional approach of theoreticians, going back to the foundation of quantum mechanics, is to run to Schrödinger's equation when confronted by a problem in atomic, molecular or solid state physics! One establishes the Hamiltonian, makes some (hopefully) sensible approximations and then proceeds to attempt to solve for the energy levels, eigenstates and so on. However, for truly complicated systems in what, these days, is much better called "condensed matter physics," this is a hopeless task; furthermore, in many ways it is not even a very sensible one! The modern attitude is, rather, that the task of the theorist is to understand what is going on and to elucidate which are the crucial features of the problem. For instance, if it is asserted that the exponent α depends on the dimensionality, d, and on the symmetry number, n, but on no other factors, then the theorist's job is to explain why this is so and subject to what provisos. If one had a large enough computer to solve Schrödinger's equation and the answers came out that way, one would still have no understanding of why this was the case! Thus the need is to gain understanding, not just numerical answers: that does not necessarily mean going back to Schrödinger's equation which, in any case, should be really regarded just as an approximation to some sort of gauge field theory. So the crucial change of emphasis of the last 20 or 30 years that distinguishes the new era from the old one is that when we look at the theory of condensed matter nowadays we inevitably talk about a "model". As a matter of fact even Schrödinger's equation and gauge field theories themselves are just models of the physical world, albeit pretty good ones as far as we can presently judge!

We should be prepared to look even at rather crude models, and, in particular, to study the relations between different models. We may well try to simplify the nature of a model to the point where it represents a "mere caricature" of reality. But notice that when one looks at a good political cartoon one can recognize the various characters even though the artist has portrayed them with but a few strokes. Those well chosen strokes tell one all one really needs to know about the individual, his expression, his intentions and his character. So, accepting Frenkel's guidance,[11] a good theoretical model of a complex system should be like a good caricature: it should emphasize those features which are most important and should downplay the inessential details. Now the only snag with this advice is that one does not really know which are the inessential details until one has understood the phenomena under study. Consequently, one should investigate a wide range of models and not stake one's life (or one's theoretical insight) on one particular model alone. Nevertheless, one model which, historically, has been of particular importance and which has given us a great deal of confidence in the phenomenological descriptions of critical exponents and scaling presented earlier deserves special attention: this is the so-called Ising model. Even today its study continues to provide us with new insights.[12]

4.2 Ising model

This model is absolutely the simplest model of a many body system! First of all we regard space as divided up into a lattice of cells of volume v_0, each represented by a single lattice point. The easiest lattice to think about in two dimensions is the square lattice but, following our resolution, we should at least look also at some other types, such as the triangular lattice (see Fig. 4.1). At each lattice site we allow just two possible microscopic states: in the language of a ferromagnet we place an Ising "spin", s, on each site. To distinguish spins on different sites I will usually label the spins with the position vectors $\underset{\sim}{R}, \underset{\sim}{R}', \ldots$ of the lattice sites or with site indices i, j, ...: thus $s_{\underset{\sim}{R}}$ or s_i, etc.

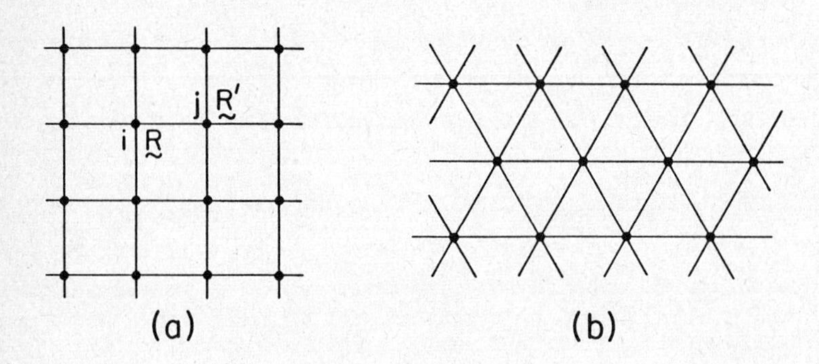

Fig. 4.1 Two dimensional lattices: (a) the plane square lattice of coordination number 4; (b) the plane triangular lattice with coordination number 6.

As implied, an Ising spin is permitted to take just two values which are expressed numerically or symbolically as

$$s_i = +1, \ (\uparrow, \text{"up"}), \qquad (4.1)$$
$$= -1, \ (\downarrow, \text{"down"}).$$

This two-valued variable is a similar but somewhat simpler entity than the quantum mechanical spin variable, $\vec{S}$, for total spin S = 1/2, whose z-component can take the two values $S^z = \pm \ ^1\!/_2$. One must notice that the Ising model also constitutes a model for a fluid, albeit the very simplest one, namely, a lattice gas. In this model we replace continuum space by the lattice of sites and suppose that the atoms or molecules can sit only on the sites. Since two atoms cannot easily be forced on top of one another, only two possibilities are contemplated at each site: either there is an atom present or there is not. Thus one can obviously establish a one-to-one correspondence between an Ising magnet and a lattice gas in which each 'down' spin, ↓, represents an occupied site and each 'up' spin, ↑, represents a vacant site: pictorially we have:

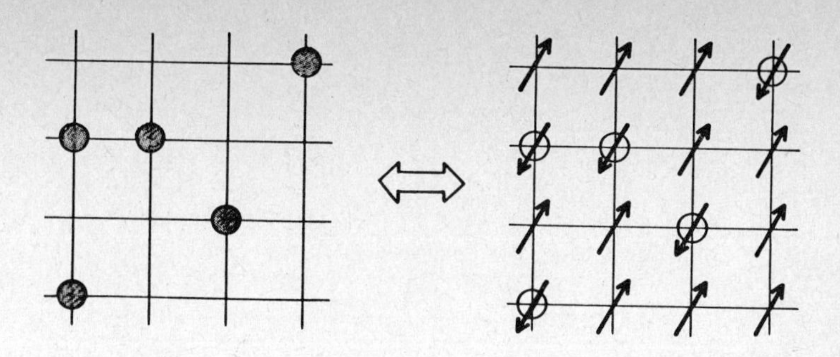

Fig. 4.2 lattice gas Ising ferromagnet

Similarly a direct correspondence exists between an Ising ferromagnet and the simplest models of a binary alloy or of a binary fluid whose composition is the important local physical variable. One can conveniently adopt the convention that an A-atom (molecule, or ion) is represented by an 'up' spin, ↑, while a B-atom then corresponds to a 'down' spin, ↓.

Mathematically all of these situations are precisely analogous: there is only one problem to solve! An Ising spin is, clearly, just a scalar and hence the Ising model is the prototype of an n=1 system.

In order to obtain any interesting behavior there must be some interactions between the spins. The standard, simplest Hamiltonian for an Ising model, given that there are N lattice sites, is

$$\mathcal{H}_N(\{s_i\}) = -H\sum_{i=1}^{N} s_i - \sum_{\langle i,j\rangle} J s_i \cdot s_j . \tag{4.2}$$

The first term in $\mathcal{H}_N$ takes account of any externally applied magnetic field, H. This term on its own would give us only a paramagnet. The second term describes the interactions between spins. For $J > 0$ it is of ferromagnetic character (approximating the so-called exchange coupling) and tends to line up the neighboring spins, s_i and s_j, in the same direction. The notation $\langle i,j\rangle$ indicates that the sum extends only over nearest neighbor pairs of lattice sites. (Sometimes the notation [i,j] is used.) When it is appropriate to consider interactions of longer range, the sum must run over all pairs (i,j) and the coupling or "exchange parameter", J, is replaced by $J_{ij} = J(\underset{\sim}{R}_i, \underset{\sim}{R}_j)$ with $J_{ij} \equiv J(\underset{\sim}{R}_j - \underset{\sim}{R}_i)$ in the normal, translationally invariant case.

4.3 Solution of the one-dimensional Ising model

The one-dimensional case of this simple model was originally solved by E. Ising in his 1925 thesis: the solution can be found in a number of text books on statistical mechanics. It will not be repeated here in full but some of the essential steps will be outlined. (They should be part of any graduate level course in statistical mechanics!) The modern way of solving the Ising model in one-dimension is first to recognize that what has to be calculated is the partition function

$$Z_N[\mathcal{H}] = \mathrm{Tr}_N\{e^{-\mathcal{H}(s)/k_BT}\} = \sum_{s_1=\pm1}\sum_{s_2=\pm1}\cdots\sum_{s_N=\pm1} e^{-\mathcal{H}(s_1,\ldots,s_N)/k_BT}. \tag{4.3}$$

Because there are so many spins the calculation is difficult. On facing a hostile army in overwhelming numbers the classical tacticians advised: "divide and conquer", or if possible, "pick them off one by one". The easy method of solving the one-dimensional Ising model (which, incidentally, is not the method Ising himself used) follows the second adage. One considers a linear lattice of N spins in which the spin summations, $\sum_{s_i=\pm1}$, have been done over all spins s_i $(i=1,2,\ldots)$ except the last one, s_N. Then one asks what happens on adding one further spin, s_{N+1}. Assuming one knows the "partial partition function" for the N-spin system, say $Z_N[\mathcal{H};s_N]$, one sees that only one more summation, i.e., over s_N, is needed to compute $Z_{N+1}[\mathcal{H};s_{N+1}]$, and so on. Finally one must take the thermodynamic limit in order to compute the bulk free energy density and see if it has any singularities that might represent phase transitions or critical behavior. Explicitly, we define the reduced free energy density via

$$f(T,H) = \lim_{N\to\infty} (Nv_0)^{-1} \ln Z_N[\mathcal{H}], \tag{4.4}$$

where v_0 is the volume (length in this d=1 case) per site. It turns out that the process of sequentially adding one spin at a time can be done very simply and directly in terms of a 2×2 matrix, which depends on T and H. Furthermore, on taking the thermodynamic limit one finds that for any boundary conditions all one needs to know about the matrix is its largest eigenvalue, say, $\Lambda_{max}(T,H)$. In terms of this maximal eigenvalue (which, since the matrix is nonnegative, has to be real) one simply has

$$f(T,H) = \ln \Lambda_{max}(T,H). \tag{4.5}$$

On deriving the appropriate 2×2 matrix and solving a quadratic equation for its eigenvalues the answer that comes out (after subtracting the harmless ground state contribution) is

$$f(T,H) = \ln[\cosh h + \sqrt{\sinh^2 h + x}\,]. \tag{4.6}$$

Here we have introduced the reduced variables

$$h = H/k_BT, \text{ and } K = J/k_BT, \tag{4.7}$$

and taken

$$x = e^{-4K} = \exp\,(-4J/k_BT). \tag{4.8}$$

The variable x is a temperature-like quantity which vanishes when $T \to 0$. For $H = 0$ and $T \to 0$ the ground state will be attained and, clearly, this will correspond to all the spins pointing in the same direction, either 'up' <u>or</u> 'down'. To reverse a particular spin then requires an energy input of 4J because of the interaction of this spin with its two neighbors. So this first, 'single-spin-flip' excited state comes with a Boltzmann factor of $\exp(-4J/k_BT)$ which demonstrates why, in Ising systems, x is the natural low temperature in terms of which one would, <u>a priori</u> expect simple, analytic behavior.

What, historically, was disappointing to the early investigators was that this model seemed not to give any phase transition whatsoever! In particular it displays no sharp specific heat anomaly at any finite (nonzero) temperature. This can be seen by setting H=0 in (4.6) so obtaining $f = \ln(1+x^{1/2})$ which is clearly a smooth function of x or T right down to the absolute zero. In fact it is now known that one-dimensional systems with quite general finite range pair interactions cannot have phase transitions at any nonzero temperature. Nevertheless, even the simplest nearest neighbor, one-dimensional Ising model does have a transition <u>at</u> $T = 0$ which can properly be regarded as a <u>critical point</u>! This can be seen, for example, by studying small x (or $T \to 0$) for which one has

$$f = \ln(1+x^{1/2}) = x^{1/2}[1 + 0\,(x^{1/2})]. \tag{4.9}$$

Since <u>a priori</u> we would have expected f to vary simply as x (as explained above) we see that the power 1/2 must really be regarded as a special sort of critical exponent (for the exponential temperature variable x).

At first sight it may seem a bit artificial to regard this effect as signifying a phase transition but, in fact, the case for doing so is strong! Consider, for example, the magnetization isotherms for the model, as sketched in Fig. 4.3. At $T = 0$ a <u>discontinuity</u> in M occurs as H passes through zero, whereas for any nonzero temperature M varies smoothly with H. Moreover, as $T \to 0$ the susceptibility, $\chi = (\partial M/\partial H)_T$ diverges very strongly: in fact, one has $\chi \sim x^{-\gamma_x}$ where $\gamma_x = 1/2$. This exponentially strong divergence of χ with $T \to 0$ should be contrasted with the simple paramagnetic behavior $\chi \sim 1/T$. Similar exponents can be defined for all other quantities of interest. Also scaling theory

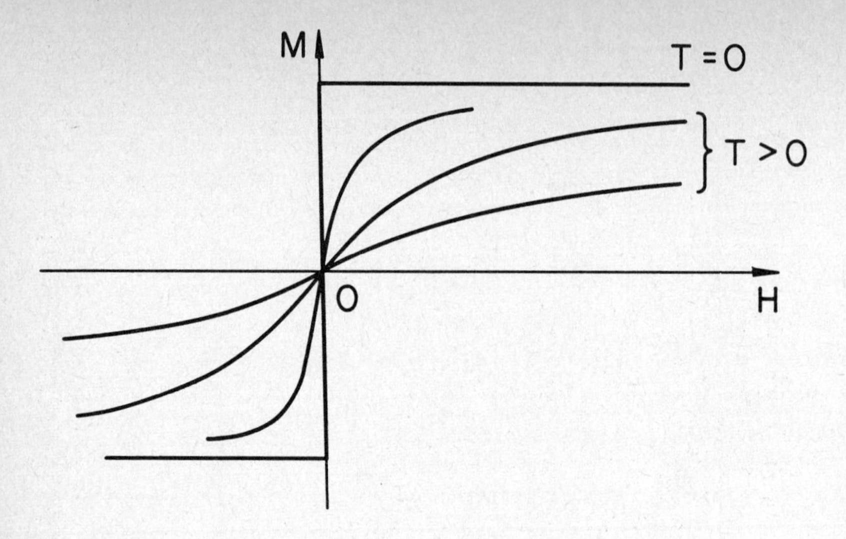

Fig. 4.3 Magnetization curves for the one-dimensional Ising model.

can be applied with $(T-T_c)$ replaced by x. (It is a good exercise for the student to check this and find the scaling functions!)

4.4 The two-dimensional Ising model

We will, here, just present some of the exact results, calculated for the two-dimensional Ising model, which reinforce our belief in the scaling hypothesis. This model is almost synonymous with the name of Lars Onsager who solved it analytically in 1944 by a generalization of the matrix method sketched above for the one-dimensional model. The first important result to emerge was that there was indeed a phase transition at $T_c > 0$. At the critical point the specific heat diverged with exponents

$$\alpha = \alpha' = 0 \quad (\log), \ (d=2, \text{ Ising}). \tag{4.10}$$

The logarithmic divergence was the first striking demonstration that the classical theory was quite wrong! Onsager also showed that

$$\beta = \frac{1}{8}, \quad (d=2, \text{ Ising}), \tag{4.11}$$

which is very different from the classical value of 1/2. (Onsager announced his result at an early stage but delayed publishing his derivation. The first published calculation for β is due to C. N. Yang.) Onsager also calculated the correlation length from which one finds

$$\nu = \nu' = 1, \ (d=2, \text{ Ising}), \tag{4.12}$$

which contrasts with the classical value 1/2. Finally he set up the calculations which lead to the demonstration that

$$\eta = \frac{1}{4}, \ (d=2, \text{ Ising}), \tag{4.13}$$

in disagreement with the classical Ornstein-Zernike prediction $\eta=0$. Later investigators showed that

$$\gamma = \gamma' = 1\frac{3}{4} \quad \text{and } \delta = 15, \quad (d=2, \text{ Ising}), \tag{4.14}$$

the classical values being $\gamma = \gamma' = 1$ and $\delta = 3$.

Note that precisely the same exponent values apply to all soluble two-dimensional Ising models, which means all those with only nearest-neighbor interactions. Furthermore, the exact values satisfy all the scaling and all the hyperscaling relations derived earlier. Unfortunately, the full scaling of the equation of state itself cannot be checked because the model has not yet been solved in the presence of a magnetic field. However, scaling of the pair correlation functions can be checked in detail. (One might remark that the exact solution of the two-dimensional Ising model <u>in a magnetic field</u> would probably teach us much more about phase transitions and critical phenomena, at this stage, than the exact solution of the three-dimensional model in zero magnetic field: in particular, it would reveal the nature of the singularities on the approach to a <u>first order</u> transition, a fascinating but subtle matter beyond the scope of these lectures.)

4.5 Ising model in three dimensions: series expansions

One thing that has been very clearly revealed by the exact analysis of two-dimensional Ising models and by comparison of the results with experiment and with the classical predictions, is that the dimensionality, d, must play a crucial role in determining the critical exponents. It is obvious, therefore, that one should also want to study the three-dimensional Ising model! This model cannot be solved analytically in the same way that Onsager solved the two-dimensional one (although there were, initially, quite a few attempts). However, answers to many of the crucial questions have been obtained to rather good precision by means of numerical "solutions". The method that has been used is the technique of exact series expansions (pioneered by Cyril Domb[13]). To see how this works, the easiest quantity to consider is the reduced susceptibility

$$\chi = \chi_T/\chi_T^{ideal}. \tag{4.15}$$

At high temperatures, as $T \to \infty$, any spin system will behave more and more like an ideal paramagnet, and so $\chi \to 1$. Thus we may seek an approximation which will approach this result in the high temperature limit but which we will, in fact, attempt to use also near in the critical region. To this end, recall that for $H = 0$ the partition function is given by

$$Z_N = \mathrm{Tr}_N\{e^{-\mathcal{H}/k_BT}\} = \sum_{\{s_i=\pm 1\}} \exp[K \sum_{\langle i,j\rangle} s_i s_j]. \tag{4.16}$$

It is a function of the single parameter $K = -J/k_BT$ which becomes small when $T \to \infty$. It is thus natural to search for an expansion of the properties of the Ising model, for arbitrary dimensionalities, in powers of K. The most direct approach is to make use of the identity (see also the more detailed discussion in Appendix C).

$$e^x = 1 + x + \frac{1}{2}x^2 + \ldots, \tag{4.17}$$

and thence obtain an expansion for Z_N in powers of K. However, to simplify the calculations it turns out to be better to introduce another temperature-like variable, namely,

$$v = \tanh K = K + O(K^3) \to 0 \text{ as } T \to \infty. \tag{4.18}$$

Then, and this is not very difficult to show for the first few terms, one finds, for example, that the expansion for χ in powers of v for the simple cubic Ising lattice is

$$\chi = 1 + 6v + 30v^2 + 150\,v^3 + 726\,v^4 + 3510\,v^5 + 16710\,v^6 + 79494\,v^7 \tag{4.19}$$
$$+ 375174\,v^8 + 1769686\,v^9 + .. + 86228667894\,v^{16} + 401225391222\,v^{17} + ..,$$

where we now know even the coefficient of v^{19} although it is too long to write here! The coefficients have a fairly simple interpretation. Starting at the origin site on a simple cubic lattice, $a_1=6$ is the number of ways to reach the first nearest neighbor sites; $a_2=30$ to reach the second nearest neighbors and so on, except that from $a_5=3510$ onwards further complications enter. The general coefficient, a_m, is, in first approximation, just the total number of distinct self-avoiding walks of m steps starting from the origin: however, this has to be corrected by allowing for a "gas of polygons" that use some of the m available lattice bonds. A great deal of effort has been expended in calculating the higher order coefficients; computers are of some assistance but they by no means make the task trivial: indeed for many years they could not compete effectively with systematic hand calculations!

Now it is quite evident that the coefficients a_m in (4.19) are increasing in a rather regular fashion with m. To study this, let us examine the ratios $\mu_m = a_m/a_{m-1}$ of successive coefficients by plotting them versus 1/m. Readers are urged to do this for themselves using the data given in (4.19). (Note that the ratios for m = 1 to 9 and for m = 17 can be plotted). Those lacking the energy or time may consult the literature.[14] For lattices like the triangular or fcc lattices that contain three-sided polygons, i.e., triangles, one finds that successive ratios fall close to a straight line of positive slope _vs_ 1/m. For 'loose-packed' lattices like the square and simple cubic lattice (containing no triangles) there is an odd-even alternation of the ratios but both sets of ratios rapidly approach a similar straight line!

What does this mean? Certainly one may conclude that, to apparently ever better approximation as m increases, one may write

$$\mu_m = \frac{a_m}{a_{m-1}} \equiv \mu_\infty [1 + \frac{g}{m} + \dots], \qquad (4.20)$$

where the dots stand for terms vanishing more rapidly than 1/m (see below). Here μ_∞ represents the (asymptotic) intercept of the line of ratios with the $m = \infty$, i.e., $1/m = 0$ axis, while g represents a dimensionless measure of the _slope_ of the plot. But again, what is this telling us?

Now it is a simple matter to see, with the aid of the binomial expansion, that the power series expansion of the function

$$A_g(v) = \frac{1}{[1-(v/v_c)]^{1+g}} = \sum_{m=0}^{\infty} \binom{-1-g}{m} \left(\frac{v}{v_c}\right)^m = 1 + a_1 v + a_2 v^2 + \dots , \qquad (4.21)$$

produces coefficients which generate ratios given exactly by

$$\mu_m = \frac{a_m}{a_{m-1}} = \frac{(g+m)}{m v_c} = \frac{1}{v_c} [1 + \frac{g}{m}], \qquad (4.22)$$

that is, which fall _exactly_ on a straight line in a plot versus 1/m. Further we see that the limiting ratio determines the point of divergence of the series as $v \to v_c-$ via

$$\mu_\infty = 1/v_c. \qquad (4.23)$$

More importantly, however, the slope, g, of the ratio plot evidently tells us the _exponent_ of divergence! (Higher order terms in a ratio plot correspond, of course, to deviations from the ideal, pure binomial form (4.20): see further below).

To apply this observation to the Ising model note the successive relations

$$\chi(v) \sim [1-(\frac{v}{v_c})]^{-(1+g)} \sim (K_c-K)^{-(1+g)} \sim 1/(T-T_c)^{1+g}. \tag{4.24}$$

So that, on recalling the exponent definition $\chi \sim 1/t^{\gamma}$, we see that the susceptibility ratio plots provide an estimate of the exponent γ via

$$\gamma = 1 + g. \tag{4.25}$$

In a similar way μ_∞, the intercept at $1/\mu = 0$, is directly related to the critical temperature. In fact one finds for the square, honeycomb and triangular lattices that the ratios μ_m extrapolate to the exactly known transition temperatures found by Onsager for these lattices to a precision of 1 in 10^4 to 10^5 or better (depending on the length of the series used). Likewise the estimates of γ for the two-dimensional lattices come out very close (to within ±0.01 to ±0.003) to the exact value γ=1.75.

For three-dimensional Ising lattices a parallel analysis yields estimates for the critical points (which, of course, depend on the lattice) of, apparently, quite comparable precision. Furthermore, to within the apparent accuracy <u>all</u> three-dimensional lattices studied (sc, bcc, fcc, and diamond) yield the same value of γ. This checks the concept of universality. In 1966, a reasonable best estimate obtained by these methods and various refinements was quoted as γ = 1.250 ± 3, where the assessed uncertainty refers to the last decimal place given. More recent work, stimulated in particular by renormalization group calculations, by the availability of longer series and by methods for studying corrections to (4.21) (see below) leads to series estimates like[15]

$$\gamma = 1.239 \pm 2. \tag{4.26}$$

which are about 1% lower. These latest estimates agree remarkably well with experimentally measured values for fluids, binary alloys, and other n=1 systems. Other exponent values obtained for three-dimensional Ising models by these techniques are:

$$\alpha \simeq 0.105 \pm 10,\ \nu \simeq 0.632 \pm 2, \text{ and } \beta \simeq 0.328 \pm 8. \tag{4.27}$$

To within the apparent uncertainties these estimates satisfy all the exponent relations including the hyperscaling relations.

Series extrapolation methods are not applicable only to critical phenomena, but can be used also in many other situations. Notice the fundamental difference in the approach from the normal truncation method of just adding up those finite number of terms of an infinite series that one has been able to calculate, and merely stopping there! Even for a convergent series this latter method must fail completely in critical phenomena since it can yield no more than a finite polynomial which could never reproduce the divergence to infinity of the susceptibility or other more subtle singular behavior that occurs at critical points.

When a large number of terms in a series have been calculated one can hope to estimate the higher order coefficients in the asymptotic form (4.20) for the ratios. The actual variation of χ for a reasonably wide range of t is, as seen earlier, given by

$$\chi = \frac{C}{t^{\gamma}} \left(1 + c_{\theta} t^{\theta} + \ldots + c_1 t + c_2 t^2 + \ldots \right). \qquad (4.28)$$

If the leading correction term is $c_1 t$ (i.e., $\theta > 1$), it turns out that c_1 can be related to the coefficient of a $1/m^2$ term in (4.20). If, however, there is a singular correction term, $c_{\theta} t^{\theta}$, with $\theta < 1$, this will show up as a dominant non-integral power $1/m^{1+\theta}$ in the expansion of μ_m. Extracting reliable information from these terms has proved difficult but, as indicated before (4.26), there has been recent progress on the problem.

In the case of the spontaneous magnetization of three-dimensional Ising models the series turn out to be much more erratic in appearance. For example, the fcc Ising lattice yields, after much labor, the low temperature expansion

$$\begin{aligned} M_0(T) = 1 &- 2x^{12} - 24x^{22} + 26x^{24} + 0+0-48x^{30} - 252x^{32} + 720x^{34} \\ &- 438x^{36} - 192x^{38} - 984x^{40} - 1008x^{42} + 12924x^{44} - 19536x^{46} \\ &+ 3062x^{48} - \ldots + 400576168x^{78} - 410287368x^{80} + \ldots . \end{aligned} \qquad (4.29)$$

The ratio method evidently fails completely for a case like this! Fortunately, however, there is another method, the so-called Padé approximant technique (propounded originally for this sort of problem by G. A. Baker, Jr. and J. L. Gammel). Padé approximant methods and their generalizations are able to handle such series and yield estimates for β and other low temperature exponents like those quoted above. In addition Padé approximants and their extensions provide efficient methods of approximate summation of series over the whole range of temperatures.

Considerable effort by various research groups has gone into these series expansion methods in an effort to calculate increasingly precise and reliable values for all the critical exponents. An important historical motivation has been the desire to check universality over a variety of models beyond the simplest Ising models. Do the exponents depend on anything besides d and n, e.g., lattice structure, quantum effects, the magnitude S of the spin, further neighbor couplings and so on? As more extensive results have become available these calculations have increasingly confirmed the surmise that for systems with interactions of finite range d and n are the only relevant quantities; other parameters embodying the fundamental constants $\hbar$, c, e etc., apparently play no role at all in this question! This was a great surprise to many of the earlier workers in the field but now, thanks to the work of Kadanoff, Wilson, and others, the reasons for this fact are much better understood. As we shall see, the renormalization group concept provides a natural explanation.

4.6 The n-vector spin models

As pointed out earlier, we need to study not just one model but, rather, a variety of models. A natural hierarchy of classical spin models is represented by the following choices of spin variables:

(a) Ising model: $s_i = \pm 1$ (n=1),

(b) XY model: $\vec{s}_i = (s_i^x, s_i^y)$, $|\vec{s}_i| = 1$ (n=2);

(c) Heisenberg model: $\vec{s}_i = (s_i^x, s_i^y, s_i^z)$, $|\vec{s}_i| = 1$ (n=3);

and finally

(d) n-vector models: $\vec{s}_i = (s_i^{(1)}, s_i^{(2)}, \ldots, s_i^{(n)})$, $|\vec{s}_i| = 1$ (general n).

In these models the spin components are simply regarded as classical variables and there are no problems associated with noncommutability, as in the more realistic quantal Heisenberg model with spin $S < \infty$ (but still with n=3). The total spin magnitudes may be normalized to unity as indicated or, as is more appropriate if one wants to consider large n, to $|\vec{s}_i|^2 = n$.

The Hamiltonian in the simplest case would be

$$\mathcal{H} = -J \sum_{\langle i,j \rangle} \vec{s}_i \cdot \vec{s}_j - H \sum_i s_i^{(1)}, \qquad (4.30)$$

where the first term is the coupling between nearest neighbor spins and the second represents the interaction with the external magnetic field which one supposes is applied in such a way that it couples only to the first component of the vectors $\vec{s}_i$. With the spin normalization set by $|\vec{s}_i|^2 = n$, the free energy density of interest is derived from the Hamiltonian via

$$f[\mathcal{H}] = \lim_{N\to\infty} - F_N/nNa^d k_B T = \lim_{N\to\infty} \frac{1}{nNa^d} \ln Z_N[\mathcal{H}], \qquad (4.31)$$

where we have expressed the cell volume v_0 in terms of the lattice spacing _a_ supposing, for simplicity, a d-dimensional hypercubic lattice. The factor of n in the denominator means that f is a free energy evaluated not only per unit volume, but _also_ per single spin component. When defined in this way one can extract sensible results even when (as proposed by H. E. Stanley) the limit $n\to\infty$ i.e., of an infinite number of spin components, is taken! It transpires that this leads to the so-called _spherical model_ invented by Mark Kac. The interest in this seemingly most artificial limit is not because the model is at all physical, but rather because it can be solved _exactly_ and it embodies important features, characteristic of models with $n > 1$ which cannot be studied in the scalar case of Ising models. The exponent values that emerge for the exact solutions are

$$\eta = 0, \text{ and } \beta = \frac{1}{2}, \text{ for all d}, \qquad (4.32)$$

and, with a striking dimensional-dependence,

$$\begin{aligned} \gamma = 2\nu &= 2/(d-2), \text{ for } d \leqslant 4, \\ &= 1 \quad , \text{ for } d \geqslant 4, \end{aligned} \qquad (4.33)$$

and, for the specific heat:

$$\alpha = \varepsilon/(d-2) \text{ where } \varepsilon = 4-d. \qquad (4.34)$$

We see here the appearance of the dimensionality parameter $\varepsilon = 4-d$; later on we will use this as a crucial expansion parameter in renormalization group theory. From a mathematical point of view it makes perfectly good sense (with a little care) to treat the spatial dimensionality, d, as a _continuous variable_ even though it is only integral values that have a direct physical meaning. In this model there are clearly two special values of d, namely, $d = 4$, called the _upper borderline_ (or _marginal) dimensionality_; and $d = 2$, called the _lower borderline dimensionality_. For $d > 4$, classical theory is seen to work. At $d = 2$ the critical temperature vanishes, $T_c = 0$, and for $d < 2$ there is no phase transition at all. As $d \to 2+$ some of the critical exponents diverge to ∞ as can be seen from the results above. It turns out that these two borderlines apply for _all_ $n > 1$. Symptomatic of another feature that appears generally at borderline dimensionalities is that for $d = 4$ one finds that the full critical behavior of the susceptibility is

$$\chi \approx \frac{C}{t} \ln t \text{ as } t \to 0+, \qquad (4.35)$$

i.e., there is a logarithmic correction factor to the leading power law behavior.

The appearance of a logarithmic factor at the borderline dimensionality $d = 4$ suggests, correctly, that simple power law scaling forms must break down somewhat at this margin. However, for $d < 4$ (including continuous values of d) the spherical model satisfies all the exponent relations and the thermodynamic and correlation functions scale completely in the standard way. Above $d = 4$ one finds that scaling again works: indeed the classical theory is, asymptotically, valid. However, that does mean that the hyperscaling relations (that involve d explicitly) must fail for $d > 4$. This observation indicates the somewhat different status that should be accorded the hyperscaling relations: below $d = 4$, however, they are precisely satisfied in the spherical model.

While the physically realizable values of d are severely limited, the values of n are much less restricted. Thus, as we have seen $n = 1,2$ and 3 are commonplace; but the case $n = 18$, for example, is of interest in describing the superfluid properties of He^3 (where the order parameter can be represented as a complex 3×3 matrix) and values as high as $n = 48$ are at least conceivable in connection with certain incommensurate phase transitions. In the opposite direction, $n = 0$ turns out to describe the pure self-avoiding walk problem or, in physical terms, the excluded volume problem for polymers in solution. Even a negative number of components may be considered! Thus the case $n = -2$ has certain attractive analytical features.[16]

4.7 Continuous spin models

In these models the spin is again regarded as an n-component classical vector, but now each component is allowed to range from $+\infty$ to $-\infty$, so we have

$$\vec{s}_i = (s_i^{(1)}, \ldots, s_i^{(n)}), \text{ with } -\infty < s_i^{(\mu)} < +\infty. \tag{4.36}$$

As regards the interactions between the spins, the Hamiltonian is just the same as before, that is

$$\mathcal{H}_{int} = -J \sum_{\langle i,j \rangle} \vec{s}_i \cdot \vec{s}_j - H \sum_i s_i^{(1)}. \tag{4.37}$$

However, there is now a further feature one must consider: this is the spin distribution. If one were not to place some sort of constraint on the magnitudes of the spins, $\vec{s}_i$, or on the way in which the components $s_i^{(\mu)}$ can be distributed then the total energy could be made indefinitely large and negative over an infinitely large region of phase space; the partition function would thus diverge while all the spins become infinitely large! We may, however, choose to regard the standard Ising model as a special class of continuous spin models, and this provides us with some guidance as to what to do to obtain a sensible model. An Ising spin s_i can be considered as a continuous, one-component spin, but with the constraint that it can take on only the values ± 1. Another way of describing this would be to say that each spin is subject to a distribution function or spin weighting function

$$e^{-w(s_i)} = \delta(s_i+1) + \delta(s_i-1), \qquad \text{(all } i). \qquad (4.38)$$

With the use of this weight function, the trace sums that are involved in calculating the partition function of an Ising spin system can be transformed into integrals so that we obtain

$$Z_N[\mathcal{H}] = \int_{-\infty}^{\infty} ds_1 \ldots \int_{-\infty}^{\infty} ds_N \, e^{-\mathcal{H}_{int}/k_B T} \prod_{i=1}^{N} e^{-w(s_i)}. \qquad (4.39)$$

(In the case of the n-vector model with $n > 1$ each integral becomes an n-fold integral over the components $s_i^{(\mu)}$.)

The simplest generalization of (4.37) that provides a genuinely continuous spin distribution, is the Gaussian model. This model (also due to Kac) is obtained by setting

$$w(\vec{s}_i) = \frac{1}{2}|\vec{s}_i|^2, \qquad \text{(all } i). \qquad (4.40)$$

The integrand in (4.38) is now just an exponential of a quadratic expression. Consequently the calculation of $Z_N[\mathcal{H}]$ reduces, after diagonalizing the quadratic form, simply to taking a product of Gaussian integrals. The model is thus exactly soluble! The Gaussian model happens to correspond precisely to the artificial limit $n = -2$ mentioned above! The exponent values that emerge from its solution are

$$\eta = 0, \text{ and } \gamma = 2\nu = 1 \text{ for all } d, \qquad (4.41)$$

while the specific heat exponent is given by

$$\begin{aligned} \alpha &= \frac{1}{2}\varepsilon \text{ for } d \leqslant 4, \qquad (4.42) \\ &= 0 \quad \text{for } d \geqslant 4. \end{aligned}$$

Unfortunately, the Gaussian model has serious shortcomings. Its worst feature is that it has no low temperature behavior! The reason for this is that the exponential decrease $e^{-|\vec{s}|^2}$ of the spin weighting function for large $|\vec{s}|$ is just not rapid enough to keep the attractive coupling terms under control when T is too small: as a consequence the integrals diverge and the model collapses!

To overcome this fatal defect we introduce a generalization of fundamental significance: this is the so-called s^4 model now often called the LGW or Landau-Ginzburg-Wilson model. In this model the spin weighting function is taken as

$$e^{-w(\vec{s})} = e^{-\frac{1}{2}|\vec{s}|^2 - \tilde{u}|\vec{s}|^4} \quad \text{with } \tilde{u} > 0. \tag{4.43}$$

The effect of the $\tilde{u}|\vec{s}|^4$ term is to pull the tails of the Gaussian weighting function down rapidly and hence give it a squarer looking shape (see Fig. 4.4). It then approximates the Ising model more closely, at least inasfar as there is little weight for large s; unphysical features of the Gaussian model below T_c are quite absent. It is widely believed that the exponent values for the scalar (n = 1) s^4

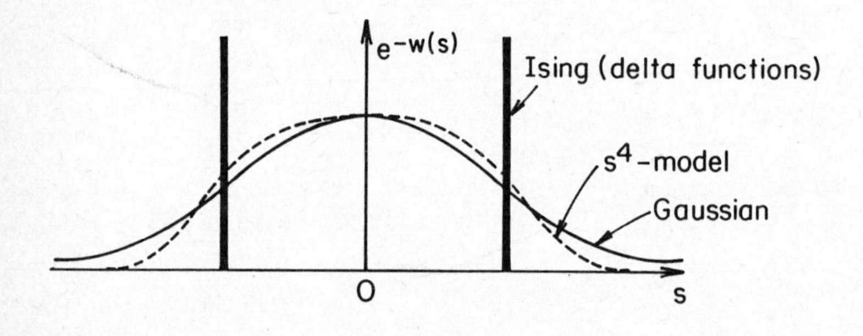

Fig. 4.4 Schematic comparison of the spin weighting functions exp[-w(s)] for the Ising, Gaussian and s^4 models.

and Ising models are exactly the same. The only sad result of including the s^4 terms in the exponential is that the integrals defining the partition function in (4.37) can now no longer be done exactly as before! This, however, is where ε-expansions, which will be discussed in a later section, have a valuable role to play.

The weighting function contributions, $w(s_i)$, are often treated as an integral part of the overall Hamiltonian: thus one writes the total reduced Hamiltonian as

$$\bar{\mathcal{H}}(s) = -\mathcal{H}_{int}(\vec{s}_1, \ldots \vec{s}_N)/k_B T - \sum_{i=1}^{N} w(\vec{s}_i), \tag{4.44}$$

and the partition function is given by

$$Z_N[\mathcal{H}] = \mathrm{Tr}_N \{e^{\bar{\mathcal{H}}(\vec{s}_1, \ldots, \vec{s}_N)}\} = \int_{-\infty}^{\infty} d^n s_1 \ldots \int_{-\infty}^{\infty} d^n s_N \, e^{\bar{\mathcal{H}}(\vec{s}_1, \ldots, \vec{s}_N)}. \tag{4.45}$$

These expressions will be our starting point in discussing the renormalization group ε expansions.

At first sight the continuous spin models with smooth weighting functions seem intrinsically different from the discrete spin or fixed-length spin models. However, as we show in Appendix A, they can in fact represent the discrete and fixed length models exactly!

5 Renormalization Group Theory

5.1 Preamble

To start with let us concentrate on the essence of renormalization group theory, putting the ideas in their simplest form. An analogy may be useful to give some perspective. In the progression from classical mechanics to a full account of quantum mechanics one starts first of all with the Bohr-Sommerfeld model or picture. Although this represents only a crude approximation, it nevertheless introduces some important ideas, such as quantization and energy levels, and it provides an explanation for the existence of discrete spectral lines and other specifically quantum-mechanical phenomena. Naturally one wants to move on from there to Schrödinger's equation and the particle-wave duality, to Bose-Einstein and Fermi-Dirac statistics, to Dirac's equation, and to quantized field theory! Nevertheless, it is instructive to start with the simplest embodiment of the most basic ideas.

In critical phenomena, the counterpart of quantization is the concept of a renormalization group transformation. The simplest such transformation which corresponds to the Bohr-Sommerfeld picture, is realized in the renormalization group treatment of the one dimensional nearest neighbor Ising model. This model can, as we have seen, be solved exactly in a fairly easy way but an analysis using a renormalization group approach still serves to introduce some important concepts. From there one hopes to progress to more subtle models. In general, the mose basic task of renormalization group theory is to explain scaling, to show us where the critical exponents come from, and to explain universality. Beyond that one would like to calculate, more-or-less explicitly, critical exponents and scaling functions. Further, the theory should tell us where the simplest scaling ideas fail and what should replace them when they do!

5.2 A renormalization group for the one-dimensional Ising model

What always enters into the partition function, as discussed previously, is the quantity $-\mathcal{H}/k_BT$, which for brevity will be called $\bar{\mathcal{H}}$. For the one-dimensional nearest neighbor Ising model we therefore have

$$\bar{\mathcal{H}} = -\mathcal{H}/k_BT = K\sum_j s_js_{j+1} + h\sum_j s_j + C\sum_j 1, \tag{5.1}$$

in which

$$K = J/k_BT, \; h = H/k_BT, \tag{5.2}$$

while an extra term, equal to CN, has been added. This has no physical consequences but turns out to serve a useful mathematical purpose in any full renormalization group treatment. It might be regarded as equal to $-E(0)/k_BT$ where E(0) is some reference energy.

This "reduced" Hamiltonian, , is evidently "equivalent" to the set of variables (K,h,C): knowledge of these three variables specifies , and therefore determines the free energy completely in the thermodynamic limit. Thus can be regarded as a point in the space of the three parameters K,h, and C. As the physical variables T and H are changed this point moves around.

One of the first approaches to renormalization group theory is to regard it merely as a special way of calculating the partition function which, in this case may be written

$$Z_N[\bar{\mathcal{H}}] = \mathrm{Tr}_N^s\{e^{\bar{\mathcal{H}}}\} = \frac{1}{2^N}\sum_{S_1=\pm 1} \cdots \sum_{S_N=\pm 1} e^{\bar{\mathcal{H}}(s)} . \tag{5.3}$$

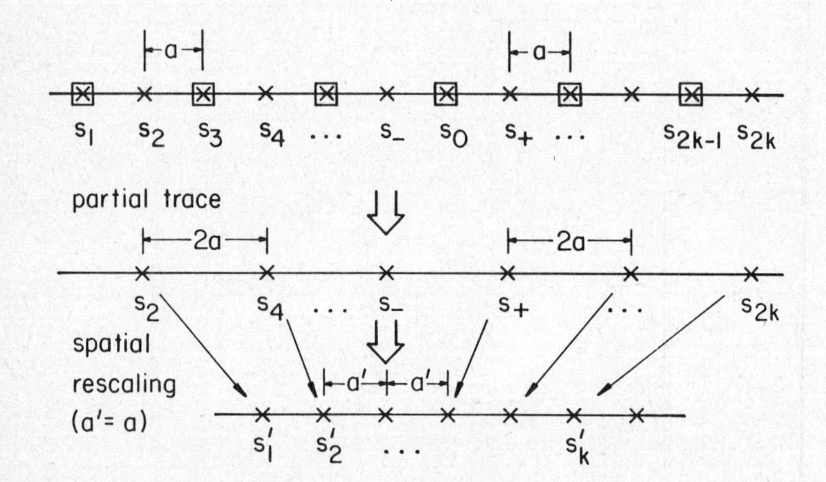

Fig. 5.1 Schematic representation of the simplest "decimation" or "dedecaration" renormalization group for a one-dimensional nearest neighbor Ising model in which a partial trace is taken by summing over alternate spins (boxed) to reduce the number of spins, followed by a spatial rescaling to restore the original appearance of the problem. (see text).

The factor $1/2^N$ has been introduced here so that Z_N is conveniently normalized to unity when $T \to \infty$. This too has no physical consequences. The free energy per spin is then

$$f[\mathcal{H}] \equiv f(K,h,C) = \frac{1}{N} \ln Z_N[\mathcal{H}]. \tag{5.4}$$

The renormalization group method of tackling the problem of evaluating $f[\mathcal{H}]$ is, like the matrix method discussed earlier, one of "divide and conquer". The idea is that instead of trying to do all the N spin summations at once, one should somehow do the summations over only some of the spins at one time, in such a way as to try to preserve the system looking as much as possible like it did before the summation, and in such a way that a spatial rescaling of the system is effected.

To achieve these ends in the present case we perform a partial trace by summing over only every second spin variable along the chain, leaving the alternate spins unaffected.[17] In this way we obtain a "renormalized" chain with only half the number of original spins as illustrated schematically in Fig. 5.1. To see what this really entails we first of all write the total Boltzmann weight in the factored form

$$e^{\mathcal{H}} = \ldots e^{Ks_-s_0 + \frac{1}{2}h(s_- + s_0) + C} e^{Ks_0s_+ + \frac{1}{2}(s_0 + s_+) + C} \ldots$$

$$= \ldots \mathcal{P}(s_-,s_0)\, \mathcal{P}(s_0,s_+) \ldots, \tag{5.5}$$

where $\mathcal{P}$ can be regarded as the Boltzmann factor for a nearest neighbor "bond", and depends only on the two spins lying at the ends of that bond. The spins s_0, s_-, and s_+ just denote one of the typical spins over which we wish to sum, together with its two nearest neighbors, respectively. The partial trace to be taken will eliminate the spin variable s_0, and result in a new Boltzmann factor, namely,

$$\mathcal{P}'(s_-,s_+) = \frac{1}{2} \sum_{s_0=\pm 1} \mathcal{P}(s_-,s_0)\, \mathcal{P}(s_0,s_+), \tag{5.6}$$

for the new "bond" connecting s_- and s_+. It is unlikely that the new Boltzmann factor will look exactly like the old one and so it has been written with a prime and is said to be "renormalized". The factor $\frac{1}{2}$ is included in this relation because with each spin eliminated one must remove a factor of $\frac{1}{2}$ from the overall normalizing factor for the partition function in (5.3). This process of eliminating spins is usually called "decimation" although "secundation" might be a more appropriate term in view of the fact it is every second spin that is "killed off" rather than every tenth one (from which Roman disciplinary procedure the word derives!) The term "dedecoration" is sometimes also used since the process is the reverse of "decorating" every bond with a new spin.

Now the renormalization group ideal is to be able to express the new bond factor, $\mathcal{P}'$, in the same basic form as the old one i.e., we would like to have

$$\mathcal{P}'(s_-,s_+) = e^{K's_-s_+ + \frac{1}{2}h'(s_- + s_+) + C'}, \tag{5.7}$$

so that the new spin chain would also be completely Ising-like. There is no reason, however, to expect that the new or renormalized parameters, K', h' and C' should take the same values as the old ones. Rather we suppose the new parameters define the renormalized Hamiltonian

$$\bar{\mathcal{H}}' \equiv (K', h', C'), \tag{5.8}$$

which will have only half as many spins. If one can indeed achieve this, one is said to have accomplished one step of a renormalization transformation. The result is written formally as

$$\bar{\mathcal{H}}' = \mathbb{R}_b[\bar{\mathcal{H}}]. \tag{5.9}$$

The important parameter b is called the spatial rescaling factor which in this case is simply equal to 2 (see Fig. 5.1). The change in the number of spins is described by b since $N \Rightarrow N' = N/2 = N/b$. In two spatial dimensions, however, one could consider decimation by knocking out alternate rows of spins and alternate columns of spins: then one would have $N \Rightarrow N' = N/4 = N/b^2$. Generally, in a d-dimensional system the spatial rescaling factor is related to the reduction in the number of degress of freedom, here simply spins, by

$$N \Rightarrow N' = N/b^d. \tag{5.10}$$

Back in one dimension, we have expressed the hope that $\mathcal{P}'(s_-,s_+)$ as obtained from (5.5) can somehow be expressed in the desired form (5.6). Now we have three variables K', h' and C' that can be adjusted in order to make this hoped-for identity true. Since s_- and s_+ can only take on the values (+1,+1), (-1,-1), (+1,-1) and (-1,+1), imposing the identity leads to four matching equations, the last two of which turn out to be identical (because the two ends of a bond are symmetrically related). It is thus an elementary exercise to show that these matching conditions are solved by

$$e^{4K'} = \frac{\cosh(2K + h)\cosh(2K - h)}{\cosh^2 h}, \tag{5.11}$$

$$e^{2h'} = e^{2h}\frac{\cosh(2K + h)}{\cosh(2K - h)}, \tag{5.12}$$

and, lastly, demonstrating why it was important to introduce the "constant" term C,

$$e^{4C'} = e^{8C}\cosh(2K + h)\ \cosh(2K - h)\ \cosh^2 h. \qquad (5.13)$$

We thus see that the proposed renormalization group transformation can be performed explicitly and exactly in this case. It has resulted in a new Hamiltonian but one retaining the same general form and it entails a reduction in the number of spins and, as we shall see shortly, an associated spatial rescaling.

5.2.1 Spatial rescaling and spin correlations

The original Ising model consisted of a chain of spins each separated from its neighbors by the lattice spacing, a. After eliminating every second spin the remaining spins are now a distance 2a apart (see Fig. 5.1). In an effort to have the renormalized model look as much like the old as possible, we rescale all lengths in such a way that the new lattice spacing, a', equals the old one. Under this scale transformation any distance R in the original lattice becomes $R' = \frac{1}{2}R$ in the new lattice when measured in units of the lattice spacing. In general we have

$$R \Rightarrow R' = R/b. \qquad (5.14)$$

This spatial rescaling is of particular importance in relation to the spin-spin correlation function, $\langle s_0 s_R \rangle$. First let us notice that it obviously makes sense to renumber the remaining spins so that their labels again run consecutively. Thus, as shown in Fig. 5.1, we take

$$s_2 \Rightarrow s_1',\ s_4 \Rightarrow s_2',\ \ldots\ , s_{2k} \Rightarrow s_k',\ \ldots, \qquad (5.15)$$

which, if we regard the labels as distance coordinates, is the same as making the identification $s'_{R'} = s_{2R'}$. Second, note that (for this renormalization transformation) since the undecimated spins retain their characters and relation to one another the renormalized correlation function $\langle s_0' s'_{R'} \rangle$ is actually equal to the original correlation function $\langle s_0 s_{2R'} \rangle$. It follows that if the renormalized correlation length is $\xi' \equiv \xi[\mathcal{H}']$ the original correlation length, $\xi \equiv \xi[\mathcal{H}]$, is just twice as long! More generally for a spatial rescaling factor b we have derived the important renormalization relation

$$\xi[\mathcal{H}] = b\xi[\mathcal{H}']. \qquad (5.16)$$

We see from this that the renormalization group procedure has the effect of shrinking the correlation length. Hence if we recall the central fact of critical phenomena, namely, that ξ becomes indefinitely large as $t \to 0$, we see that a renormalization transformation has the effect of driving a system <u>away from criticality</u>. It transpires that this is, perhaps, the most crucial feature of the method, the one that enables us to focus on critical points!

5.2.2 Unitarity

Another crucial aspect that a renormalization transformation should embody may be called unitarity (although the term is not here being used in the sense familiar in matrix theory or quantum mechanics). Recall that the renormalized Hamiltonian arose in the process of carrying out a partial trace over some of the original spin degrees of freedom. If one now simply completes the trace operation by summing over the remaining spins as coupled through the renormalized Hamiltonian to obtain the renormalized partition function, the end result must be the same as if one had performed the entire trace operation in one go. In mathematical terms, we have, first, the partial trace

$$e^{\bar{\mathcal{H}}'(s')} = \mathrm{Tr}_{N''}^{s''}\{e^{\bar{\mathcal{H}}(s)}\}, \tag{5.17}$$

where s'' stands for the $N'' = N-N'$ spins over which the decimation trace was taken, and then we compute

$$Z_{N'}[\bar{\mathcal{H}}'] \equiv \mathrm{Tr}_{N'}^{s'}\{e^{\bar{\mathcal{H}}'(s')}\}$$

$$= \mathrm{Tr}_{N'}^{s'}\{\mathrm{Tr}_{N''}^{s''}\{e^{\bar{\mathcal{H}}(s)}\}\} = \mathrm{Tr}_{N}^{s}\{e^{\bar{\mathcal{H}}(s)}\}$$

$$= Z_N[\bar{\mathcal{H}}]. \tag{5.18}$$

In other words the partition function is preserved under renormalization or, equivalently, renormalized by the simple factor unity!

This central result yields the law of renormalization for the free energy itself as follows:

$$f[\bar{\mathcal{H}}] = \frac{1}{N} \ln Z_N[\bar{\mathcal{H}}]$$

$$= \left(\frac{N'}{N}\right) \frac{1}{N'} \ln Z_{N'}[\bar{\mathcal{H}}'] = b^{-d} f[\bar{\mathcal{H}}']. \tag{5.19}$$

Note that for the sake of generality we have used (5.10) which applies for any renormalization group.

Having seen how to construct a renormalization group explicitly (in at least one case!), and having identified a number of important general properties, let us enquire into how it may be used to elucidate the nature of a critical point. To this end, we will leave aside the particular algebraic forms that appear in analysing the one-dimensional Ising model, and focus instead upon the more abstract features which they illustrate.

5.3 Flow equations, recursion relations, and fixed points

A renormalization transformation, as we have just seen, fundamentally changes a given problem into a new one, which, however, still contains the same essential information as the original one. If we rewrite (5.19) and (5.17) [which entails (5.11)-(5.13)] we may describe the renormalization procedure by a set of flow equations which describe the motion of a point describing the reduced Hamiltonian, $\bar{\mathcal{H}}$, in the appropriate space of parameters, which for our Ising chain are the variables K, h, and C. First, from (5.19) we obtain

$$f[K,h,C] = b^{-d} f[K',h',C'], \qquad (5.20)$$

for the free energy, and then we have

$$K' = \mathcal{R}_K(K,h), \qquad (5.21)$$

$$h' = \mathcal{R}_h(K,h), \qquad (5.22)$$

and, finally,

$$C' = b^d C + \mathcal{R}_0(K,h), \qquad (5.23)$$

for the "coupling constants" or "thermodynamic fields" specifying $\bar{\mathcal{H}}$. The last three flow equations are sometimes also called the recursion relations for the coupling constants: of course, for the Ising chain they are just (5.11)-(5.13) written in a more abstract form.

Now since the temperature, T, is built into the parameter K, these relations also imply a flow equation for T. To explore the significance of this let us, for simplicity, first suppose that the magnetic field on our model vanishes, i.e., $H = 0$, so that this flow equation can be written simply as

$$T \Rightarrow T' = \mathcal{R}(T), \tag{5.24}$$

where $\mathcal{R}(T)$ is the appropriate function of T. We will also suppose, without justification at this point, that $\mathcal{R}(T)$ has the form shown in Fig. 5.2, the important feature being that the plot crosses the line $T' = T$. We may call the temperature, T^*, at the crossing a <u>fixed point</u> because it clearly satisfies the relation

$$T^* = \mathcal{R}(T^*), \tag{5.25}$$

which means that when the recursion relation (5.24) is <u>iterated</u> the temperature T does not change if its initial value is set at $T = T^*$ i.e., it remains "fixed" at T^*. Subsequently we will see that T^*, in fact, represents the <u>critical point</u> (in this simplified, "Bohr-Sommerfeld" description).

Now if one starts with a temperature $T_1 > T^*$ then, as is easily seen from Fig. 5.2 one finds that $T_1' = \mathcal{R}(T_1) > T_1$, so that if the renormalization transformation is iterated, it drives the temperature, T, further and further from T^*. The same thing applies if the starting temperature, say T_0, is below T^* as also illustrated in the figure. Consequently we see that T^* is an <u>unstable fixed point</u>: the temperature always moves <u>away</u> from it under successive renormalizations. Of course, there is, in the figure another fixed point at $T = 0$ which is stable, but this turns out to be only of limited interest. In practice a similar stable fixed point also occurs at $T = \infty$ as we might have guessed. These totally stable fixed points are usually referred to as "trivial" fixed points.

As we have seen, the flow equation for the correlation length, is

$$\xi(T) = b\xi(T'). \tag{5.26}$$

At a fixed point therefore we must have

$$\xi(T^*) = b\xi(T^*), \tag{5.27}$$

but since $b > 1$, this equation has only two possible solutions, namely,

(i) $\xi(T^*) = \infty$, which evidently characterizes a critical point, and

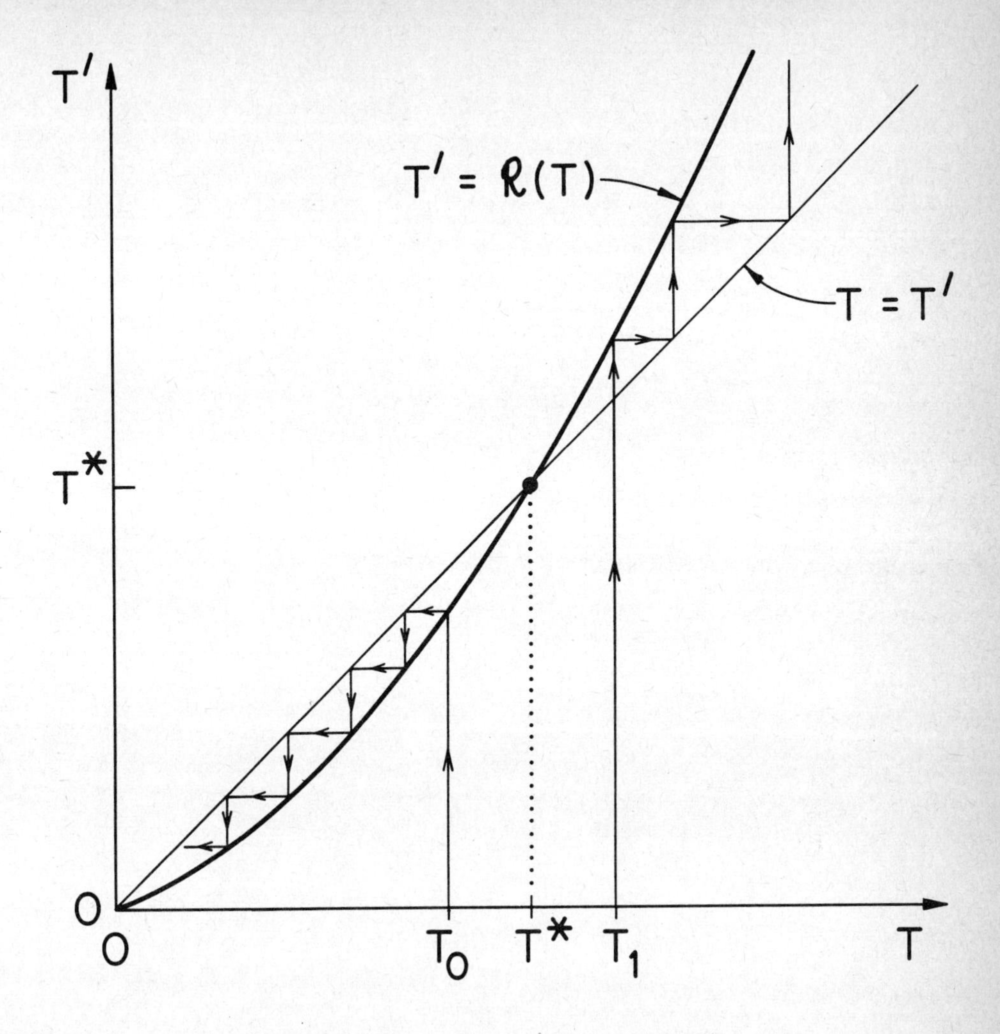

Fig. 5.2 Plot of the renormalization function or recursion relation, $T' = \mathcal{R}(T)$ for the temperature, showing the line $T' = T$, the fixed point at $T = T^*$, and successive renormalization "flows" resulting from iterating the recursion relation from two starting temperatures T_0 and T_1.

(ii) $\xi(T^*) = 0$, which corresponds to a trivial fixed point.

The vanishing of the correlation length at infinite temperature where the spins are totally uncoupled, or at zero temperature where they are frozen in a ground state is of little physical interest here and we shall ignore it. However, we have clearly made good our promise to prove

$$T_C = T^*, \tag{5.28}$$

i.e., to show that the critical temperature is located at the fixed point. We will now show that the properties of the renormalization group in the vicinity of the fixed point determine the values of the critical exponents.

5.3.1 Linearization about a fixed point

A renormalization transformation is, in general, a non-linear transformation as evident, for example from (5.11)-(5.13), but in the close vicinity of a fixed point we should be able to linearize it on the assumption that it behaves sufficiently smoothly. In the present context this merely means replacing the curved plot of $\mathcal{R}(T)$ near T^* by its tangent at T^*. Writing, as before

$$t = \frac{(T-T_c)}{T_c} = \frac{(T-T^*)}{T^*}, \tag{5.29}$$

it follows that after renormalization the temperature deviation will be given by

$$t' \equiv t^{(1)} \approx \Lambda_1(b)t, \tag{5.30}$$

for small enough t, where $\Lambda_1(b)$ is the slope of the tangent, which, as has been indicated will depend explicitly on the spatial rescaling factor b. To see this suppose one iterates twice so obtaining

$$t'' \equiv t^{(2)} \approx \Lambda_1(b)\Lambda_1(b)t. \tag{5.31}$$

Clearly this should be quite equivalent to transforming with a spatial rescaling factor b^2. Thus we conclude that one must also have

$$t^{(2)} \approx \Lambda_1(b^2)t, \tag{5.32}$$

from which we see that

$$\Lambda_1(b)\Lambda_1(b) = \Lambda_1(b^2). \tag{5.33}$$

We learn from this that Λ_1 must depend on b in a rather special way, namely as

$$\Lambda_1(b) = b^{\lambda_1}, \tag{5.33a}$$

where λ_1 is some constant (independent of b). If the renormalization transformation is iterated ℓ times one clearly obtains

$$t^{(\ell)} \approx \Lambda_1^{\ell} t = b^{\lambda_1 \ell} t, \tag{5.34}$$

while the effect on the correlation length follows from (5.26) as

$$\xi(t) \approx b^{\ell}\xi(b^{\lambda_1 \ell} t). \tag{5.35}$$

We have been explicit in these equations about the fact that the behavior stated really holds only asymptotically close to the critical point within the regime where the linearization represents a good approximation.

Now ℓ, the number of iterations, is quite arbitrary and so we may select its value in a way which procures a major simplification. Specifically if we choose ℓ to satisfy

$$b^{\ell} = t^{-1/\lambda_1}, \tag{5.36}$$

the flow relation for the correlation length becomes

$$\xi(t) \approx t^{-1/\lambda_1} \xi(1) = \frac{\text{const.}}{t^{1/\lambda_1}}. \tag{5.37}$$

This evidently matches the power law behavior, $\xi \approx 1/t^{\nu}$, which we expect to see near a critical point! On comparing exponents we make the identification

$$\nu = \frac{1}{\lambda_1}. \tag{5.38}$$

One can carry out a precisely similar analysis, based on (5.20), for the free energy. The result that emerges (on choosing C so that f vanishes at the fixed point) is

$$f(t) \approx b^{-d\ell} f(b^{\lambda_1 \ell} t), \tag{5.39}$$

from which, again choosing b to satisfy (5.36), one obtains

$$f(t) \approx t^{d/\lambda_1} f(1) = t^{d\nu} f(1). \tag{5.40}$$

This we may compare with the standard critical behavior for the free energy which we recall from (3.25) is $f \approx A_0 t^{2-\alpha}$. Hence we deduce the hyperscaling relation

$$2-\alpha = \frac{d}{\lambda_1} = d\nu, \tag{5.41}$$

first introduced heuristically in Sec. 3.8.

5.3.2 A second variable and scaling

If the magnetic field H is no longer constrained to vanish, the renormalization group operator, $\mathbb{R}_b$, acts in a more complicated non-linear fashion to generate T' and H' from T and H. We can express this fact either in terms of the pair of coupled recursion relations

$$T' = \mathcal{R}_T(T,H), \tag{5.42}$$

$$H' = \mathcal{R}_H(T,H), \tag{5.43}$$

or as the "vector" recursion relation

$$\binom{T}{H} \Rightarrow \binom{T}{H}' = \mathbb{R}_b \binom{T}{H}. \tag{5.44}$$

Note that we may ignore the "constant" term C because its flow, while depending on T and H, cannot itself have any influence on T and H, since it merely represents an additive contribution to the Hamiltonian and, thence, to the free energy but does not affect the coupling or spin configurations in any way. [This can, of course, be seen explicitly in (5.11)-(5.13)]. On the other hand, in neglecting other possible variables and focusing just on T and H we are presenting what, in our quantum-mechanics analogy, might be termed only a "single-particle picture" rather than a many-particle theory which, in quantum mechanics, would entail discussion of Fermi and Bose statistics, and so on. To the extent that the nearest-neighbor one-dimensional Ising model can be treated correctly within this limited context it can be regarded as the "hydrogen atom" of critical phenomena; however, as in most of chemistry and physics, it will prove essential to move beyond the hydrogen atom to approach the most interesting problems!

With these provisos in mind, let us, as before, assume the existence of a nontrivial fixed point (T^*, H^*) which, from the symmetry of the magnetic

Hamiltonian, should occur at $H^* = 0$. By the previous arguments, this fixed point will again prove to be the critical point since we still obtain

$$\xi[\mathcal{H}^*] \equiv \xi^* = b\xi^* = \infty. \tag{5.45}$$

Near this fixed point we may linearize the recursion relations which yields

$$\binom{\Delta T}{\Delta H}' \approx \underset{\sim}{L} \binom{\Delta T}{\Delta H}, \tag{5.46}$$

where the linear operator, $\underset{\sim}{L}$ is now the matrix

$$\underset{\sim}{L} = \begin{bmatrix} \dfrac{\partial \mathcal{R}_T}{\partial T} & \dfrac{\partial \mathcal{R}_H}{\partial T} \\ \dfrac{\partial \mathcal{R}_T}{\partial H} & \dfrac{\partial \mathcal{R}_H}{\partial H} \end{bmatrix}^*, \tag{5.47}$$

the derivatives being evaluated at the fixed point, while $\Delta T = T-T^*$ and $\Delta H = H-H^*$ denote the deviations of T and H from their fixed point values.

Now the 2×2 matrix $\underset{\sim}{L}$ will have two eigenvalues

$$\Lambda_1 = b^{\lambda_1} \quad \text{and} \quad \Lambda_2 = b^{\lambda_2}, \tag{5.48}$$

with associated eigenvectors say, $\underset{\sim}{q}_1$ and $\underset{\sim}{q}_2$ in terms of which we may expand the deviation vector as

$$\binom{\Delta T}{\Delta H} = h_1 \underset{\sim}{q}_1 + h_2 \underset{\sim}{q}_2 = h_1 \binom{q_{11}}{q_{12}} + h_2 \binom{q_{21}}{q_{22}}. \tag{5.49}$$

The coefficients, h_1 and h_2, can evidently stand in for ΔT and ΔH which, in turn, represent the deviations from criticality since, in this simplified "Bohr-Sommerfeld" treatment, the critical point is <u>at</u> the fixed point. The parameters h_1 and h_2 are therefore called the <u>critical fields</u> or, in a somewhat more general context, the <u>linear scaling fields.</u> In general we must expect to find, by solving (5.49), that h_1 and h_2 are linear combinations of ΔT and ΔH: this would, for example, be the case at the critical point of a fluid where H_1 in particular, must be replaced by a particular combination of $\Delta p = (p-p_c)$, the pressure deviation, and $\Delta T = (T-T_c) \propto t$. In the case of a simple ferromagnet, such as we have in mind, symmetry under $H \Rightarrow -H$ dictates that $\underset{\sim}{L}$ is a diagonal matrix and hence that we have

$$h_1 = t \text{ and } h_2 = H, \tag{5.50}$$

where we have chosen a convenient normalization for the eigenvectors.

On iterating the linearized renormalization group transformation (5.46) ℓ times we obtain

$$t^{(\ell)} \approx b^{\lambda_1 \ell}\, t \qquad \text{and} \qquad H^{(\ell)} \approx b^{\lambda_2 \ell} H, \tag{5.51}$$

provided that $T^{(\ell)}$ and $H^{(\ell)}$ remain in the linear region. If the iterations are repeated too often then, ultimately, the flow of T and H will become non-linear and the forms (5.51) break down. This can be seen explicitly by using the recursion relations derived for the one-dimensional Ising model for which the overall flow pattern is shown in Fig. 5.3. Notice that this actually displays a whole line of trivial fixed points! The nontrivial fixed point of interest here occurs at $T^* = 0$ corresponding to the fact, discussed in Sec. 4.3, that critical point behavior occurs at $T_c = H_c = 0$.

In the linear region near the critical point the flow of the free energy is now given by

$$f(t,h) \approx b^{-d\ell}\, f(b^{\lambda_1 \ell} t\, ,\, b^{\lambda_2 \ell} H), \tag{5.52}$$

(where, again, C has been chosen so that f vanishes at the fixed point). On making use of the freedom to choose the value of ℓ we may set

$$b^{\ell} = \left(\frac{t}{t^{\dagger}}\right)^{-1/\lambda_1}, \tag{5.53}$$

where $t^{\dagger}$ is a suitably small, fixed reference temperature, selected to keep the iterations within the linear regime. Thus we obtain the relation

$$f(t,h) \approx \left(\frac{t}{t^{\dagger}}\right)^{d/\lambda_1} f\left(t^{\dagger}, \frac{H}{(t/t^{\dagger})^{\lambda_2/\lambda_1}}\right). \tag{5.54}$$

But since $t^{\dagger}$ is now just a fixed parameter, this result corresponds exactly to our original scaling ansatz for the free energy, namely

$$f(t,h) \approx A_0 t^{2-\alpha}\, Y\left(D\, \frac{H}{t^{\Delta}}\right). \tag{5.55}$$

Comparison yields the exponent identification

$$2-\alpha = \frac{d}{\lambda_1}, \qquad \text{and} \qquad \Delta = \frac{\lambda_2}{\lambda_1}. \tag{5.56}$$

Thus we see how the renormalization group eigenvalues, λ_1 and λ_2, at the appropriate nontrivial fixed point determine the critical exponents! At the same time scaling is implied just by the form of the transformation. The non-universal amplitudes, A_0 and D, are also easily expressed in terms of $t^{\dagger}$, λ_1 and λ_2, while the scaling function itself is given formally by

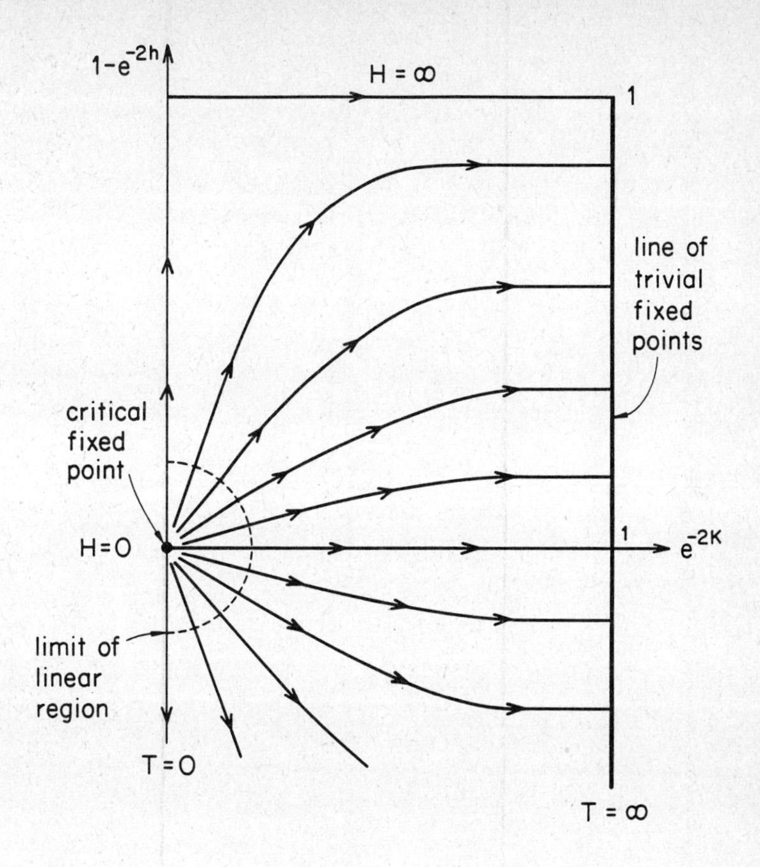

Fig. 5.3 Overall flow pattern in the (T,H) plane for the decimation or dedecoration renormalization group for the one-dimensional Ising model (based on Nelson and Fisher (1975) loc. cit.). Recall tha $K = J/k_BT$ and $h = H/k_BT$. The dashed curve delimits, approximately, the region over which a linearization of the renormalization group (in this case in the variables $x = \exp(-4J/k_BT)$ and H) is justifiable.

$$Y(y) = f(t^{\dagger}, y), \tag{5.57}$$

where $t^{\dagger}$ must evidently be chosen sufficiently small to eliminate the effects of the neglected nonlinearities. Sometimes the scaling function may be calculated by a "matching" procedure which involves iterating $\mathbb{R}$ sufficiently many times that a noncritical region is reached where the renormalized free energy can be matched, to sufficient accuracy, onto results obtained from some other theory, such as mean field theory or perturbation theory which can be regarded as valid away from criticality.

At this point the reader will find it a very instructive exercise to return to the exact recursion relations (5.11)-(5.13) for the linear Ising chain and work explicitly through the chain of reasoning leading first to the zero-field fixed point and evaluation of the critical exponents α, and ν and then through the two-variable situation to the scaling behavior (5.54) and the gap exponent Δ. It will be found that the nontrivial fixed point occurs at $T^* = H^* = 0$. Because this is a zero-temperature fixed point one finds it appropriate to use the variable $x = e^{-4K}$ in place of T, in terms of which the recursion relations are readily linearized. In this way all the critical features derived in Sec. 4.3 are recaptured correctly without the need of solving exactly for the full free energy: that, of course, is what the renormalization group is all about! Details will be found in Nelson and Fisher[17] but we quote the exponents

$$2 - \alpha_x = \nu_x = \frac{1}{\lambda_x} = \frac{1}{2} \qquad \text{and} \qquad \Delta_x = \frac{\lambda_h}{\lambda_x} = \frac{1}{2}, \tag{5.58}$$

where the subscript x denotes the use of x in place of t in the exponent definitions and scaling forms, while λ_h is merely an alternative notation for the eigenvalue λ_2. The scaling function is found to be simply

$$Y(y) = (1 + y^2)^{\frac{1}{2}}, \tag{5.59}$$

the nonuniversal amplitudes then being $A_0 = D = 1$.

5.4 General Renormalization Groups

We have used the one-dimensional Ising model to introduce some of the most important aspects of renormalization group theory at an initial level. However, we have presented no explanation of the observed universality of critical phenomena. Nor have we shown how one might construct a renormalization group transformation, $\mathbb{R}$, for systems of higher dimensionality or with other types of local variables than the simplest Ising spins. Neither have we seen how to calculate explicitly for more complex systems. Accordingly, we will now resurvey the terrain, but from a general perspective, presenting, as it were the Schrödinger picture of critical phenomena.

5.4.1 The space of Hamiltonians

Let us start by listing some of the essential attributes of, and some of the important assumptions we will make (or that should be proved) about an effective renormalization group transformation. A crucially important point is the need, in formulating $\mathbb{R}$, for a "large" space, $\mathbb{H}$, of Hamiltonians. Historically, this aspect was rather late in being generally recognized; it was K. G. Wilson who first emphasized it strongly. A simple example serves to illustrate why and how this need arises. Consider the double Ising chain or two-layer lattice which, pictorially, constitutes a ladder. What is the effect on the corresponding Hamiltonian of a b=2 decimation which eliminates alternate pairs of spins? In the absence of a

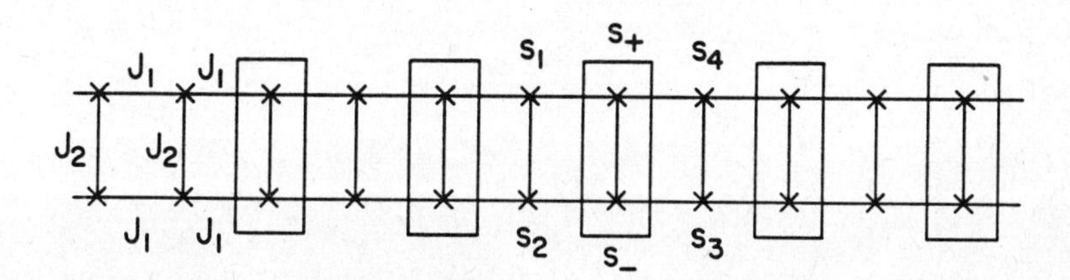

Fig. 5.4 A double-chain or two-layer Ising model ladder with nearest neighbor interactions of strength J_1 along the chains and J_2 between the chains. The boxed pairs of spins are summed over and thence eliminated in a b=2 decimation transformation.

magnetic field the original Hamiltonian is specified, as shown in Fig. 5.4, by three interaction parameters, namely, J_1, the coupling between nearest neighbors along one chain, J_2, the cross-chain coupling between adjacent spins on opposite chains, and by C, the additional "constant" parameter that was introduced earlier in treating the simple Ising chain. Carrying out the partial trace clearly results in a new double chain, and we need to match the Hamiltonians for this renormalized system and for the original system on sets of four untransformed spins like s_1, s_2, s_3 and s_4 in Fig. 5.4. Now there are 16 possible configurations of these four spins and these give rise to 16 matching equations. For H=0 a number of these equations turn out to be equivalent. Nevertheless, as each reader should convince him or herself, it is quite impossible to achieve matching using only three renormalized coupling constants C', J_1', and J_2' as would be expected for a simple ladder. Rather it proves essential to introduce two more new parameters, namely, J_3', for pair couplings like s_1 s_3 and s_2 s_4, and J_4', for a quartic coupling term s_1 s_2 s_3 s_4 as illustrated in Fig. 5.5. Thus we establish the need for an enlarged space, $\mathbb{H}$,

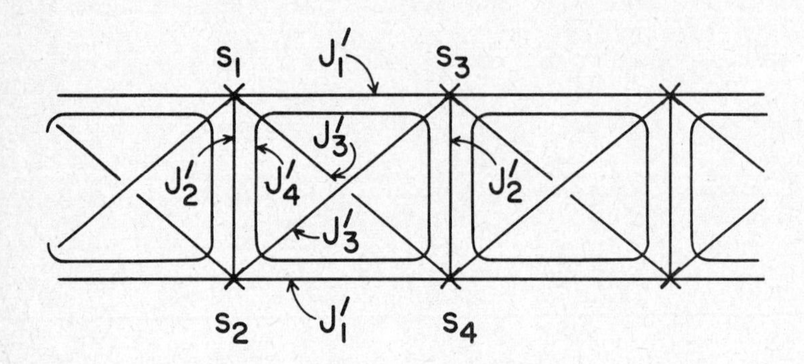

Fig. 5.5 A "braced ladder" of Ising spins with four-spin couplings resulting from renormalizing the simple Ising ladder of Fig. 5.4 by a decimation transformation.

of in this case five-parameter Hamiltonians. In the "initial", physically given Hamiltonian two of these, J_3 and J_4, just "happen" to vanish identically! After renormalization, however, they necessarily appear.

In general, then, one must allow for an indefinitely large space $\mathbb{H}$ of Hamiltonians $\mathcal{H} \equiv (C, K_1, K_2, K_3, \ldots)$, in order to provide a reasonable chance for a useful renormalization group to exist. The Hamiltonians will be characterized by coupling parameters C, $K_1 \equiv J_1/k_BT$, $K_2 \equiv J_2/k_BT$, etc. which, in general, will be infinite in number. For this reason renormalization group problems tend to be difficult, and as yet, there are not many that have been solved exactly or analyzed by rigorous methods.

5.4.2 Renormalization group desiderata

A renormalization group $\mathbb{R}_b$ for a space $\mathbb{H}$ of Hamiltonians should satisfy the following requirements:

A. Existence. There should, in the first place, clearly be a well-defined transformation, or mapping,

$$\bar{\mathcal{H}} \Rightarrow \bar{\mathcal{H}}' = \mathbb{R}_b[\bar{\mathcal{H}}], \tag{5.60}$$

which, in particular, remains unambiguous and well-defined <u>in</u> the thermodynamic limit $N \to \infty$. (See also below.)

B. <u>Elimination</u>. In the process of making the transformation there should be a reduction in the number (or density) of the original degrees of freedom, so that we can write, as before,

$$N \Rightarrow N' = N/b^d, \tag{5.61}$$

where b is the spatial rescaling factor.

C. <u>Spatial locality</u>. The transformation should not be so drastic that it mixes up the local degrees of freedom, the spins, in a hopelessly haphazard way! More concretely, one should be able to identify the same regions of space and associated local variables before and after the transformation, although, of course, spatial distances will have been changed: two regions of space originally separated by a distance $\underset{\sim}{x}$ will be brought closer together by a factor b after the transformation,

$$\underset{\sim}{x} \Rightarrow \underset{\sim}{x}' = \frac{\underset{\sim}{x}}{b}, \tag{5.62}$$

thus preserving the overall density of degrees of freedom (which is what, basically, fixes distance scales).

Hand in hand with this rescaling of space goes the transformation of the correlation length according to

$$\xi \Rightarrow \xi' \approx \xi/b. \tag{5.63}$$

However, this particular relation must, in general, be regarded as mainly heuristic since the true transformation for ξ, especially away from criticality, must depend on the details of the particular definition of correlation length which is adopted.[18]

In momentum space the effect of $\mathbb{R}_b$ is to enlarge wave-vectors by a factor b, so that

$$\underset{\sim}{q} \Rightarrow \underset{\sim}{q}' = b\underset{\sim}{q}. \tag{5.64}$$

D. <u>Unitarity</u>. Thermodynamics should be preserved by a renormalization group transformation. In other words, the two Hamiltonians should give rise to the equivalent thermodynamic functions under proper transformations of the thermodynamic fields or the couplings. In particular the unitarity relation

$$Z_N[\bar{\mathcal{H}}] = Z_{N'}[\bar{\mathcal{H}}'], \qquad (5.65)$$

preserves the total partition function and from this we obtain the flow equation or recursion relation for the overall free energy density, namely,

$$f[\bar{\mathcal{H}}] = b^{-d} f[\bar{\mathcal{H}}'], \qquad (5.66)$$

as already demonstrated in (5.19).

E. Smoothness and uniformity. In employing a renormalization group transformation it is normally essential to assume that the transformation is smooth in the sense that if $\mathcal{H} \Rightarrow \mathcal{H}'$ and $\mathcal{H} + \delta\mathcal{H} \Rightarrow \mathcal{H}' + \delta\mathcal{H}'$ then as $\delta\mathcal{H} \to 0$ one has $\delta\mathcal{H}' \to 0$; or, more strongly, that $\delta\mathcal{H}'$ becomes proportional to $\delta\mathcal{H}$ so that a first derivative exists; and so on for one or more higher derivatives. Again, such smoothness is normally assumed to hold uniformly over interesting regions of the Hamiltonian space $\mathbb{H}$ and to apply, in particular, to the flow equation for the free energy where, furthermore, one trusts that one is entitled to neglect the differences $f_N[\mathcal{H}] - f_\infty[\mathcal{H}]$ and their derivatives, etc., in the thermodynamic limit $N \to \infty$. These properties are not obviously guaranteed and really need thought and justification as the specific cases arise. (Indeed, smoothness has been seriously questioned by Griffith's and Pearce[19] for certain types of renormalization group transformation.)

F. Aptness or focusability. For any given Hamiltonian or class of Hamiltonians there is not just one renormalization group - "the renormalization group" as some people say - but rather there are many that might be introduced, and one must question, for example, whether the process is best carried out in real space or momentum space and so on. A "good" renormalization group must be "apt" or appropriate for the problem at hand, and it must, in particular, "focus" properly on the critical phenomena of interest. To this end it is sometimes necessary to introduce additional devices to make the renormalization group work satisfactorily. An important instance is provided by spin rescaling in which the (continuous) spin variables undergo the transformation

$$\vec{s} \Rightarrow \vec{s}' = \vec{s}/c, \qquad (5.67)$$

where c, the spin rescaling factor, may have to be chosen appropriately as some function $c = c(b)$ of the rescaling factor b (or, even, as $c[\bar{\mathcal{H}};b]$). We will see concretely how this need arises in Section 6 when the momentum shell renormalization group is used to generate the $\varepsilon = 4-d$ expansion for critical exponents. However, the significance of c can be seen more generally in the context of an important special class of renormalization groups which we characterize as quasi linear.

5.4.3 Quasi-linear renormalization groups

Spin rescaling has an intimate connection with the correlation functions as we now show. The spin-spin correlation function for two points, say $\underset{\sim}{o}$ and $\underset{\sim}{x}$, will depend on $\underset{\sim}{x}$ and also on the magnetic field and interaction terms in the Hamiltonian, so we may write quite generally

$$\langle s_{\underset{\sim}{o}}\, s_{\underset{\sim}{x}} \rangle = G[\underset{\sim}{x};\ \bar{\mathcal{H}}\ ]. \qquad (5.68)$$

Now a renormalization group transformation not only changes $\underset{\sim}{x}$ and $\bar{\mathcal{H}}$ but also involves some definition of the renormalized spin variables and their relation to the original spin variables. In the decimation transformation this simply amounted to a re-identification (or relabelling) of the original spins. More generally, however, the relation between $s_{\underset{\sim}{x}}$ and $s'_{\underset{\sim}{x}'}$ may be, and usually will be more complex. Consequently the transformation law for $G[\underset{\sim}{x};\ \bar{\mathcal{H}}\ ]$ is not necessarily simple. In the case of a quasi linear renormalization group, however, an identification such as (5.67) holds so that, in particular, the pair spin correlation function has the transformation law

$$G[\underset{\sim}{x};\ \bar{\mathcal{H}}\ ] = c^2 G[\underset{\sim}{x}';\ \bar{\mathcal{H}}\,']. \qquad (5.69)$$

The factor c^2 appears simply because each spin in the definition (5.68) is to be rescaled.

Now, granting such a relation, consider the situation at a nontrivial fixed point which, by definition satisfies

$$(\bar{\mathcal{H}}^*)' = \mathbb{R}_b[\bar{\mathcal{H}}^*] = \bar{\mathcal{H}}^*, \qquad (5.70)$$

so that (5.69) yields, for the fixed point correlations

$$G^*(\underset{\sim}{x}) = (c^*)^2 G^*(\underset{\sim}{x}/b). \qquad (5.71)$$

Since b is essentially arbitrary this represents a functional equation for G^* which has the unique solution

$$G^*(x) \approx \frac{D}{x^{2\omega}}, \text{ with } c^* = c[\mathcal{H}^*] = b^{-\omega}, \qquad (5.72)$$

where D and ω are constants (independent of b). However, such power law behavior is just what is to be expected at a fixed point which represents the critical point of a system. Thus we can make the identification

$$\omega = \frac{1}{2}(d-2+\eta), \tag{5.73}$$

where η is the critical point decay exponent introduced in (3.48).

This conclusion can be restated in another way: in order to obtain a nontrivial fixed point of appropriate critical character, it is necessary to adjust c (at least close to $\bar{\mathcal{H}} = \bar{\mathcal{H}}^*$) to satisfy (5.73), where η need not be known a priori. This is somewhat analogous to the adjustment of the energy in a Schrödinger equation for a stationary state so that the wave function satisfies proper boundary conditions, and the energy then yields the desired eigenvalue.

If one now returns to our exact decimation solution of the one-dimensional Ising model one sees that, without raising the question, we implicitly took a spin rescaling factor $c = 1$ or $\omega = 0$ in (5.72). Further, as mentioned, the quasilinear criterion, (5.69), was indeed satisfied. The fact that we then obtained a sensible fixed point for one-dimensional Ising criticality at $T = 0$ was thus really a result of the "accident" that $\eta = 1$ describes the scaling behavior of the Ising correlations via (3.52) as $T \to 0$: this, happily, agrees with (5.73) since $2\omega-d+2 = 0-1+2 = 1$! It is evident, however, that if one suspects that η is not simply equal to (2-d) it is inappropriate to use a quasilinear renormalization group unless one allows for a spin rescaling factor.

5.5 Flows, universality and scaling

The assumption of smoothness means that systems represented by Hamiltonians corresponding to nearby points in our multidimensional parameter space, $\mathbb{H}$, flow under renormalization to other points which also lie relatively close together. Let us apply this observation to the set of Hamiltonians representing a single physical system, say for concreteness the ferromagnet pure nickel, in the vicinity of its critical point. We will enquire into the flow trajectories generated by iteration of the renormalization group and see how this leads naturally to a concept of universality.

5.5.1 Universality, relevance and irrelevance

Consider Fig. 5.6 which presents a visualization of the space, $\mathbb{H}$, of Hamiltonians and, in particular, exhibits a "physical manifold" described by the

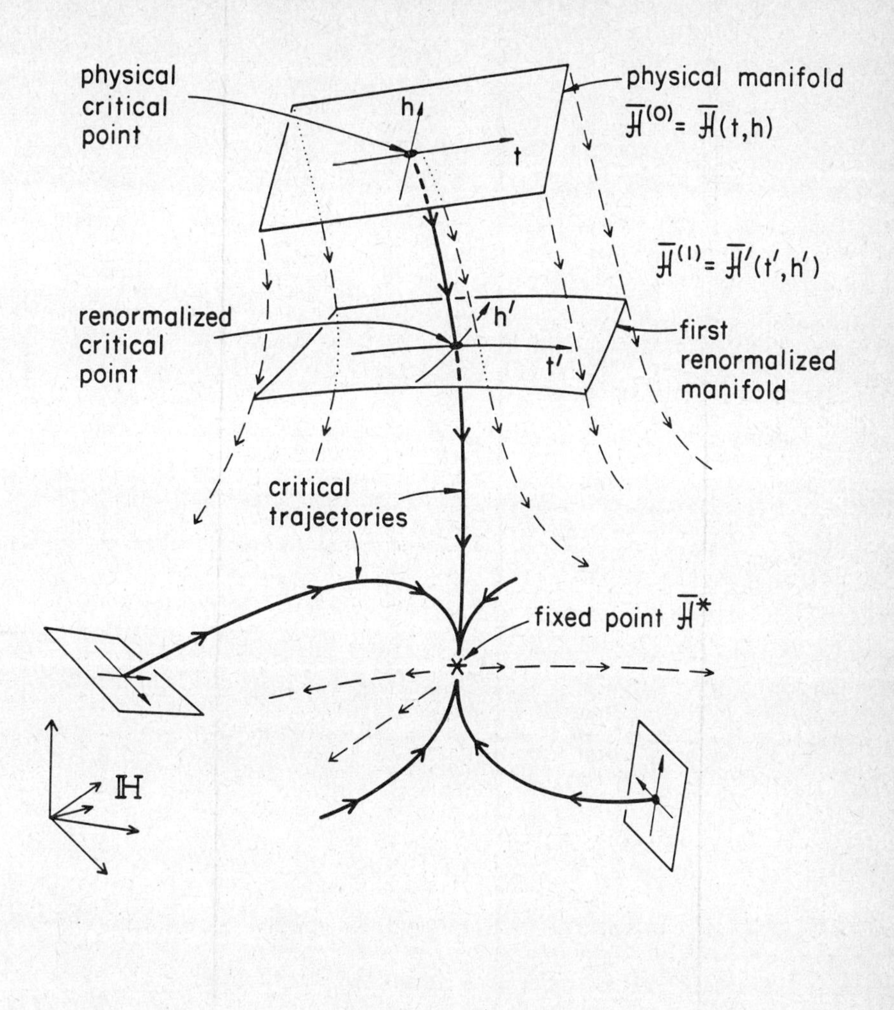

Fig. 5.6 A representation of the space of Hamiltonians, $\mathbb{H}$, showing initial or physical manifolds and subsequent renormalization group flows. Critical trajectories are shown bold: they terminate on the fixed point $\overline{\mathcal{H}}^{*}$.

initial, unrenormalized Hamiltonians, $\mathcal{H}^{(0)}(t,h)$, corresponding, as we agreed, to nickel near its ferromagnetic critical point. At the critical point itself, when the physical fields t and h vanish, we have $\xi = \infty$ because of the characteristic slow decay of the correlations. However, the critical Hamiltonian $\mathcal{H}_c^{(0)} = \bar{\mathcal{H}}^{(0)}(0,0)$ is not, in general, a fixed point! After one operation of the renormalization group transformation we obtain a new manifold, representing the first-renormalized Hamiltonians $\bar{\mathcal{H}}'(t',h')$, in which is embedded the renormalized critical Hamiltonian, $\bar{\mathcal{H}}_c' = \mathbb{R}_b[\bar{\mathcal{H}}_c]$. It is crucial to realize that this also will be a critical Hamiltonian [and, hence, equal to $\bar{\mathcal{H}}'(t'=0,h'=0)$]: the reason for this is simply that the flow equation (5.63) for the correlation length tells us that $\xi_c \Rightarrow \xi_c' = \xi_c/b = \infty/b = \infty$! Thus under successive renormalization a line or trajectory of critical points is generated. In principle this critical trajectory might eventually fly off to infinity or it might wander around in $\mathbb{H}$ forever, even perhaps in some sort of turbulent or chaotic motion! Nevertheless in the light of the previous examples (and further calculations to be performed) it is also very plausible to suppose that the critical trajectory eventually terminates at some fixed point $\bar{\mathcal{H}}^*$ at which, of course, further renormalization produces no further motion. The critical trajectory starting from the critical point of nickel and proceeding through a sequence of critical points of renormalized forms of nickel, lies on the stable critical manifold of the fixed point $\bar{\mathcal{H}}^*$ i.e., the set of all points in $\mathbb{H}$ which are ultimately carried by the renormalization group flows into $\bar{\mathcal{H}}^*$. Evidently all points on this stable critical manifold, including the fixed point itself, correspond to systems at criticality.

Now one might start in quite a different region of parameter space corresponding, say, to iron or gadolinium, as suggested by the other initial, physical manifold indicated in Fig. 5.6. Then, perhaps, under renormalization the critical point Hamiltonians for iron and gadolinium flow to the same fixed point as before! If it happens this way, then Ni, Fe and Gd must all lie on the same critical manifold. The universality of their critical behavior then follows from this fact! To demonstrate this point consider what happens to the free energy under ℓ successive renormalization group iterations: by (5.66) we have

$$f[\bar{\mathcal{H}}] = b^{-d\ell} f[\bar{\mathcal{H}}^{(\ell)}]. \tag{5.74}$$

We see that the behavior of f(t,h) for any $\bar{\mathcal{H}} \equiv (t,h)$ which lies near a critical Hamiltonian is determined by the behavior of $f[\bar{\mathcal{H}}]$ for a multiply renormalized Hamiltonian which will lie close to the fixed point. Thus the critical behavior for Ni, Fe, and Gd, and for any other systems whose critical points lie on the same manifold, will be essentially identical. In particular, because of the smoothness of the mapping all will display the same critical exponents and, furthermore, all will be described by the same scaling functions. It is only as regards the various non-universal amplitudes that the various systems will differ.

Not all systems, of course, are expected to have critical Hamiltonians which flow to the same fixed point. For instance, suppose iron is placed under a uniaxial stress. The initial parameters, and hence the initial physical manifold, are now slightly altered. If the critical point Hamiltonian were still to flow to the same fixed point as before, we would call this uniaxial stress an _irrelevant perturbation_ since it does not change any of the essential asymptotic critical properties. On physical grounds, however, in this particular case we suspect strongly that the critical behavior will change since the uniaxial stress should enhance parallel spin fluctuations but tend to suppress transverse fluctuations. Thus the new flow should carry $\bar{\mathcal{H}}_c$ to a _different fixed point_ which may be described in terms of a single-component, scalar or $n = 1$ Ising-like order parameter, whereas the original fixed point for unstressed iron is expected to correspond to an isotropic Heisenberg-like or $(n = 3)$-component order parameter. Now there will be another manifold of Hamiltonians that all flow to the new Ising-like fixed point; the critical properties of systems described by these Hamiltonians will be different from those of the former set. In such a case we say the uniaxial stress represents a _relevant perturbation_ since it causes the critical Hamiltonian to flow to a distinct, new fixed point.

The flow picture brings out clearly the idea of various universality _classes_. Systems which belong to the same universality class have critical Hamiltonians which flow into the same (or equivalent) fixed points. The corresponding critical manifolds can be regarded as the _catchment areas_ or _basins of attraction_ of the different fixed points.

5.5.2 Continuous flows

As we have seen, one utilizes a renormalization group transformation by iterating it, obtaining successive renormalized Hamiltonians. Accordingly it usually proves convenient to introduce a discrete flow variable ℓ, which counts the iterations. It can clearly be thought of as a time-like renormalization or rescaling variable which parameterizes the flow trajectories. To this end we rewrite the spatial rescaling factor as $b = e^{\ell}$, and recall that the renormalization transformation is parametrized by b as $\bar{\mathcal{H}}' = \mathbb{R}_b[\bar{\mathcal{H}}]$. Quite often, however, there arise situations in which ℓ can be regarded as a _continuous_, truly time-like flow variable. When this is so, the renormalization group equations can be written more directly as differential flow equations. Thus for the Hamiltonian itself the transformation is represented by

$$\frac{d\bar{\mathcal{H}}}{d\ell} = \mathbb{G}[\bar{\mathcal{H}}], \qquad (5.75)$$

where the $\mathbb{G}$ is the <u>infinitesimal generator</u> for $\mathbb{R}$. It can thus be expressed via a limit operation as

$$\mathbb{G} = \lim_{b \to 1+} \left(\frac{\mathbb{R}_b - 1}{b - 1}\right). \tag{5.76}$$

Since $\bar{\mathcal{H}}$ merely represents the point $(C, K_1, K_2, \ldots)$, the flow can also be written as a set of simultaneous differential equations of the form

$$\frac{dC}{d\ell} = \mathcal{G}_0(C, K_1, K_2, \ldots),$$

$$\frac{dK_1}{d\ell} = \mathcal{G}_1(C, K_1, K_2, \ldots), \tag{5.77}$$

for the parameters C, K_1, K_2, etc. (If, as before, C represents the "constant" term in $\bar{\mathcal{H}}$, it will not actually enter in the $\mathcal{G}_i$ for $i > 0$.)

5.5.3 The fixed point spectrum

In order to use a renormalization group to describe critical phenomena we must assume that there is an appropriate fixed point $\bar{\mathcal{H}}^*$. This assumption is backed up in many cases by various more-or-less detailed calculations. A few, like those for the one-dimensional Ising model, are exact but most are at best systematic approximations. However, if we follow the assumption through, powerful general conclusions follow: conversely if no proper fixed point exists we may expect scaling and other consequences to fail.

The first step, as we have seen, is linearization. To implement the procedure we take $\bar{\mathcal{H}}$ close to the fixed point and write

$$\bar{\mathcal{H}} = \bar{\mathcal{H}}^* + g\mathcal{Q}, \tag{5.78}$$

where g is small and $\mathcal{Q}$ is some "operator", i.e., a partial Hamiltonian. On operating with $\mathbb{R}_b$ and invoking the smoothness assumption, we obtain

$$\bar{\mathcal{H}}' = \mathbb{R}_b[\bar{\mathcal{H}}] = \bar{\mathcal{H}}^* + g\mathbb{L}_b\mathcal{Q} + O(g^2), \tag{5.79}$$

where $\mathbb{L}_b = (\delta\bar{\mathcal{H}}'/\delta\bar{\mathcal{H}})$ is a linearized renormalization group operator. As a linear operator, it can be expected to have a spectrum of eigenvalues $\Lambda_j(b)$ and associated "eigenvectors", $\mathcal{Q}_j$, which are "operators" or partial Hamiltonians. Sometimes the $\mathcal{Q}_j$ are called <u>critical densities</u> or <u>scaling operators</u>, etc. They are determined by the eigenvalue equation

$$\mathbb{L}_b \mathcal{Q}_j = \Lambda_j(b)\, \mathcal{Q}_j. \tag{5.80}$$

Each of the eigenvalues should be expressible in the form

$$\Lambda_j(b) = b^{\lambda_j}, \tag{5.81}$$

where the individual λ_j's are independent of b. (This reflects the semigroup property, $\mathbb{R}_{b_1 b_2} = \mathbb{R}_{b_1} \mathbb{R}_{b_2}$, of the renormalization group transformations.) Typically, one can make the indentifications

$$\mathcal{Q}_1 = \mathcal{E}, \qquad \mathcal{Q}_2 \equiv \Psi, \tag{5.82}$$

where Ψ denotes the order parameter and $\mathcal{E}$ the energy (see also below).

If we assume the eigenvectors form a complete set, or at least a sufficiently complete basis in some asymptotic sense, we may expand $\bar{\mathcal{H}}$ in terms of them as

$$\bar{\mathcal{H}} = \bar{\mathcal{H}}^* + \sum_j g_j \mathcal{Q}_j + \ldots . \tag{5.83}$$

Acting on $\bar{\mathcal{H}}$ with $\mathbb{R}_b$ then yields

$$\bar{\mathcal{H}}' = \bar{\mathcal{H}}^* + \sum_j g_j \Lambda_j \mathcal{Q}_j + O(g_j^2, g_i g_j), \tag{5.84}$$

and on iterating ℓ times we find

$$\bar{\mathcal{H}}^{(\ell)} = \bar{\mathcal{H}}^* + \sum_j g_j \Lambda_j^{\ell} \mathcal{Q}_j + O(g_j^{\,2}, g_i g_j). \tag{5.85}$$

The g_j are called <u>critical fields</u> or <u>linear scaling fields</u>. Evidently we may express the ℓ-renormalized field as

$$g_j^{(\ell)} \approx g_j \Lambda_j^{\ell} = b^{\ell \lambda_j} g_j, \tag{5.86}$$

where we must write $\approx$ ("asymptotically equals") in place of = because we are neglecting the higher order terms in (5.84) and (5.85). Now as ℓ increases there are three possible courses for $g_j^{(\ell)}$:

(a) If $\lambda_j > 0$ then we have $\Lambda_j > 1$ and $g_j^{(\ell)}$ grows rapidly larger, carrying the system away from the fixed point and, hence, away from the corresponding criticality. In accord with our previous discussion, such g_j are called <u>relevant fields</u> and the associated $\mathcal{Q}_j$ are called <u>relevant operators</u>. At a normal critical point we know that criticality is destroyed by varying the temperature, which couples to the energy, $\mathcal{E}$, from T_c or by changing the ordering field, h, which couples to Ψ, from its critical value, h = 0. Thus we expect to find <u>two</u> relevant scaling fields which, reflecting (5.82), may be identified as $g_1 \equiv t$ and $g_2 \equiv h$.

(b) If $\lambda_k < 0$ one has $\Lambda_k < 1$ and $g_k^{(\ell)}$ shrinks steadily to zero. Ultimately, therefore, it should be possible to ignore such fields. For this reason, again in concordance with the earlier discussion, g_k is called an irrelevant variable or field in such a case and the associated $\mathcal{Q}_k$ is an irrelevant operator. If the relevant fields are all set to zero the flows will take $\bar{\mathcal{H}}$ to the fixed point, so it must then lie *on* the critical manifold. Thus another way of stating universality is to note that the fixed point is insensitive to the irrelevant variables, so that systems differing from one another only with respect to irrelevant variables belong to the same universality class and are "governed" or "controlled" by the same fixed point.

(c) Finally there is the borderline where $\lambda_m = 0$ so that $\Lambda_m = 1$. The corresponding g_m are called marginal variables, and neither grow nor shrink very rapidly. Rather the flow of a marginal variable must be described by

$$\frac{dg_m}{d\ell} = 0 + O(g_i g_j, g_j^2), \tag{5.87}$$

and so is determined by terms quadratic in the fields. Thus a marginal variable varies only relatively slowly with ℓ. On following through an analysis in which marginal variables feature (see e.g. in Sec. 6) one finds there are various special things that can happen, which violate the simplest scaling precept. One of the typical effects is the appearance of logarithmic correction factors, such as $(\ln|t|)^{\tilde{\nu}}$, multiplying the usual critical power laws. The ability to identify and predict such departures from straight forward scaling represents one of the powers of the renormalization group approach.

5.5.4 Scaling of the free energy and hyperscaling

Let us now express the free energy in terms of the set of scaling fields g_1, g_2, g_3, etc. The flow equation for the free energy then takes on the simple asymptotic form

$$f(t, h, g_3, \ldots) \approx b^{-d\ell} f(b^{\ell\lambda_1}t, b^{\ell\lambda_2}h, \ldots, b^{\ell\lambda_j}g_j, \ldots), \tag{5.88}$$

where we have made use of the identifications $g_1 \equiv t$ and $g_2 \equiv h$.

Now we can make the previous choice for b or ℓ by setting

$$b^{\ell\lambda_1} = 1/t, \tag{5.89}$$

which yields the general scaling result

$$f(t,h, \ldots, g_j, \ldots) \approx t^{2-\alpha} f(1, \frac{h}{t^{\Delta}}, \ldots, \frac{g_j}{t^{\phi_j}}, \ldots), \tag{5.90}$$

where the standard thermodynamic exponents are given, as before, by

$$2-\alpha = d/\lambda_1 \quad \text{and} \quad \Delta = \lambda_2/\lambda_1, \tag{5.91}$$

while the "crossover exponent" for the scaling field g_j is given by

$$\phi_j = \lambda_j/\lambda_1. \tag{5.92}$$

Now if $\phi_j > 0$ for some j, the scaled combination g_j/t^{ϕ_j} becomes large as $t \to 0$ and so it clearly cannot be ignored: in other words g_j is another relevant variable and its presence will normally lead to crossover to different critical behavior (or, perhaps, to noncriticality as for t and h). On the other hand, when ϕ_k is negative one has

$$g_k/t^{\phi_k} \equiv g_k t^{\theta_k} \to 0 \text{ as } t \to 0, \tag{5.93}$$

and so g_k becomes inconsequential: it is an irrelevant variable. By expanding (5.90) in terms of the scaled combination g_k/t^{ϕ_k}, if this is allowed, we see that such irrelevant variables can contribute "corrections-to-scaling" i.e., correction factors to leading power laws of the form $[1 + c_j\, t^{\theta_j} + \ldots]$. At some slight risk therefore (in case the g_k enter in a "dangerous" way), one can thus discard all the irrelevant variables and worry only about the relevant ones. Finally, this justifies the postulate of asymptotic scaling near a critical point in terms of only a few important variables. For a standard critical point with only two relevant variables we thus recapture the scaling form

$$f(t,h,g_3,g_4, \ldots) \approx t^{2-\alpha}\, Y\!\left(\frac{h}{t^{\Delta}}\right), \tag{5.94}$$

in considerable generality.

It is worth mentioning that in addition to the singular corrections to this asymptotic scaling form which arise from the irrelevant variables and their exponents as factors $(1+ct^{\theta}+\ldots)$, one must always expect farther analytic corrections to scaling which will appear as t and h depart increasingly from criticality. At the most trivial level the "harmless" change from $t = (T-T_c)/T_c$ to $t' = 1-(T_c/T)$, which is often useful theoretically and experimentally, introduces such correction terms since one has $t' = t - t^2 + t^3 + \ldots$. More generally, on solving the recursion relations near a fixed point beyond linear order one finds that the scaling fields $t, h, \ldots, g_j, \ldots$ in the scaling relation (5.90) should, for greater accuracy, be replaced by non-linear scaling fields, $\tilde{t}, \tilde{h}, \ldots, \tilde{g}_j, \ldots$ which, in quadratic and higher order can couple the original fields together so that, for example, one has $\tilde{t} = t + a_1t^2 + a_2h^2 + a_3th^2 + \dots$, where symmetry may dictate that certain terms are absent although the coefficients a_i here are nonuniversal. Clearly further analytic corrections to asymptotic scaling arise from this source and can be significant in practice.

Finally, let us appeal to the locality assumption for the renormalization group and recall (5.63) to obtain, for $h = 0$,

$$\xi(t,0,\dots) \approx b\xi(b^{\ell\lambda_1}t,0,\dots) \approx t^{-1/\lambda_1}\xi(1,0,0,\dots), \tag{5.95}$$

where we have used (5.89) and allowed the irrelevant variables to go to zero. Comparing this with the definition $\xi(T) \sim t^{-\nu}$ as $t \to 0$ $(h=0)$ yields again (see (5.38)) the identification

$$\nu = 1/\lambda_1, \tag{5.96}$$

which may, with good reason, be regarded as the most fundamental of the renormalization group exponent relations. However, the provisos explained after (5.63) and in Sec. 5.4.3 must be borne in mind and, more properly, one should work with the correlation flow equation (5.69) or its analogue for nonquasilinear renormalization groups.

If we combine (5.96) with (5.91) we immediately obtain the <u>hyperscaling relation</u>, $d\nu = 2 - \alpha$, first introduced in Sec. 3.8 on heuristic grounds [see (3.57) to (3.59)]. From this and the previous d-independent scaling relations, follow all the other hyperscaling relations such as (3.60), which relates η and δ, (3.65) and (3.72). It is clear at this stage that hyperscaling is "built into" renormalization group theory in a rather intimate and deep way. Nevertheless hyperscaling <u>fails</u>, as mentioned previously, for classical or mean field theory (unless one has $d = 4$); but we have already seen evidence, most concretely through the exact results for spherical models, that the classical exponent values <u>are valid</u> for $d > 4$; furthermore, this is confirmed generally by the explicit renormalization group ε-expansion analysis presented below in Sec. 6! Thus we are faced with the paradox that hyperscaling seems to be predicted very generally by renormalization group analysis but, nonetheless, fails strongly for $d > 4$: this issue is discussed further in Appendix D where it is resolved in a consistent way in terms of the properties of <u>dangerous irrelevant variables</u>.

This is also an appropriate place to caution the reader that one can encounter, in the critical spectrum of operators, certain so-called <u>redundant operators</u>: these appear formally in the specification of the Hamitonian $\bar{\mathcal{H}}$ and its flow under renormalization but the associated scaling fields have <u>no effect</u> on the free energy or other observable properties! As discussed by Wegner,[20] redundant operators may be envisaged as describing, in a continuous spin system, for example, a mere change in the origin or scale of the spin variables (which, since all spins are eventually integrated out, cannot have an effect if all couplings, fields, etc. are changed in a covariant way as specified by the corresponding scaling field). In well-controlled practical calculations redundant operators do not normally cause problems and the reader will find little reference to them in the literature!

5.6 The construction of renormalization groups

The actual process of explicitly constructing a useful renormalization group is not trivial. We will only consider briefly a few particular renormalization groups, and then delve a little more deeply into one of them. A renormalization group typically involves going over from one set of local variables or spins to another set, $\{s\}_N \Rightarrow \{s\}_{N'}$. A rather general form for $\mathbb{R}$ can be expressed via

$e^{\bar{\mathcal{H}}\{s\}} \Rightarrow e^{\bar{\mathcal{H}}'\{s'\}}$, where the renormalized Boltzmann factor, which is what really matters, is defined by

$$e^{\bar{\mathcal{H}}'(s')} = \mathrm{Tr}_N^s \{\mathcal{R}_{N',N}(s',s)\, e^{\bar{\mathcal{H}}(s)}\}, \qquad (5.97)$$

in which the kernel $\mathcal{R}_{N',N}(s',s)$ has N original or unrenormalized variables, s, but a smaller number N' of renormalized variables s'. [The rescaling factor b is defined as usual via (5.61).] Now in order to meet the unitarity requirement (5.65), this kernel must satisfy the condition

$$\mathrm{Tr}_{N'}^{s'} \{\mathcal{R}_{N',N}(s's)\} = 1 \text{ for all s.} \qquad (5.98)$$

Two of the simplest and more fashionable renormalization groups can then be specified as follows:-

5.6.1 Kadanoff's block spin renormalization group

Kadanoff[21] was the first person to expose the intimate connection between the idea of a rescaled "block" or "cell" spin and the scaling properties of a critical point, thereby prefiguring Wilson's development of the general renormalization group approach. He was also the first to bring this particular approach to the point of being a practical computational scheme. (It might be mentioned, however, that the idea of block spins as a way of approaching critical phenomena had been proposed independently a year or so earlier by M. J. Buckingham[22] at one of the first conferences to explicitly recognize the unity and universality of diverse critical phenomena.) The simplest way of picturing Kadanoff's construction is to consider a

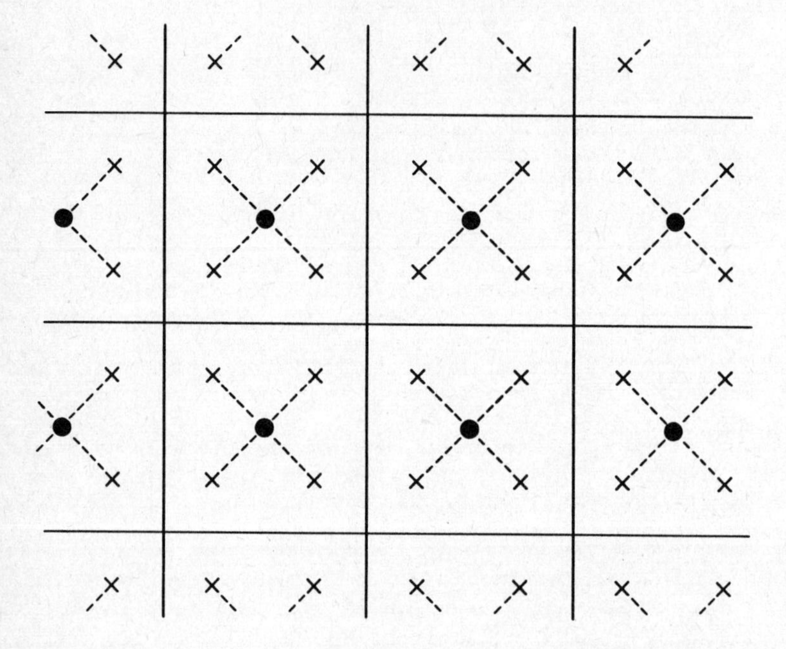

Fig. 5.7 An illustration of Kadanoff's block spin renormalization scheme for a square lattice Ising model. The original spins are denoted by crosses; the renormalized block spins are shown by solid circles. The spatial rescaling factor here is $b = 2$.

two-dimensional Ising model on a square lattice as illustrated in Fig. 5.7. The lattice is divided into blocks or cells each containing $2 \times 2 = 4$ spins. On renormalization each cell of spins is replaced by a single block or renormalized spin. Thus for this renormalization group we have $b = 2$. There are several algebraic ways in which this replacement can be effected.

One might, for example, imagine a pair coupling of strength J_0 between the block spins and the original spins in addition to the given couplings, say J_1, J_2, etc., between the original spins. A full trace is then taken over the original spins leaving a square lattice of block spins.

A crucial feature that appears directly one sums over the original spins in such a "real space" renormalization group for a lattice of dimensionality $d > 1$ is that couplings now apppear not only between first neighbor block spins but also between second neighbors, third neighbors, fourth neighbors, and so on. Worse in fact, since one is actually forced at the first step to go over to a space of Hamiltonians in which there are an infinite number of coupling constants not only between all pairs of spins but also between all triplets, all quartets, etc. Despite this inescapable complexity one can write down a formally exact expression for $\mathbb{R}$. In the simplest general case this is expressed by the kernel

$$\mathcal{R}_{N',N} = \prod_{\substack{\text{cells} \\ \underset{\sim}{x}'}}^{N'} \frac{1}{2} [1 + c s_{\underset{\sim}{x}'}' \sum_{\substack{\underset{\sim}{x} \text{ in} \\ \text{cell } \underset{\sim}{x}'}}^{b^d} s_{\underset{\sim}{x}}], \tag{5.99}$$

where c serves as a spin rescaling factor. The product runs over all the blocks or cells; the sum runs over all the b^d spins in a given cell. One can also check that this yields a quasilinear transformation (as discussed in Sec. 5.4.3).

Although (5.99) is a neat closed formula, it certainly does not mean that the problem is solved! In fact the best that can be done (unless d=1) in order to actually implement this renormalization group, i.e., to relate the new renormalized couplings explicitly to the old ones, is to invoke some approximation scheme. Unfortunately, the methods of approximation normally used entail truncating the number of interactions at each stage of renormalization to some finite number of more-or-less short range coupling terms. This is a fairly uncontrollable method of approximation, and really useful new results from this renormalization scheme have not been very plentiful.

5.6.2 Niemeijer and van Leeuwen's majority rule

This method is most frequently applied to plane triangular lattices but can be adapted to other geometries and dimensionalities. In the simplest case triangles of adjacent spins, s, are formed into blocks and associated with a block spin, s'. Now in any block of three Ising spins, s_1, s_2, and s_3 say, at least two will always be pointing in the same direction. The transformation rule then states that the corresponding block spin, s', points in the direction of the majority! The corresponding kernel can be written, with rescaling factor given by $b^d = 3$, as

$$\mathcal{R}_{N',N} = \prod_{\substack{\text{cells}\\ \underset{\sim}{x}'}} \frac{1}{2}\left[1 + c_0 s'_{\underset{\sim}{x}'} \sum_{\underset{\sim}{x}} s_{\underset{\sim}{x}} + c_1 s'_{\underset{\sim}{x}'} \prod_{\underset{\sim}{x}} s_{\underset{\sim}{x}}\right], \tag{5.100}$$

where the sum and second product run over the three spins in the cell $\underset{\sim}{x}'$; the coefficients are fixed by $c_0 = -c_1 = \frac{1}{2}$. It is easily seen that this transformation is not quasilinear so that it is not necessary to adjust c_0 or c_1 to achieve a condition such as (5.72) and (5.73). One may be tempted, however, to try different values for c_0 and c_1 and to try to optimize their choice in some way. However, the exact transformation again necessarily involves an infinite number of coupling constants so some sort of approximation scheme must be used. Many possibilities arise and it is hard to find a truly systematic procedure since no small parameter presents itself. Variational criteria for choosing the optimal c_0 and c_1 have been explored but they cannot be relied on to yield correct final results.

5.6.3 Wilson's momentum shell integration

This renormalization group, which is particularly important since it turns out to allow a systematic expansion procedure, is designed for or, perhaps more fairly, requires continuous spins, $\vec{s}_{\underset{\sim}{x}} \equiv (s^{\mu}_{\underset{\sim}{x}})$ with $-\infty < s^{\mu}_{\underset{\sim}{x}} < \infty$ as discussed in Sec. 4.7. If, for simplicity, one considers a d-dimensional hypercubic lattice of spacing a (i.e., a square lattice for d = 2, simple cubic for d = 3, etc.) one can introduce the associated Fourier transformed spin variables

$$\hat{s}_{\underset{\sim}{q}} = \sum_{\underset{\sim}{x}} e^{i\underset{\sim}{q}\cdot\underset{\sim}{x}} \vec{s}_{\underset{\sim}{x}}, \tag{5.101}$$

where the wave vector $\underset{\sim}{q}$ runs over the appropriate first Brillouin zone of the reciprocal space lattice: this can be expressed by

$$|q_x|,\ |q_y|,\ \ldots \leqslant q_\Lambda = \pi/a, \tag{5.102}$$

where q_Λ represents a momentum space cutoff which, of course, simply reflects the underlying lattice structure. The situation for a square lattice is illustrated in Fig. 5.8. For the original spins in real space one has, reciprocally,

$$\vec{s}_{\underset{\sim}{x}} = \frac{1}{N}\sum_{\underset{\sim}{q}} e^{-i\underset{\sim}{q}\cdot\underset{\sim}{x}} \hat{s}_{\underset{\sim}{q}}. \tag{5.103}$$

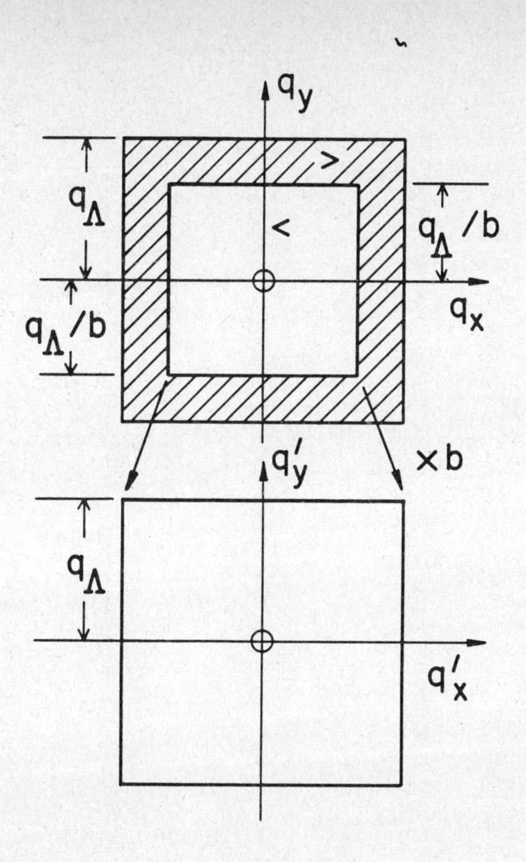

Fig. 5.8 Momentum space for a square lattice illustrating the construction of an inner zone, marked <, and an outer zone or shell marked > (upper part of figure). After integrating over spin variables with momenta in the shell, the inner zone is expanded by a factor b to form the new, renormalized Brillouin zone.

In this way any reduced Hamiltonian $\bar{\mathcal{H}}(\vec{s}_{\underset{\sim}{x}})$, expressed in terms of the real space (or lattice) spins, can be re-expressed precisely in terms of the Fourier spins, $\hat{s}_{\underset{\sim}{q}}$ as $\bar{\mathcal{H}}(\hat{s}_{\underset{\sim}{q}})$. Likewise the trace operation

$$\mathrm{Tr}_N^{s_{\underset{\sim}{x}}} = \prod_{\underset{\sim}{x}} \prod_{\mu=1}^{n} \int_{\infty}^{\infty} ds_{\underset{\sim}{x}}^{\mu}, \tag{5.104}$$

becomes simply

$$\mathrm{Tr}_N^{\hat{s}_{\underset{\sim}{q}}} = \prod_{\underset{\sim}{q}} \prod_{\mu=1} \int_{\infty}^{\infty} d\hat{s}_{\underset{\sim}{q}}^{\mu}. \tag{5.105}$$

Now, motivated by the idea that it is the low momentum or long wavelength fluctuations that are of most importance for critical phenomena, while the short wavelength, high momentum fluctuations are less crucial, Wilson divides the Brillouin zone into two regions as shown in Fig. 5.8. In the inner region, which we will indicate by a superscript $<$, the wavevectors $\underset{\sim}{q}$ satisfy

$$|q_x|, |q_y|, \ldots \leqslant q_{\Lambda}/b, \tag{5.106}$$

while the remaining, outer region constitutes a "momentum shell" of thickness $\Delta q = (1-b^{-1})\pi/a$ which can, if convenient, be chosen infinitesimal. Now the original Hamiltonian is a function of spin variables $\hat{s}_{\underset{\sim}{q}}$ with wavevectors distributed uniformly throughout the whole zone. Let us partition these into a set, $\{\hat{s}_{\underset{\sim}{q}}^{<}\}$, of all those $N' = N/b^d$ spins with wavevectors in the inner zone, and into the remaining set, $\{\hat{s}_{\underset{\sim}{q}}^{>}\}$, of the (N-N') spins in the shell. We can then write the reduced Hamiltonians as

$$\bar{\mathcal{H}}(\hat{s}_{\underset{\sim}{q}}) = \bar{\mathcal{H}}(\hat{s}_{\underset{\sim}{q}}^{<}, \hat{s}_{\underset{\sim}{q}}^{>}). \tag{5.107}$$

Evaluation of the partition function requires an integration over all these spins as implied by (5.105). Instead of doing this in one step, Wilson proposes that the integration be performed in stages, starting with an integration over only the spin variables $\{\hat{s}_q^{>}\}$ in the outer zone or shell. This procedure clearly embodies the physical idea that the high momentum variables play a smaller role in the critical behavior and hence may reasonably be eliminated first. (It should be stressed, however, that it is a serious over-simplification to assert that all the critical behavior occurs only at low momentum: this is not the case and is not assumed in the renormalization group approach which, on the contrary, allows properly for all contributions.)

This renormalization procedure yields a new Hamiltonian $\bar{\mathcal{H}}'$ given by

$$e^{\bar{\mathcal{H}}'(\hat{s}'_{\underset{\sim}{q}'})} = \mathrm{Tr}^{\hat{s}^{>}_{\underset{\sim}{q}}}_{N,N'} \{ e^{\bar{\mathcal{H}}(\hat{s}^{<}_{\underset{\sim}{q}}, \hat{s}^{>}_{\underset{\sim}{q}})} \}. \tag{5.108}$$

In this expression we have also allowed for spin and spatial rescaling. The latter proceeds simply in accord with (5.63) and (5.64). As illustrated in Fig. 5.8, the rescaling

$$\underset{\sim}{q} \Rightarrow \underset{\sim}{q}' = b\underset{\sim}{q}, \tag{5.109}$$

of the wavevectors corresponds to an expansion of the inner region of the original Brillouin zone to fill out the new, renormalized zone back to the size of the original zone.

A spin rescaling is needed since it is not hard to see, by examining the transformation of the Fourier space spin correlation functions $\hat{G}(\underset{\sim}{q}) = \langle \hat{s}_{\underset{\sim}{q}} \hat{s}_{-\underset{\sim}{q}} \rangle$, that the renormalization group defined by (5.108) is quasilinear. Accordingly, the renormalized spins are defined via

$$\hat{s}_{\underset{\sim}{q}} \Rightarrow \hat{s}'_{\underset{\sim}{q}'} = \hat{s}_{\underset{\sim}{q}}/\hat{c}, \tag{5.110}$$

where, in comparing with (5.67), we have the relation

$$\hat{c} = b^{d} c. \tag{5.111}$$

It follows by the previous arguments that at a fixed point the Fourier spin rescaling factor is related to the exponent η via

$$\hat{c}^{*} = b^{(d + 2 - \eta)/2}. \tag{5.112}$$

Of course, other critical exponents must come from an analysis of the fixed point spectrum.

Naturally one cannot, in general, implement this momentum shell transformation exactly. Nor can one be necessarily assured of smoothness, locality and aptness. However, in the same way that the one-dimensional Ising model can be treated exactly by the decimation or a block spin renormalization group - thus constituting an analogue to the quantum mechanical "particle in the box" problem - so can Gaussian models, as described in Sec. 4.7, be treated exactly by the momentum shell transformation. One might, indeed, regard the Gaussian model as the "hydrogen atom" of critical phenomena: unfortunately, however, in itself it is of distinctly less <u>direct</u> physical relevance than the hydrogen atom. Even so, as we shall show in the next section, a solution of the Gaussian model via the momentum shell renormalization group provides a foundation on which can be built a systematic expansion procedure for solving more realistic and challenging models!

6. Dimensionality Expansions

On the face of it, the Wilson momentum space renormalization group seems to suffer from the same afflictions as the previously-described real-space renormalization groups; but it turns out to have an overwhelming advantage in that the unavoidable approximations can now be made in a systematic and controlled way. It has thus proved possible to make many useful, novel, and incisive calculations with the momentum shell integration technique and, even in rather low orders of approximation, quite accurate numerical results have been obtained.

As explained, the momentum shell renormalization group requires the use of continuous spins with, say, n-components so that $\vec{s}_{\underset{\sim}{x}} = (s_{\underset{\sim}{x}}^{(\mu)})_{\mu=1,2,\ldots,n}$ with $-\infty < s^{(\mu)} < \infty$. At first sight this precludes its application to discrete spin systems, like the spin 1/2 Ising model, or to systems with fixed length spins like the classical Heisenberg model with, say, $|\vec{s}_{\underset{\sim}{x}}| = 1$. However, this view proves too naive since, via a Kac-Hubbard-Stratonovich transformation, such models can be transformed <u>exactly</u> into thermodynamically equivalent continuous spin models with definite spin weighting functions of the general sort discussed in Sec. 4.7. How this works is explained in Appendix A. Here we will assume that a continuous spin model is given and we start by transforming it into a Fourier space representation suitable for application of the momentum shell procedure.

6.1 Transformation of the Hamiltonian

Following (4.44) we consider the total reduced Hamiltonian expressed in real space variables as

$$\bar{\mathcal{H}} = -\mathcal{H}_{int}(\vec{s}_{\underset{\sim}{x}})/k_B T - \sum_{\underset{\sim}{x}} w(\vec{s}_{\underset{\sim}{x}}), \tag{6.1}$$

where the interaction Hamiltonian (or, for true spins, the "exchange" Hamiltonian) is given by

$$\mathcal{H}_{int}(\vec{s}_{\underset{\sim}{x}}) = -\frac{1}{2}\sum_{\underset{\sim}{x}_1}\sum_{\underset{\sim}{x}_2} J(\underset{\sim}{x}_1-\underset{\sim}{x}_2)\, \vec{s}_{\underset{\sim}{x}_1}\cdot\vec{s}_{\underset{\sim}{x}_2}, \tag{6.2}$$

while the site vectors $\underset{\sim}{x}$ range over a d-dimensional hyper-cubic lattice of spacing $\underline{a}$. The single-spin weighting function is expanded as

$$w(\vec{s}) = \frac{1}{2}|\vec{s}|^2 + \tilde{u}|\vec{s}|^4 + \tilde{v}|\vec{s}|^6 + \ldots . \tag{6.3}$$

It is worth recalling at this point that, as mentioned in Sec. 4.7 and demonstrated in Appendix A, the discrete variable spin $\frac{1}{2}$ Ising model and the classical, fixed-length, n-vector models can all be cast <u>exactly</u> in the form of continuous spin models as considered here with no approximation. Now we introduce momentum space variables $\hat{s}_{\underset{\sim}{q}}$ via (5.103) with the inverse relation (5.101). It is then straightforward to transform (6.1) into the form

$$\bar{\mathcal{H}} = -\frac{1}{2}\frac{1}{N}\sum_{\underset{\sim}{q}}[1-\hat{K}(\underset{\sim}{q})]\hat{s}_{\underset{\sim}{q}}\cdot\hat{s}_{-\underset{\sim}{q}}$$

$$-\tilde{u}\frac{1}{N^3}\sum_{\underset{\sim}{q}_1}\sum_{\underset{\sim}{q}_2}\sum_{\underset{\sim}{q}_3}(\hat{s}_{\underset{\sim}{q}_1}\cdot\hat{s}_{\underset{\sim}{q}_2})(\hat{s}_{\underset{\sim}{q}_3}\cdot\hat{s}_{\underset{\sim}{q}_4}) + \dots , \tag{6.4}$$

where the wave vectors appearing in the multiple sums are restricted by

$$\underset{\sim}{q}_1 + \underset{\sim}{q}_2 + \underset{\sim}{q}_3 + \underset{\sim}{q}_4 = 0, \underset{\sim}{Q}, \tag{6.5}$$

for the fourth order term, and similarly in higher orders, where $\underset{\sim}{Q}$ is any reciprocal lattice vector. The interactions now appear via the Fourier transform

$$1 - \hat{K}(\underset{\sim}{q}) = 1 - \sum_{\underset{\sim}{x}} e^{i\underset{\sim}{q}\cdot\underset{\sim}{x}}\,\frac{J(\underset{\sim}{x})}{k_BT}. \tag{6.6}$$

If the couplings are of reasonably short range it is possible to expand $\hat{K}(\underset{\sim}{q})$ in a power series in q in which, for symmetry reasons the linear term vanishes and the quadratic term is proportional to $|\underset{\sim}{q}|^2$. The result can be written in the form

$$1 - \hat{K}(\underset{\sim}{q}) = \frac{T-T_0}{T} + \frac{\hat{J}(0)}{k_BT}R_0^2q^2 + O(\underset{\sim}{q}^4), \tag{6.7}$$

where the mean field critical temperature, T_0, has been introduced via

$$k_BT_0 = \hat{J}(0) = \sum_{\underset{\sim}{x}} J(\underset{\sim}{x}), \tag{6.8}$$

while R_0 measures the range of the interactions.

Of course we will be interested in the thermodynamic limit $N \to \infty$. The wavevector sums then become integrals and, to simplify formulae, we will employ the shorthand notation

$$\frac{a^{-d}}{N}\sum_{\underset{\sim}{q}} \quad \int\frac{d^dq}{(2\pi)^d} \equiv \int_{\underset{\sim}{q}}. \tag{6.9}$$

A rescaling of the spin variables by the substitution

$$\hat{s}_{\underset{\sim}{q}} = \vec{\sigma}_{\underset{\sim}{q}}\left(\frac{T}{T_0}\right)^{1/2}\frac{1}{R_0a^{d/2}}, \tag{6.10}$$

transforms $\bar{\mathcal{H}}$ into the more convenient standard form

$$\bar{\mathcal{H}} \simeq -\frac{1}{2}\int_{\underset{\sim}{q}}(r + eq^2)\,\vec{\sigma}_{\underset{\sim}{q}}\cdot\vec{\sigma}_{-\underset{\sim}{q}} \tag{6.11}$$

$$-u\int_{\underset{\sim}{q}_1}\int_{\underset{\sim}{q}_2}\int_{\underset{\sim}{q}_3}\sum_{\mu,\nu=1}^{n}\sigma^\mu_{\underset{\sim}{q}_1}\sigma^\mu_{\underset{\sim}{q}_2}\sigma^\nu_{\underset{\sim}{q}_3}\sigma^\nu_{\underset{\sim}{q}_4},$$

with

$$g_1 + g_2 + g_3 + g_4 = 0, \tag{6.12}$$

which is known as the Landau-Ginzburg-Wilson (LGW) reduced Hamiltonian or as a field-theoretic Hamiltonian (or "action"). The coefficient e has been introduced in the quadratic term but, at this stage, it is simply equal to unity. The leading coefficient, r, now stands in for the temperature since one has

$$r = \frac{T-T_0}{T_0 R_0^2} = \frac{t_0}{R_0^2}, \tag{6.13}$$

where t_0 measures the deviation from mean field criticality. Finally, the coefficient of the fourth order term becomes

$$u = \tilde{u}\left(\frac{T}{T_0}\right)^2 \frac{a^{d-4}}{(R_0/a)^4}, \tag{6.14}$$

which reveals, for the first time, how the deviation in dimensionality

$$\varepsilon = 4 - d, \tag{6.15}$$

arises naturally. Note that "umklapp" processes, with $\underset{\sim}{Q} \neq 0$, have been ignored, q^4, q^6, ... terms have been dropped and sixth and higher order terms in $\sigma_{\underset{\sim}{q}}$ have been neglected: in the end one can (and should) return to check that all of these contributions represent irrelevant variables in the domain of interest. As usual, $\bar{\mathcal{H}}$ can be regarded as a point, (r,e,u,v, ...) in the space of Hamiltonians where v represents the coefficient of the sixth order terms and so on.

6.2 Computing with continuous spins

Since we have continuous spin variables, computing the trace of the Boltzmann factor involves multiple integrals over the spins as indicated in (5.104) and (5.105). The general Hamiltonian may be written

$$\bar{\mathcal{H}} = \bar{\mathcal{H}}_2 - u\bar{\mathcal{H}}_4 - v\bar{\mathcal{H}}_6 + \dots, \tag{6.16}$$

where the first term is quadratic in the $s_{\underset{\sim}{x}}$ (and, hence, in the $\hat{s}_{\underset{\sim}{q}}$ and $\vec{\sigma}_{\underset{\sim}{q}}$), while the second is quartic, and so on. If the higher order terms could be dropped, leaving only a quadratic or "free-field" Hamiltonian, the Boltzmann factor $\exp(\bar{\mathcal{H}}_2)$ would decompose into a product of Gaussian functions of the individual $\vec{\sigma}_{\underset{\sim}{q}}$ variables. The trace integrations would then be trivial! The most obvious way of handling the higher order terms is thus to treat them as a perturbation and to attempt an expansion of the free energy in powers of u, v, etc. Confining ourselves to the quartic term we would then expand as

$$e^{\bar{\mathcal{H}}_2 + u\bar{\mathcal{H}}_4} = e^{\bar{\mathcal{H}}_2}\left(1 - u\,\bar{\mathcal{H}}_4 + \frac{1}{2!}u^2\,\bar{\mathcal{H}}_4^{\,2} + \ldots\right), \tag{6.17}$$

which represents a Gaussian function times polynomials in the $\vec{\sigma}_{q}$. One is now confronted by various combinations of products of integrals, all of the same basic type, namely,

$$\int_{-\infty}^{\infty} d\sigma_{q}^{\mu}\, e^{-\frac{1}{2}(r + eq^2)|\sigma_{q}^{\mu}|^2}\, |\sigma_{q}^{\mu}|^k = \frac{I_k}{(r+eq^2)^{(k+1)/2}}, \tag{6.18}$$

which integrate out as shown, I_k being a constant which vanishes for k odd.

Since the wavevectors, q, form a quasicontinuum, the products referred to become infinite products in the thermodynamic limit and the process of taking the logarithm of the overall trace to obtain an expression for the free energy, thus yields momentum integrals of the form

$$\mathcal{J}_k(d) = \int \frac{d^d q}{(r+q^2)^{k/2}}. \tag{6.19}$$

If we procede straight ahead in a perturbation theoretic spirit we now confront a major problem, namely, the so-called <u>infra-red divergences</u>. In the absence of the perturbation we have a Gaussian model which becomes critical as $r \to 0$. With the quartic term present we actualy expect the critical point to be depressed to <u>negative</u> r but, in any case, if we want to study the critical region we must at least consider $r \to 0$. However, in that limit all the integrals will diverge for large enough k owing to the singularity of the integrand as $q \to 0$ (whence the terminology "infra-red"). Specifically, counting powers of momentum shows that $\mathcal{J}_k(d)$ diverges as $k \to 0$ whenever $d \leqslant k$. Since even the leading term in (6.17) involves $k = 4$ we see that the naive perturbation method fails immediately for $d \leqslant 4$!

The Wilson approach circumvents this basic problem by never actually integrating over momenta beneath the reduced cutoff q_Λ/b (see Fig. 5.8). Thus no divergences are encountered.

6.3 Implementation of momentum shell renormalization

In order to implement Wilson's momentum shell renormalization group in a perturbative manner we split the original Hamiltonian (6.11) in the form

$$\bar{\mathcal{H}} = \bar{\mathcal{H}}^{<} + \bar{\mathcal{H}}_2^{>} - u\bar{\mathcal{H}}_4^{>}, \tag{6.20}$$

where $\bar{\mathcal{H}}^{<}$ includes all those parts of $\bar{\mathcal{H}}$ which contain only spins, $\vec{\sigma}^{<}_{\underset{\sim}{q}}$, $\bar{\mathcal{H}}^{>}_{2}$ is the Gaussian, free-field or quadratic part of the total Hamiltonian with spins $\vec{\sigma}^{>}_{\underset{\sim}{q}}$ with momenta lying in the outer shell (see Fig. 5.8). Finally $u\bar{\mathcal{H}}_4$ consists of all the remaining terms which involve spins $\vec{\sigma}^{>}_{\underset{\sim}{q}}$ (as well as, in general, some $\vec{\sigma}^{<}_{\underset{\sim}{q}}$).

The trace operation

$$\mathrm{Tr}^{\vec{\sigma}^{>}}_{N-N'}\{e^{\bar{\mathcal{H}}(\vec{\sigma}^{<},\,\vec{\sigma}^{>})}\}, \tag{6.21}$$

which we want to carry out, can be expressed conveniently in a perturbation series if we make use of the notation

$$\langle X\rangle_{>} = \frac{\mathrm{Tr}^{\vec{\sigma}^{>}}_{N-N'}\{Xe^{\bar{\mathcal{H}}^{>}_{2}}\}}{\mathrm{Tr}^{\vec{\sigma}^{>}}_{N-N'}\{e^{\bar{\mathcal{H}}^{>}_{2}}\}}, \tag{6.22}$$

which represents averaging with the free-field Hamiltonian over only the higher momentum fluctuations, i.e., those with $\underset{\sim}{q}$ in the shell. Now the renormalized Hamiltonian can be written

$$\bar{\mathcal{H}}' = \left[\ln\left(\mathrm{Tr}^{>}\{e^{\bar{\mathcal{H}}^{<}}\, e^{\bar{\mathcal{H}}^{>}_{2}}\, e^{-u\bar{\mathcal{H}}^{>}_{4}}\}\right)\right]_{\vec{\sigma}_{\underset{\sim}{q}}\Rightarrow\vec{\sigma}'_{\underset{\sim}{q}'}}. \tag{6.23}$$

Notice first that the factor $e^{\bar{\mathcal{H}}^{<}}$ commutes with the trace operation since it involves only spins $\vec{\sigma}^{<}_{\underset{\sim}{q}}$. Then, using the notation (6.22), one readily establishes the expansion

$$\begin{aligned}
\bar{\mathcal{H}}' = [\,&\bar{\mathcal{H}}^{<} + \ln(\mathrm{Tr}^{>}\{e^{\bar{\mathcal{H}}^{>}_{2}}\}) \\
&+ \ln(1 - u\langle\bar{\mathcal{H}}^{>}_{4}\rangle_{>} + \tfrac{1}{2}u^{2}\langle(\bar{\mathcal{H}}^{>}_{4})^{2}\rangle_{>} + \ldots)]_{\vec{\sigma}_{\underset{\sim}{q}}\Rightarrow\vec{\sigma}'_{\underset{\sim}{q}'}}, \\
= [\,&\bar{\mathcal{H}}^{<} + \ln(\mathrm{Tr}^{>}\{e^{\bar{\mathcal{H}}^{>}_{2}}\}) - u\langle\bar{\mathcal{H}}^{>}_{4}\rangle_{>} \\
&+ \tfrac{1}{2}u^{2}(\langle(\bar{\mathcal{H}}^{>}_{4})^{2}\rangle_{>} - \langle\bar{\mathcal{H}}^{>}_{4}\rangle^{2}_{>}) + 0(u^{3})]_{\vec{\sigma}_{\underset{\sim}{q}}\Rightarrow\vec{\sigma}'_{\underset{\sim}{q}'}}
\end{aligned} \tag{6.24}$$

which we have performed to order u^2. We may now set about calculating more explicitly the recursion relations

$$r' = \mathcal{R}_r(r,e,u, \ldots),$$
$$e' = \mathcal{R}_e(r,e,u, \ldots), \qquad (6.25)$$
$$u' = \mathcal{R}_u(r,e,u, \ldots),$$

to successive powers of u. (We will neglect the constant term since it cannot enter the recursion relations for r, e, u, etc. and cannot play a role in determining critical exponents. However, it would be needed for studying the full free energy.) The details of the derivation are presented more fully in Appendix B. In the lowest orders one finds just

$$r' = \hat{c}^2 b^{-d}[r + 2u \; (\quad) + O\;(u^2)], \qquad (6.26)$$

$$e' = \hat{c}^2 b^{-d-2}[e + 2u \; (\quad) + O\;(u^2)], \qquad (6.27)$$

$$u' = \hat{c}^4 b^{-3d} u \; [1 - \tfrac{1}{2} u \; (\quad) + O\;(u^2)]. \qquad (6.28)$$

The origin of the factors $\hat{c}$ here is easy to understand. The spin rescaling (5.110) introduces a factor of $\hat{c}$ for each unrenormalized spin component $\sigma^\mu_{\underset{\sim}{q}}$, and so r' and e', which are associated with the $\sigma^\mu_{\underset{\sim}{q}} \sigma^\mu_{-\underset{\sim}{q}}$ terms in $\bar{\mathcal{H}}'$, acquire factors of $\hat{c}^2$. Likewise u, which is associated with the quartic term, acquires a factor $\hat{c}^4$. The factors of b come from the spatial rescaling (5.109). Since the momentum integrals in (6.11) transform as

$$\int_{\underset{\sim}{q}} (r + eq^2) \times \ldots \Rightarrow \int b^{-d} \frac{d^d q'}{(2\pi)^d} \; [\; r + e\;(q'/b)^2] \times \ldots \;, \qquad (6.29)$$

the expression for r' acquires a factor of b^{-d}, and that for e' a factor b^{-d-2}. Similarly, since there are integrations over three different momenta involved in the quartic spin term, a factor of b^{-3d} enters for u. The reader should check these statements carefully: although they involve only dimensional analysis they turn out to be a most crucial ingredient!

Now it is clear that the choice of an overall scale factor remains at our disposal [as used in writing (6.10)]. This freedom can be used to fix one of the parameters: following (6.11) we will choose to maintain the constraint (or normalization)

$$e' = e = 1. \qquad (6.30)$$

The reason for this choice is that the q^2 term (or, in real space, the gradient squared term) is the one that sets the physical length scales. It follows that the spin rescaling factors are determined (to leading order) by

$$\hat{c}^2 = b^{d+2} \text{ and } c^2 = \hat{c}^2/b^{2d} = b^{-d+2}. \qquad (6.31)$$

If we find a nontrivial fixed point under these conditions it must, via (5.112) or (5.73), mean that

$$\eta = 0 + O(u). \qquad (6.32)$$

The other two recursion relations then give us, in leading i.e., zeroth order, the results

$$r' = b^2 r, \qquad (6.33)$$

$$u' = b^{4-d}\, u = b^{\varepsilon} u. \qquad (6.34)$$

The appearance of the factor b^{ε} in this last equation is a vital feature. Note that it simply reflects the <u>canonical dimensions</u> of u (in terms of lengths) as revealed in (6.14).

6.4 The Gaussian fixed point

The only fixed point that exists in the zeroth order approximation developed above is the Gaussian fixed point given by

$$r_G^* = 0, \; u_G^* = 0. \qquad (6.35)$$

With u = 0 the recursion relations are now diagonal as they stand, and the Gaussian eigenvalues are evidently

$$\Lambda_1 \equiv \Lambda_r = b^2, \text{ so } \lambda_1 = \lambda_{\mathcal{E}} = 2, \qquad (6.36)$$

and

$$\Lambda_u = b^{\varepsilon}, \text{ so } \quad \lambda_u = \varepsilon. \qquad (6.37)$$

We see that the parameter u changes from being irrelevant at large d to being relevant at the <u>border-line dimensionality</u> $\varepsilon = 0$, i.e., $d = 4$. Thus for $d > 4$ the quartic spin terms prove to be <u>irrelevant</u>, and $u \to 0$ under renormalization. The first, relevant eigenvalue must clearly be identified as the <u>thermal</u> eigenvalue. Through (5.38) or (5.96) we thus find

$$\nu = \frac{1}{2}, \qquad (d > 4), \tag{6.38}$$

which is the classical value! At $d = 4$ we see that u is marginal and, to this order, does not shift under renormalization. Finally, u is relevant for $d < 4$ and flows away from the Gaussian fixed point. The crucial question is: "Where to?"

Before answering that question, for which we must study the recursion relations in higher order, note that we may still consider the full scaling form (5.90) for the free energy around the Gaussian fixed point. Evidently, crossover away from Gaussian) critical behavior (for which all other exponents also prove to be classical) is controlled by the scaled combination

$$u/r^{\phi_u} \quad \text{with } \phi_u = \frac{\lambda_u}{\lambda_1} = \frac{1}{2}\varepsilon. \tag{6.39}$$

But notice now, from (6.13) and (6.14), that both r and u depend inversely on the range of the forces R_0. It follows that the range enters in the combination $a^d/R_0{}^d t_0{}^{\varepsilon/2}$. Hence, if R_0/a is large the Gaussian fixed point should describe the critical point (which will then look classical) until $R_0 t^{\varepsilon/2d}/a$ becomes small. Since the exponent here is comparatively small (being 1/6 for $d = 3$) the crossover to nonclassical behavior may take place rather slowly. The exponent we have found[23] for the long range crossover agrees with that following from the Ginzburg criterion for the validity of classical theory.[24-27] Its small value serves to explain, for example, why the BCS theory of superconductivity, which is a classical or mean field theory, works so well in practice; the ratio R_0/a is there measured by T_F/T_c, where T_F is the Fermi temperature and T_c is the superconducting transition temperature. (Of course the BCS theory is quantum-mechanical in nature: the word "classical" here, as elsewhere, refers only to the neglect of fluctuations in the statistical mechanical treatment.)

6.5 The renormalization group to order ε

If we are to obtain useful results for $d < 4$, the unstable flow from the Gaussian fixed point must, for at least one direction of flow, terminate at some new nontrivial fixed point. Since (6.34), the zeroth order recursion relation for u, is linear this is impossible unless we carry the perturbation calculation explicitly to at least the next order. To do this, a diagrammatic formulation, modeled on field theory, is helpful as sketched in Appendix B: the requisite analysis serves to fill the blanks in (6.26) to (6.28) and yields,

$$r' = \hat{c}^2 b^{-d}\left[r + 4u\,(n+2)\int_{\underset{\sim}{q}}^{>} \frac{1}{r+q^2} + O\,(u^2)\right], \tag{6.40}$$

$$e' = \hat{c}^2 b^{-d-2}\ [1 + 0 + 0\ (u^2)], \tag{6.41}$$

and

$$u' = \hat{c}^4 b^{-3d} u\ [1 - \frac{1}{2}\ u.8(n+8)\int_{\underset{\sim}{q}}^{>} \frac{1}{(r+q^2)^2} + 0\ (u^2)]. \tag{6.42}$$

We see from the second relation that in order to maintain $e' = e = 1$, we must set

$$\hat{c}^2 = b^{d+2+0(u^2)}, \tag{6.43}$$

This implies that if there is a nontrivial fixed point at some $u = u \neq 0$, then the critical point decay exponent satisfies

$$\eta = 0(u^{*2}). \tag{6.44}$$

Although this is not an explicit formula it gives us some understanding of why η is so small in most physical systems relative to the deviations of other exponents from their classical values.

Now if we substitute with (6.43) in (6.42) the prefactor b^{ε} appears again. Let us, then, invoke the idea of <u>continuous dimensionality</u> and enquire as to what happens if $\varepsilon = 4 - d$ is small! First we can write

$$b^{\varepsilon} = 1 + \varepsilon \ln b + 0(\varepsilon^2). \tag{6.45}$$

Then, keeping only terms linear in ε and u, the recursion relation for u can be rewritten as

$$u' - u = u[\varepsilon \ln b - 4\ (n+8)\ u \int_{\underset{\sim}{q}}^{>} \frac{1}{(r+q^2)^2}], \tag{6.46}$$

where the integral is now to be evaluated <u>at</u> $d = 4$. Evidently this recursion relation has a new fixed point at $u = u^* \propto \varepsilon/(n+8)$. Since, by supposition, ε is small, u^* is also small and therefore we can neglect the $0(u^2)$ corrections since, in the neighborhood of this fixed point they will be of order ε^2. This is the crux of the ε-expansion idea: by expanding in powers of ε we may utilize the field-theoretic perturbation theory in powers of u in a systematic way. We rely on the renormalization group framework since although the initial, physical value of u may well be 'large', i.e., of order unity, the flow of the critical trajectory to the fixed point allows us to calculate only for u small, of order ε.

Clearly the integrals require a little further thought. Note, first, that if we approximate the momentum shell by a hyper<u>sphere</u> (instead of using a hyper<u>cube</u>) we have

$$\int_{\underset{\sim}{q}}^{>} \equiv \int_{q_\Lambda/b}^{q_\Lambda} \hat{C}_d q^{d-1} dq, \tag{6.47}$$

where the area of the unit hypersphere is, as before,

$$C_d = (2\pi)^d \hat{C}_d = 2\pi^{d/2}/\Gamma(\tfrac{1}{2}d). \tag{6.48}$$

Since the gamma function is an analytic function of its argument, we certainly can see that any spherically symmetric integrals, such as involved here, can be extended to continuous dimension. (As indicated in Appendix C one can even extend hypercubic lattices to continuous d.) Accepting the hyperspherical approximation we can perform both the needed integrals in the critical region, i.e., for small r. One finds

$$\int_{\underset{\sim}{q}}^{>} \frac{1}{(r+q^2)^2} = \int_{q_\Lambda/b}^{q_\Lambda} \frac{1}{8\pi^2} \frac{q^3 dq}{q^4} + O(\varepsilon, r),$$

$$= K_2 \ln b + O(\varepsilon, r), \tag{6.49}$$

with $K_2 = 1/8\pi^2$ and, similarly,

$$\int_{\underset{\sim}{q}}^{>} \frac{1}{(r+q^2)} = K_1(1-b^{-2}) - K_2\, r \ln b + O(\varepsilon, r^2). \tag{6.50}$$

The constant K_1, has the value $q_\Lambda{}^2/16\pi^2$ in the spherical approximation but its actual value proves to be immaterial as regards all universal quantities; conversely the value of K_2 in these two equations is <u>independent</u> of the approximation, as is easily seen by more careful analysis.

We can now write the recursion relations correct to relative order ε, u and ur as

$$r' = b^2 r\, [1-4(n+2)K_2 u \ln b] + 4(n+2)K_1(b^2-1)u, \tag{6.51}$$

$$u' = u + u \ln b\, [\varepsilon - 4(n+8)K_2 u]. \tag{6.52}$$

Before analyzing these relations it is worth noting that they can be cast in differential form, as in (5.75) to (5.77), by putting

$$b = e^{\delta\ell} = 1 + \delta\ell + O(\delta\ell^2), \tag{6.53}$$

and taking the limit $\delta\ell \to 0$. Thus one obtains

$$\frac{dr}{d\ell} = [2-4(n+2)K_2u]r + 8(n+2)K_1u, \tag{6.54}$$

$$\frac{du}{d\ell} = u[\varepsilon - 4(n+8)K_2u]. \tag{6.55}$$

6.6 The n-vector fixed point

Now we can investigate the new fixed point.[28] From (6.52) or (6.55) we have

$$u^* = \frac{\varepsilon}{4(n+8)K_2} = \frac{2\pi^2\varepsilon}{(n+8)}, \tag{6.56}$$

which is, of course, only correct to order ε. Then from (6.51) or (6.54) we find

$$r^* = -\frac{K_1}{K_2}\left(\frac{n+2}{n+8}\right)\varepsilon, \tag{6.57}$$

where the ratio K_1/K_2 has the nonuniversal value $\frac{1}{2}q_\Lambda^2$ in the spherical approximation. Evidently this new fixed point "breaks off" from the Gaussian fixed point as the dimensionality falls below the borderline $d = 4$. (Actually, it exists also for $d > 4$ but at negative u^* where it is unstable and plays no role since $u \geq 0$ is needed if the partition function is to be well-defined, at least in the absence of any higher order stabilizing terms.) It is easy to determine the flows in the (r,u) plane near the fixed points. Their appearance for $\varepsilon > 0$, i.e., $d < 4$ is shown in Fig. 6.1.

To determine the critical exponents and test the new fixed point for its stability we must linearize about (r^*, u^*). If we work with the discrete recursion relations (6.51) and (6.52), we may write

$$\Delta r = r - r^* \text{ and } \Delta u = u - u^*, \tag{6.58}$$

and so obtain the matrix form

$$\begin{bmatrix}\Delta r\\ \Delta u\end{bmatrix} = \begin{bmatrix} b^2(1-\frac{n+2}{n+8}\varepsilon\ \ln b) & 4(n+2)K_1(b^2-1)\\ 0 & 1-\varepsilon\ \ln b\end{bmatrix}\begin{bmatrix}\Delta r\\ \Delta u\end{bmatrix}. \tag{6.59}$$

Note that K_2 has cancelled out! The eigenvalues follows at once as

$$\Lambda_1 \approx b^2\left(1-\frac{n+2}{n+8}\varepsilon\ \ln b\right) \approx b^{2+\frac{n+2}{n+8}\varepsilon}, \tag{6.60}$$

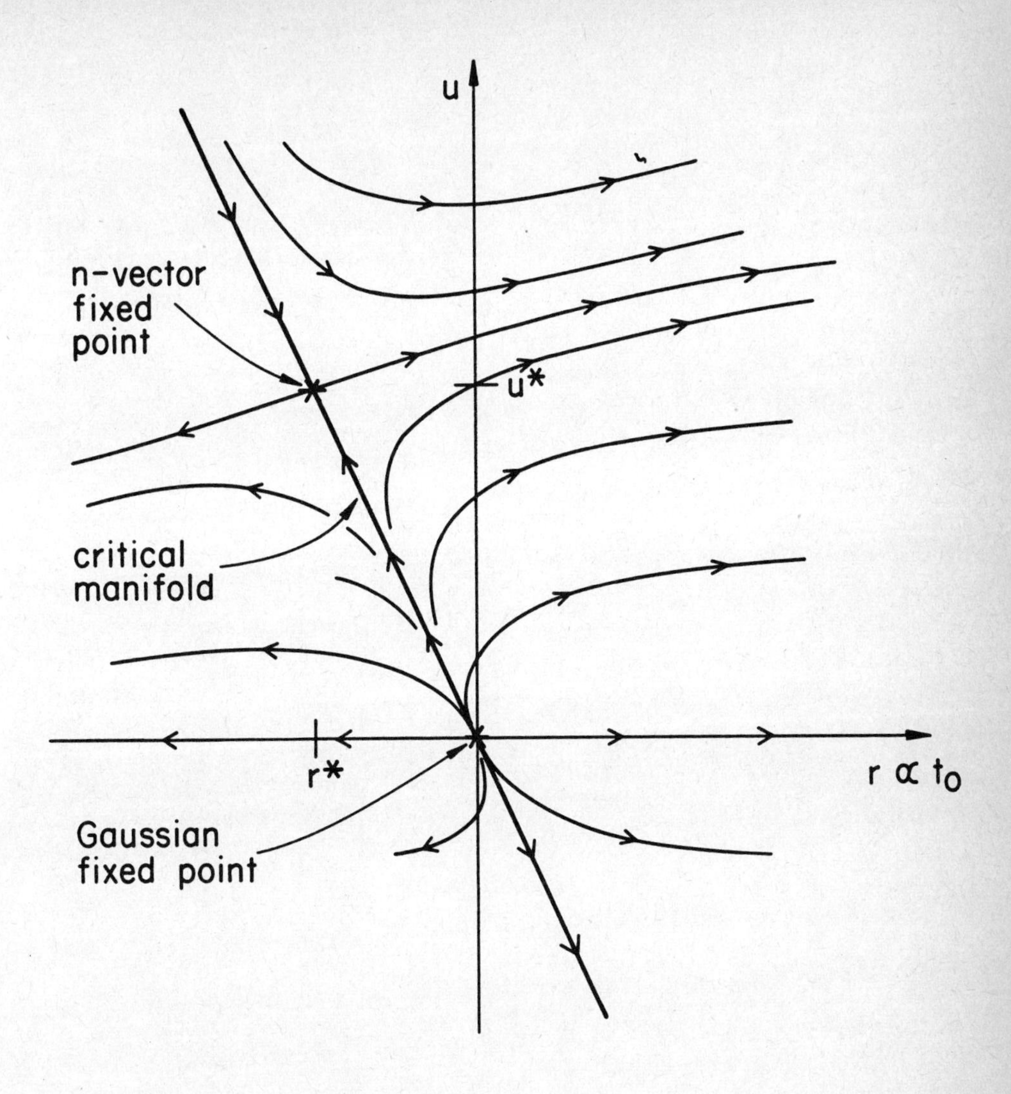

Fig. 6.1 Sketch of the renormalization group flows in the (r,u) plane for small $\varepsilon = 4 - d$. Note that the critical manifold (or trajectory) is straight only to order ε.

$$\Lambda_u \approx 1 - \varepsilon \ln b \approx b^{-\varepsilon}, \tag{6.61}$$

so that we have

$$\lambda_1 \equiv \lambda_{\mathcal{E}} = 2 - \frac{n+2}{n+8}\varepsilon + O(\varepsilon^2), \tag{6.62}$$

$$\lambda_u = -\varepsilon + O(\varepsilon^2). \tag{6.63}$$

From the last result we see that u represents an <u>irrelevant</u> variable about the new fixed point when $\varepsilon > 0$: in other words this fixed point is "stable" when $d < 4$ and hence "controls" the flow in place of the Gaussian fixed point (see Fig. 6.1) which, as we found, is now unstable. [Strictly we should say "stable (or unstable) <u>on</u> the critical manifold" but the restriction is left unstated in practical terminology since the expected, relevant, unstable directions are always understood.)

Finally, we may use the renormalization group eigenvalues to compute the critical exponents. Thus for the correlation length we find from (5.96)

$$\nu = \frac{1}{\lambda_1} = \frac{1}{2} + \frac{n+2}{4(n+8)}\varepsilon + O(\varepsilon^2). \tag{6.64}$$

If we recall our result for the correlation decay exponent, namely,

$$\eta = O(u^{*2}) = 0 + O(\varepsilon^2), \tag{6.65}$$

we may use a scaling relation (which may be verified independently by more detailed calculations) to find

$$\gamma = (2-\eta)\nu = [2 + O(\varepsilon^2)]/\lambda_1,$$

$$= 1 + \frac{n+2}{2(n+8)}\varepsilon + O(\varepsilon^2). \tag{6.66}$$

Likewise the hyperscaling relations yield α from $2 - \alpha = d\nu$ and

$$\beta = \frac{1}{2}(d-2+\eta)\nu$$
$$= \frac{1}{2} - \frac{3}{2(n+8)}\varepsilon + O(\varepsilon^2). \tag{6.67}$$

The other thermodynamic exponents follow similarly. In addition we obtain something new, namely, the leading correction-to-scaling exponent [see (5.93) <u>et seq.</u>] which is associated with u and hence given by

$$\theta = -\phi_u = -\frac{\lambda_u}{\lambda_1} = \frac{1}{2}\varepsilon + O(\varepsilon^2). \qquad (6.68)$$

Historically, the first theoretical predictions for the value of θ come from the renormalization group ε expansion.

6.7 Some numerics

It is natural to enquire how well, if at all, the ε expansion works! To answer, recall first that the limit $n \to \infty$ should reproduce the spherical model (see Sec. 4.6). The exact results for the spherical model include $\beta = 1/2$ which certainly agrees with (6.67) and, more interestingly,

$$\gamma = \frac{2}{d-2} = 1 + \frac{1}{2}\varepsilon + \frac{1}{4}\varepsilon^2 + \frac{1}{8}\varepsilon^3 + \ldots, \qquad (6.69)$$

which confirms (6.66) precisely! In this limit we see, in fact, that the ε expansions for the exponents represent convergent power series with a radius of convergence $\varepsilon_c = 2$.

For finite n it seems more likely that the ε expansion is only asymptotic (but, probably, "Borrel summable"). Nevertheless, we may, optimistically, hope that $\varepsilon = 1$ (for $d = 3$) is relatively "small" in that it is only halfway to the undoubted breakdown around $\varepsilon_c = 2$. This optimism turns out to be surprisingly well justified. Indeed, even the first order expansions yield values in much better agreement with bulk (d=3)-dimensional experiments than does classical theory. Thus from (6.66) and (6.67) we find

$$\gamma(d = 3, n = 1) \simeq 1\tfrac{1}{6} \simeq 1.67, \quad \text{while} \quad \gamma_{expt} \simeq 1.24,$$
$$\beta(d = 3, n = 1) \simeq \tfrac{1}{3}, \quad \text{while} \quad \beta_{expt} \simeq 0.32 - 0.33.$$

In second order the results are even more encouraging. Thus from

$$\alpha = \frac{4-n}{2(n+8)}\varepsilon - \frac{(n+2)^2(n+28)}{4(n+8)^2}\varepsilon^2 + O(\varepsilon^3), \qquad (6.70)$$

one obtains

for $\alpha\ (d=3) \simeq 0.08,\ -0.02,$ and $-0.10,$
$n = 1,\ 2,$ and $3,$

respectively, which correlates well with the observed values

$$\alpha_{expt} \simeq 0.11, \quad -0.02, \quad \text{and } -0.14.$$

The trends with n are clearly reproduced and the divergence of the specific heat for $n \geq 2$, but not for $n \leq 2$, is also predicted.

By working harder one can calculate further terms in the expansion. Thus correct to fourth order one knows[29]

$$\eta = \frac{(n+2)\varepsilon^2}{2(n+8)^2}\{1 + \frac{(-n^2+56n+272)}{4(n+8)^2}\varepsilon$$

$$+ [\frac{(-5n^4-230n^2+1124n^2+17920n+46144}{16(n+8)^2} - \frac{24\zeta(3)(5n+22)}{(n+8)}]\varepsilon^2$$

$$+ O(\varepsilon^3)\}, \qquad (6.71)$$

which is interesting because of the appearance of the Riemann zeta function $\zeta(s) = \sum_1^\infty n^{-s}$ in fourth order. All the other exponents are now also known to this order.[30] At this stage, however, one does not obtain better numerical results if one merely truncates the expansion: however, with suitable methods of summation[30] rather satisfactory results are obtained which, for the most part, appear to be accurate to within two or three parts in the third decimal place!

6.8 Further developments in brief

Having explained the concepts of scaling and universality, and having laid the foundations of renormalization group theory, these lectures must end. In a more extended course we would, at this point, proceed to survey some of the many significant applications of renormalization group theory. First we might demonstrate the appearance of factors like ln t to special powers at the borderline dimensionality. Then, following the historical developments, we might consider the effect of _long range_ forces[31] with a spin coupling decaying as

$$J(\underset{\sim}{R}) \sim 1/R^{d+\sigma} \quad \text{as } R \to \infty, \qquad (6.72)$$

with $\sigma > 0$. These are somewhat artificial but for $\sigma < 2$ we would discover new critical behavior with a new borderline dimensionality at $d = 2\sigma < 4$ about which we could construct a modified ε-expansion in powers of $\varepsilon_\sigma = 2\sigma - d$.

Next it would be logical to examine the crossover from, say, Heisenberg (n=3) critical behavior to Ising (n=1) or XY (n=2) behavior induced by anisotropy in the spin-spin couplings and, hence, in the quadratic part of the LGW Hamiltonian. This could lead to a discussion of _bicritical points_ as observed in many antiferromagnets.[32] Spin anisotropies of higher symmetry (induced physially by coupling to the lattice) in particular those entering as cubic symmetry breaking terms in the quartic spin terms, can lead to quite new sorts of critical behavior: so also do long-range forces of dipole-dipole character which must be examined for real ferromagnets,[33] although for $n \geq 2$ one finds that the numerical values of the exponents hardly change! For uniaxial, Ising-like dipolar ferromagnets, on the other hand, a striking new phenomenon occurs: the critical fluctuations at low momenta are suppressed and the borderline dimensionality drops from $d=4$ to $d=3$.[34,35] Thus, except for subtle logarithmic correction factors, classical theory becomes correct for a real bulk system!

A borderline dimensionality $d=3$ arises also in the description of tricritical points,[36] which is accomplished within a single-component ($n=1$) model by allowing the coefficients of the s^4 term to become negative but, as is needed for stability, retaining a term $- vs^6$ with $v > 0$. Tricritical points are observed in multicomponent fluid mixtures, in antiferromagnets, in superfluid helium three-four mixtures, etc.

Then one would want to describe the expansions in powers of $1/n$ developed by Abe[37] (without explicit reference to the renormalization group) and those about the lower borderline dimensionality, putting $d=2+\epsilon$, devised to Polyakov.[38] The Kosterlitz-Thouless[39-41] theory of XY-like or $n=2$ systems at the borderline dimensionality $d=2$, which describes thin superfluid helium films, and the subsequent Halperin-Nelson-Young[42,43] theories of two-dimensional melting would be tempting topics -- and so on! Even then, we would not have touched on the development and application of real space renormalization group techniques[44] including the versatile approximate renormalization group scheme of Migdal[45] and Kadanoff.[46] No mention would have been made of dynamical critical phenomena[47] and the application of renormalization groups in that context. Nor would we have discussed methods for calculating equations of state, or correlation functions, or crossover scaling functions,[48,49] or have described Wilson's method for solving the Kondo problem,[50] or applications to polymers, to liquid crystals, and more.

It is evident that to present even a sketchy account of all these topics would require much more time and space. Indeed, our task would grow to resemble that of giving a full account of the applications of quantum mechanics! Truly the renormalization group approach, and the associated ideas of scaling and universality, have become basic tools of the condensed matter theorist and are constantly being applied to new and more challenging problems. Happily, however, for the reader who wishes to enquire further there are now a selection of reviews and text books at various levels in which to browse and dig deeper. Some of these have already been mentioned in passing but for convenience these and a few more have been gathered together in the Bibliography. Note that the Bibliography makes no claim to completeness: indeed, we should add that in these lectures we have been somewhat cavalier in mentioning individual scientists and in making reference to the original literature. Accordingly, apologies are offered here to any who feel unjustifiably unmentioned or otherwise slighted. The reader, however, should have no difficulty in entering the literature through the sources cited in the Bibliography: please do so with best wishes for stimulating study and fruitful discovery!

ACKNOWLEDGMENTS

The author is grateful to the CSIR of South Africa and Dr. G. Heymann for the invitation to lecture at the Advanced Course on Critical Phenomena at Stellenbosch. The hospitality of the Merensky Institute for Physics and its faculty, especially Professor Fritz J. W. Hahne and Professor Chris A. Engelbrecht, is much appreciated. The author is particularly indebted to Dr. Arthur G. Every for his expert and patient assistance in preparing a draft manuscript of the lectures and for seeing the full text through to final form. The ongoing support of the National Science Foundation, in part through the Materials Science Center at Cornell University, is gratefully acknowledged. Friends, students and colleagues too numerous to mention have helped the author learn and understand much of what is retold here: special thanks, however, are due to Cyril Domb, David M. Jasnow, Leo P. Kadanoff, Martin F. Sykes, Ben Widom and Kenneth G. Wilson.

A.G.E. gratefully acknowledges the hospitality extended to him at the University of Illinois and the CSIR for providing a travel grant. Thanks also to 'Tish' Watts for typing the manuscript.

APPENDIX A The Kac-Hubbard-Stratonovich Transformation

The Kac-Hubbard-Stratonovich transformation is a way of turning one model into another. This device has played an increasingly valuable role in the theory of critical phenomena. The main theoretical factor suggesting that various different models might be expected to transform into one-another is that of universality. To belong to the same universality class different models must, somehow, be mathematically equivalent, at least in their critical regions, even though they have quite different physical interpretations and contrasting mathematical formulations.

To illustrate how the transformation is carried out we will consider the simplest example, namely, a spin $\frac{1}{2}$ Ising model; however, the approach to be described can be readily extended. (The interested reader should work through the case of the fixed-length n-vector model.)

Consider the general Ising model partition function

$$Z_N(\underset{\sim}{K}) = \mathrm{Tr}_N^{\sigma} \{\exp [\sum_{(i,j)} K_{ij}\sigma_i\sigma_j]\}, \qquad (A1)$$

where the interactions satisfy

$$K_{ij} = J_{ij}/k_B T = K_{ji}, \qquad (\text{with } K_{ii} \equiv 0), \qquad (A2)$$

and the sum in (A1) runs over all distinct pairs (i,j). The Ising spins σ_i in this expression can take on only the two values ±1, and so the operation of taking the trace over any spin means

$$\mathrm{Tr}^{\sigma} = \frac{1}{2} \sum_{\sigma=\pm 1} . \qquad (A3)$$

The factor of 1/2 is incorporated in order to normalize the trace, i.e., so that $\mathrm{Tr}^{\sigma}\{1\} = 1$. The aim now is to turn (A1) into a form which looks somewhat similar but involves a new set of <u>continuous spin variables</u>, s_i instead of the discrete σ_i. The result which we will obtain is

$$Z_N(\underset{\sim}{K}) = e^{f_0(K)} \int_{-\infty}^{\infty} \prod_{i=1}^{N} ds_i \; \exp [-\sum_{(i,j)} Q_{ij}s_i s_j] \; \exp[-\sum_i w(s_i)], \qquad (A4)$$

where $f_0(K)$ is a smooth, analytic function of K and just provides a background free energy with (in general) no interesting critical behavior. The integrations are performed over all the continuous spin variables, s_i, of which there are as many as there were original Ising spins; the limits for each integration are $-\infty$ and $+\infty$. The spin weighting function comes out to be

$$-w(s_i) = \ln(\cosh s_i) - \frac{1}{2} Q_{ii}s_i^2 = -\frac{1}{2}(Q_{ii}-1)s_i^2 - \frac{1}{12} s_i^4 + \dots . \qquad (A5)$$

A weighting function is, of course, necessary because without it the integrals would diverge: in fact, Q_{ii}, which will be defined below, must be positive and sufficiently large. Our object now is to derive equations relating the original interactions K_{ij} to the new spin-spin interactions, $-Q_{ij}$, and provide a justification for the particular form assumed by the weighting function $w(s_i)$. (The fixed-length n-vector model yields a different but qualitatively similar set of weighting functions depending on n.)

A general reason for going over to continuous spin variables is that they are easier to deal with mathematically. In particular, in order to carry out the spatial Fourier transformations on the set of spin variables, which play such a vital role in the renormalization group ε-expansion theory, the spins have to be continuous, unbounded variables.

For simplicity of exposition we will consider the fully ferromagnetic case $K_{ij} = K_{ji} \geq 0$ $(K_{ii} \equiv 0)$. If we then make use of the inequality

$$\frac{1}{2}(\sigma_i + \sigma_j)^2 = \frac{1}{2}\sigma_i^2 + \frac{1}{2}\sigma_j^2 + \sigma_i\sigma_j = 1 + \sigma_i\sigma_j \geq 0, \tag{A6}$$

we can rewrite the interactions as

$$\sum_{(i,j)} K_{ij}\,\sigma_i\sigma_j = -\frac{1}{2}NP_0 + \frac{1}{2}\sum_{i=1}^{N}\sum_{j=1}^{N} P_{ij}\sigma_i\sigma_j, \tag{A7}$$

where the symmetric, N × N matrix

$$\underset{\sim}{P} \equiv [P_{ij}] = P_0\underset{\sim}{I} + \underset{\sim}{K} = [P_0\delta_{ij} + K_{ij}], \tag{A8}$$

will be positive definite if P_0 is chosen positive and sufficiently large: specifically, by (A6) it suffices to choose

$$P_0 \geq \max_i \,[\sum_{\ell} K_{i\ell}]. \tag{A9}$$

If $\underset{\sim}{\sigma}^T = [\sigma_1, \ldots, \sigma_N]$ is a row vector and $\underset{\sim}{\sigma}$ the corresponding (transposed) column vector, we can thus rewrite the partition function as

$$Z_N(\underset{\sim}{K}) = e^{-\frac{1}{2}NP_0}\,\mathrm{Tr}_N^{\sigma}\{e^{\frac{1}{2}\underset{\sim}{\sigma}^T \underset{\sim}{P}\,\underset{\sim}{\sigma}}\}. \tag{A10}$$

Now consider another quadratic form in N continuous variables y_i, namely,

$$Q(\underset{\sim}{y}) = \sum_{i=1}^{N}\sum_{j=1}^{N} Q_{ij}y_iy_j, \tag{A11}$$

where $\underset{\sim}{Q} = [Q_{ij}]$ is a symmetric, positive definite matrix: As such, $\underset{\sim}{Q}$ may be diagonalized by an orthogonal transformation with matrix $\underset{\sim}{O}$. Explicitly, in terms of the new variables

$$\underset{\sim}{x} = \underset{\sim}{O}\,\underset{\sim}{y}, \tag{A12}$$

the quadratic form becomes

$$Q(\underset{\sim}{y}) = \sum_{r=1}^{N} \lambda_r x_r^2, \tag{A13}$$

where the λ_r (r=1, ..., N) are the real, positive eigenvalues of $\underset{\sim}{Q}$. It follows similarly that the determinant of $\underset{\sim}{Q}$ is given by

$$|\underset{\sim}{Q}| = \prod_{r=1}^{N} \lambda_r. \tag{A14}$$

The partition-function-like expression

$$I(\underset{\sim}{Q}) = \int_{-\infty}^{\infty} d^N y\, e^{-\frac{1}{2}\underset{\sim}{y}^T \underset{\approx}{Q}\, \underset{\sim}{y}}, \tag{A15}$$

may now be evaluated by changing variables from $\underset{\sim}{y}$ to $\underset{\sim}{x}$ and noticing that the Jacobian of the transformation is +1 since $\underset{\sim}{O}$ is orthogonal. The integrals are then just Gaussian and we obtain

$$I(Q) = \prod_{r=1}^{N} \int_{-\infty}^{\infty} dx_r\, e^{-\frac{1}{2}\lambda_r x_r^2} = \prod_{r=1}^{N} \sqrt{\left(\frac{2\pi}{\lambda_r}\right)} = \frac{(2\pi)^{N/2}}{\sqrt{|\underset{\sim}{Q}|}}. \tag{A16}$$

Next let us make a simple shift in the variables $\underset{\sim}{y}$ according to

$$\underset{\sim}{y} = \underset{\sim}{s} + \underset{\sim}{Q}^{-1}\underset{\sim}{\sigma} \quad \text{with} \quad \underset{\sim}{y}^T = \underset{\sim}{s}^T + \underset{\sim}{\sigma}^T \underset{\sim}{Q}^{-1}, \tag{A17}$$

where $\underset{\sim}{s}$ represents a new set of variables, which will eventually be identified as the continuous spin variables, while the $\underset{\sim}{\sigma}$ here represent only fixed shift parameters! In terms of the new variables the quadratic form Q becomes

$$Q(\underset{\sim}{y}) = \underset{\sim}{s}^T \underset{\approx}{Q}\underset{\sim}{s} + \underset{\sim}{s}^T \underset{\sim}{\sigma} + \underset{\sim}{\sigma}^T \underset{\sim}{s} + \underset{\sim}{\sigma}^T \underset{\sim}{Q}^{-1} \underset{\sim}{\sigma} = \underset{\sim}{s}^T \underset{\approx}{Q}\underset{\sim}{s} + 2\sum_{i=1}^{N} s_i \sigma_i + \underset{\sim}{\sigma}^T \underset{\sim}{Q}^{-1} \underset{\sim}{\sigma}. \tag{A18}$$

If we now choose the matrix $\underset{\sim}{Q}$ so that

$$\underset{\sim}{Q} = \underset{\sim}{P}^{-1} = 1/(P_0 \underset{\sim}{I} + \underset{\sim}{K}), \tag{A19}$$

we see quickly how (A15) and (A16) apply to the problem in hand since we obtain the identity

$$I(\underset{\sim}{Q}) = e^{-\frac{1}{2}\underset{\sim}{\sigma}^T \underset{\approx}{P}\underset{\sim}{\sigma}} \int_{-\infty}^{\infty} d^N s\, \exp\left[-\frac{1}{2}\underset{\sim}{s}^T \underset{\approx}{Q}\underset{\sim}{s} - \sum_{i=1}^{N} s_i \sigma_i\right],$$

$$\text{i.e.,}\quad I(Q) = \frac{(2\pi)^{N/2}}{|\underset{\sim}{Q}|} = (2\pi)^{N/2}\sqrt{|\underset{\sim}{P}|}. \tag{A20}$$

Note that in the first line, the spin variables σ_i appear like nonuniform or random external fields acting upon the s variables. If we solve for the factor involving $\underset{\sim}{P}$ and substitute in (A10) we can write the result as

$$Z_N(\underset{\sim}{K}) = \frac{e^{-\frac{1}{2}NP_0}}{(2\pi)^{N/2}\sqrt{|\underset{\sim}{P}|}}\, \mathrm{Tr}_N^{\sigma}\left\{\int_{-\infty}^{\infty} d^N s\, \exp[-\sum_{(i,j)} Q_{ij}s_i s_j - \tfrac{1}{2}\sum_i Q_{ii}s_i^2]\, e^{-\sum_i s_i\sigma_i}\right. . \tag{A21}$$

Note that the diagonal terms of the expression $\frac{1}{2}\underset{\sim}{s}^T\underset{\sim}{Q}\underset{\sim}{s}$ have been separated off while what remains has been written as a sum over pairs of spins (i,j).

At this point we make the crucial observation that the trace operation on the σ_i commutes with the integration over the s_i and affects only the last exponential factor in (A21). To perform the trace we use the simple result

$$\mathrm{Tr}^{\sigma_i}\{e^{-s_i\sigma_i}\} = \frac{1}{2}(e^{-s_i} + e^{+s_i}) = \cosh s_i, \tag{A22}$$

so that, finally, we obtain

$$Z_N(\underset{\sim}{K}) = \frac{e^{-\frac{1}{2}NP_0}}{(2\pi)^{N/2}\sqrt{|\underset{\sim}{P}|}}\int_{-\infty}^{\infty} d^N s\, \exp\Big[-\sum_{(i,j)}^{N} Q_{ij}s_i s_j - \sum_i(\tfrac{1}{2}Q_{ii}s_i^2 - \ln\cosh s_i)\Big]. \tag{A23}$$

This is clearly in the anticipated form (A4) with the spin weighting function given by (A5) while the new interactions, $-Q_{ij}$, etc. follow from (A19) and (A8). Note that because $\underset{\sim}{P}$ is positive definite by construction, so is $\underset{\sim}{Q}$; it then follows, since ln cosh s varies only as $|s|$ for large s, that the Q_{ii} coefficients are positive and sufficiently large to ensure covergence of the integrals over the s_i.

We have thus achieved an exact transformation of the discrete spin Ising model with couplings, K_{ij}, into a continuous spin model with new couplings, $-Q_{ij}$, defined via the inverse matrix $(P_0\underset{\sim}{I} + \underset{\sim}{K})^{-1}$. It is clearly of interest to gain some idea of the nature of these new interactions. To that end, let us suppose that $\underset{\sim}{K}$ describes only nearest neighbor ferromagnetic couplings. A little thought then shows that $\underset{\sim}{K}^2$ describes next-nearest neighbor couplings (plus some self-coupling), that $\underset{\sim}{K}^3$ describes third-neighbor couplings (plus some further first neighbor couplings) and so on. Thus the identity

$$-P_0^2\underset{\sim}{Q} = -P_0\underset{\sim}{I} + \underset{\sim}{K} - P_0^{-1}\underset{\sim}{K}^2 + P_0^{-2}\underset{\sim}{K}^3 - \dots, \tag{A24}$$

which is valid when P_0 satisfies (A9), shows that the couplings $-Q_{ij}$ are, in first approximation, the same as the K_{ij} but scaled by a factor P_0^{-2}. In higher approximation second-neighbor antiferromagnetic couplings appear but they are weaker by a factor $1/P_0$, and so on. Thus the new couplings are no longer of pure nearest-neighbor character: however, they are of 'short range' in the sense that they decay exponentially with distance (and, evidently, they are oscillatory in sign although predominantly ferromagnetic in effect).

If the lattice is translationally invariant it is advantageous, as seen in Sec. 6.1, etc., to transform to continuous Fourier space spin variables, $\hat{s}_{\underset{\sim}{q}}$. The couplings are then directly expressed, as in (6.4), in terms of the Fourier transform

$$\hat{Q}(q) = \sum_{\underset{\sim}{x}} e^{i\underset{\sim}{q}\cdot\underset{\sim}{x}} Q(\underset{\sim}{x}) \quad \text{with} \quad Q(\underset{\sim}{x}_i - \underset{\sim}{x}_j) \equiv Q_{ij}. \tag{A25}$$

This in turn is related to the corresponding transform $\hat{K}(\underset{\sim}{q})$ of $K_{ij} \equiv K(\underset{\sim}{x}_i - \underset{\sim}{x}_j)$ through

$$-\hat{Q}(\underset{\sim}{q}) = \frac{-1}{P_0 + \hat{K}(\underset{\sim}{q})} = \frac{-1}{P_0 + \hat{K}(\underset{\sim}{0})} + \frac{\hat{K}(\underset{\sim}{q}) - \hat{K}(\underset{\sim}{0})}{[P_0 + \hat{K}(\underset{\sim}{0})]^2} - \dots . \tag{A26}$$

This form is illuminating since in the case of predominantly ferromagnetic couplings of the σ spins one has, for rapidly decaying interactions in a large system,

$$\hat{K}(\underset{\sim}{q}) = \hat{K}(\underset{\sim}{0})[1 - R_0^2 q^2 + O(q^4)], \tag{A27}$$

where $\hat{K}(\underset{\sim}{0}) > 0$, while the (real) length R_0 measures the range of the interactions. By substitution in (A26) we see that the couplings of the s spins are likewise ferromagnetic with a comparable finite range. Finally, note that in the thermodynamic limit one has

$$Q_{ii} = Q(\underset{\sim}{0}) = \int \frac{d\underset{\sim}{q}}{(2\pi)^d} \frac{1}{P_0 + \hat{K}(\underset{\sim}{q})}, \tag{A28}$$

which is necessarily positive as required for a sensible weighting factor.

APPENDIX B Details of the ε-expansion calculation

In this appendix we examine the derivation of the recursion relations (6.26)-(6.27) and (6.40)-(6.42) for the perturbation-theoretic expansion of the Wilson momentum shell renormalization group near d=4 dimensions and introduce the diagrammatic language that facilitates the caluclations. We are concerned here with the LGW reduced Hamiltonian (6.4) which we rewrite, as in (6.11), in the form

$$\bar{\mathcal{H}} \simeq -\frac{1}{2}\int_{\underset{\sim}{q}} \sum_{\mu=1}^{n} (r + eq^2)\, \sigma^{\mu}_{\underset{\sim}{q}} \sigma^{\mu}_{-\underset{\sim}{q}} - u \int_{\underset{\sim}{q}_1}\int_{\underset{\sim}{q}_2}\int_{\underset{\sim}{q}_3} \sum_{\mu,\nu=1}^{n} \sigma^{\mu}_{\underset{\sim}{q}_1} \sigma^{\mu}_{\underset{\sim}{q}_2} \sigma^{\nu}_{\underset{\sim}{q}_3} \sigma^{\nu}_{\underset{\sim}{q}_4}, \tag{B1}$$

with the wavevectors restricted by

$$g_1 + g_2 + g_3 + g_4 = 0, \tag{B2}$$

so that 'umklapp' processes are neglected. Likewise, we suppose that the sixth and higher order terms may be neglected initially. The validity of both these approximations, to leading order in ε, may be checked by computing their effects by the same techniques. However, it is important to note that even if such terms are rigorously absent in the initial, physical Hamiltonian they may be generated and normally will be generated, in the process of successive renormalization. The coefficient e of the q^2 term will be constrained throughout (by spin rescaling) to be equal to unity. Thus $\mathcal{H}$ is, effectively, determined only by the two parameters r and u.

As discussed in the main text, the first step is to split up $\mathcal{H}$ as follows

$$\bar{\mathcal{H}} = \bar{\mathcal{H}}^{<} + \bar{\mathcal{H}}_2^{>} - u\, \bar{\mathcal{H}}_4^{>}. \tag{B3}$$

Then we must compute the renormalized Hamiltonian, $\bar{\mathcal{H}}'$, which is given to second order in u by

$$\bar{\mathcal{H}}' \simeq \{ \bar{\mathcal{H}}^{<} + \ln(\mathrm{Tr}^{>} [e^{\bar{\mathcal{H}}_2^{>}}]) - u \langle \bar{\mathcal{H}}_4^{>} \rangle_{>}$$

$$+ \frac{1}{2} u^2 [\langle\langle (\bar{\mathcal{H}}_4^{>})^2 \rangle_{>} - \langle \mathcal{H}_4^{>} \rangle_{>}^{2}] + \ldots \}_{\vec{\sigma}_g \Rightarrow \vec{\sigma}'_{g'}} . \tag{B4}$$

Let us start by examining the lower order terms. First, note that $\bar{\mathcal{H}}^{<}$ is of the same basic form as the original LGW Hamiltonian except that the momentum integrals are limited to the inner region, <, of momentum space which contains only $N' = N/b^d$ spins. However, the spatial rescaling restores the original domain of integration through the transformations

$$g \Rightarrow g'/b: \quad \int_g^{<} \equiv \int^{<} \frac{d^d q}{(2\pi)^d} = \int \frac{d^d(q'/b)}{(2\pi)^d} \equiv b^{-d} \int_{g'}, \tag{B5}$$

where uninflected momentum integrals run over the full zone. Recalling the spin rescaling, $\vec{\sigma}_g \Rightarrow \hat{c}\, \vec{\sigma}'_{g'}$, we see that the quadratic part of $\bar{\mathcal{H}}^{<}$ transforms as

$$-\frac{1}{2}\int_g^{<} (r + eq^2)\, \vec{\sigma}_g \cdot \vec{\sigma}_{-g} = -\frac{1}{2}\int_{g'} b^{-d}(r + eq'^2 b^{-2})\, \hat{c}^2\, \vec{\sigma}'_{g'} \cdot \vec{\sigma}'_{-g'}$$

$$= -\frac{1}{2}\int_{g'} [(\hat{c}^2 b^{-d} r) + (\hat{c}^2 b^{-d-2} e) q'^2] \vec{\sigma}'_{g'} \cdot \vec{\sigma}'_{-g'} , \tag{B6}$$

so that, ignoring possible contributions from higher order terms, etc., the parameters r and e are simply renormalized by factors $\hat{c}^2 b^{-d}$ and $\hat{c}^2 b^{-d-2}$, respectively. Notice that the momenta, g', and spins, $\vec{\sigma}'_g$, in (B6) are really 'dummy variables', in the sense that they are to be integrated over, so we may actually drop all the primes in the final expression. Evidently the quartic term similarly generates a leading renormalization factor $\hat{c}^4 b^{-3d}$ for u.

In the second term in (B4) the term $\exp(\bar{\mathcal{H}}_2^>)$ depends only on spins $\sigma^>$, in the outer zone and so when the operation $\mathrm{Tr}^>$ is applied to it, the result is just a constant contribution which can be ignored for our present purposes since it cannot further affect other interactions. (Nevertheless, if we were concerned to calculate the free energy itself, we would have to retain this constant term in $\bar{\mathcal{H}}'$.) This completes the calculation of the recursion relations to zeroth order in u: see (6.26) to (6.28).

The evaluation of the first order term, $u\langle\bar{\mathcal{H}}_4^>\rangle_>$, requires closer attention. First note that in the expression for $\bar{\mathcal{H}}_4^>$, which for finite N is just the multiple sum,

$$\bar{\mathcal{H}}_4^> = \left(\frac{a^{-d}}{N}\right)^3 \sum_{q_1}\sum_{q_2}\sum_{q_3}{}^{>} \sum_{\mu,\nu=1}^{n} \sigma^\mu_{q_1}\sigma^\mu_{q_2}\sigma^\nu_{q_3}\sigma^\nu_{q_4}, \tag{B7}$$

at least one of the four momentum labels q_1, q_2, q_3 or $q_4 = -q_1 - q_2 - q_3$ must lie in the outer region, >. We must then evaluate the momentum shell average

$$\langle\bar{\mathcal{H}}_4^>\rangle_> = \frac{\mathrm{Tr}^>\{\bar{\mathcal{H}}_4^>\, e^{\bar{\mathcal{H}}_2^>}\}}{\mathrm{Tr}^>\{e^{\bar{\mathcal{H}}_2^>}\}}, \tag{B8}$$

where $\bar{\mathcal{H}}_2^>$ can similarly be written as a sum, namely,

$$\bar{\mathcal{H}}_2^> = -\frac{1}{2}\frac{a^{-d}}{N}\sum_{q}{}^{>}\sum_{\mu=1}^{n}(r + eq^2)\,\sigma^\mu_q\sigma^\mu_{-q}, \tag{B9}$$

the momenta being all restricted to the outer zone. The trace operation has the explicit form

$$\mathrm{Tr}^> = \prod_q{}^{>}\prod_\mu\int_{-\infty}^{\infty} d\sigma^\mu_q, \tag{B10}$$

which signifies a multiple integral over all the spin components σ^μ_q with momenta in the outer region, >. In the thermodynamic limit, $N \to \infty$, this operation becomes a functional integral. However, we may avoid this concept and the question of its proper definition by keeping N finite and doing the n(N-N') integrals over all the $\sigma^>$ components before taking the thermodynamic limit. (Note that in the spatial rescaling steps in (B5) and (B6) it was advantageous to take the thermodynamic limit, as we did implicitly, at an early stage.)

Each term in (B7) can be processed separately through (B8) and the results then added together. Consider the typical term $\sigma^\mu_{q_1}\sigma^\mu_{q_2}\sigma^\nu_{q_3}\sigma^\nu_{q_4}$. If all four of the q_j lie in the > zone then the result will be a constant and hence is of no further direct interest to us here (although it contributes to the constant term in $\bar{\mathcal{H}}'$). If either one or three of the q_j lie in the > zone the result must vanish. To see this note that the integrals involved in (B8) have the symmetric Gaussian form shown in (6.18) and so vanish by symmetry if an odd power of σ^{μ}_{q} is contributed by $\bar{\mathcal{H}}^{>}_{4}$. (See further below.) At least one of the $\sigma^{>}$ integrals must be odd in the case posed. Thus the only terms that need be considered are those where two of the q_j are equal in magnitude but opposite in sign and both lie in the > zone, with the associated spin component indices being the same. This generates a term $\sigma^{\mu}_{q}\sigma^{\mu}_{-q} = [\mathrm{Re}(\sigma^{\mu}_{q})]^2 + [\mathrm{Im}(\sigma^{\mu}_{q})]^2$. The other two q_j, then belong to the inner zone and will, perforce, also be equal in magnitude and opposite in sign; their associated spin component indices, ν, will likewise match. Taking into account all possible combinations that satisfy these criteria, leads to

$$\langle\bar{\mathcal{H}}^{>}_{4}\rangle_{>} = \left(\frac{a^{-d}}{N}\right)^3 \sum_{q_<}^{<} \sum_{q_>}^{>} \sum_{\mu,\nu=1}^{n} \{\langle\sigma^{\mu}_{q_<}\sigma^{\mu}_{-q_<}\sigma^{\nu}_{q_>}\sigma^{\nu}_{-q_>}\rangle_{>}$$

$$+ \langle\sigma^{\mu}_{q_>}\sigma^{\mu}_{-q_>}\sigma^{\nu}_{q_<}\sigma^{\nu}_{-q_<}\rangle_{>} + \langle\sigma^{\mu}_{q_<}\sigma^{\mu}_{q_>}\sigma^{\nu}_{-q_<}\sigma^{\nu}_{-q_>}\rangle_{>}\delta_{\mu\nu}$$

$$+ \langle\sigma^{\mu}_{q_<}\sigma^{\mu}_{q_>}\sigma^{\nu}_{-q_>}\sigma^{\nu}_{-q_>}\rangle_{>}\delta_{\mu\nu} + \langle\sigma^{\mu}_{q_>}\sigma^{\mu}_{q_<}\sigma^{\nu}_{-q_<}\sigma^{\nu}_{-q_>}\rangle_{>}\delta_{\mu\nu}$$

$$+ \langle\sigma^{\mu}_{q_>}\sigma^{\mu}_{q_<}\sigma^{\nu}_{-q_>}\sigma^{\nu}_{-q_<}\rangle_{>}\delta_{\mu\nu}\}. \tag{B11}$$

Now all the spin variables commute. Further, the spins $\sigma^{<}$ are not affected by the $\mathrm{Tr}^{>}$ operation and can thus be removed from under the angular brackets. In addition, because of the equivalence of the different components for each spin we have

$$\sum_{\nu=1}^{n} \langle\sigma^{\nu}_{q}\sigma^{\nu}_{-q}\rangle_{>} = n\,\langle\sigma^{\nu}_{q}\sigma^{\nu}_{-q}\rangle_{>}, \tag{B12}$$

and hence find that

$$\langle\bar{\mathcal{H}}^{>}_{4}\rangle_{>} = \left(\frac{a^{-d}}{N}\right)^3 \left[(2n+4)\sum_{q_>}^{>}\langle\sigma^{\nu}_{q_>}\sigma^{\nu}_{-q_>}\rangle_{>}\right] \sum_{q}^{<}\sum_{\mu=1}^{n}\sigma^{\mu}_{q}\sigma^{\mu}_{-q}, \tag{B13}$$

where we have dropped the < subscripts on q since they are no longer essential. Note that the combinatorial factor (2n+4) is of central importance to the final answers!

The next step is to calculate $\langle\sigma^{\nu}_{q_>}\sigma^{\nu}_{-q_>}\rangle_{>}$. It is equal to a product of integrals over all the $\sigma^{>}$, divided by a similar product. Cancellation occurs for all these integrals except for those over $\sigma^{\nu}_{q_>}$ and $\sigma^{\nu}_{-q_>}$; hence the result is

$$\left\langle \sigma^{\nu}_{g_>}\sigma^{\nu}_{-g_>}\right\rangle_> = \frac{\iint d\sigma^{\nu}_{g_>} d\sigma^{\nu}_{-g_>} \sigma^{\nu}_{g_>}\sigma^{\nu}_{-g_>} \exp[-\frac{1}{2}(\frac{a^{-d}}{N})(r+eq^2_>)(\sigma^{\nu}_{g_>}\sigma^{\nu}_{-g_>} + \sigma^{\nu}_{-g_>}\sigma^{\nu}_{g_>})]}{\iint d\sigma^{\nu}_{g_>} d\sigma^{\nu}_{-g_>} \exp[-\frac{1}{2}(\frac{a^{-d}}{N})(r+eq^2_>)(\sigma^{\nu}_{g_>}\sigma^{\nu}_{-g_>} + \sigma^{\nu}_{-g_>}\sigma^{\nu}_{g_>})]} . \quad (B14)$$

The integration over pairs of complex conjugate spins $\sigma^{\nu}_{g_>}$ and $\sigma^{\nu}_{-g_>} = (\sigma^{\nu}_{g_>})^*$ is carried out by making use of the equivalence

$$\int_{-\infty}^{\infty}\int_{-\infty}^{\infty} d\sigma^{\nu}_{g}\, d\sigma^{\nu}_{-g} \equiv \int_{-\infty}^{\infty}\int_{-\infty}^{\infty} d(\mathrm{Re}\sigma^{\nu}_{g})d(\mathrm{Im}\sigma^{\nu}_{g}). \quad (B15)$$

By this means we arrive at a product of two separate Gaussian integrals over real variables divided by two other similar integrals: all are readily evaluated and yield

$$\left\langle \sigma^{\nu}_{g}\sigma^{\nu}_{-g}\right\rangle_> = \frac{Na^d}{(r+eq^2)} . \quad (B16)$$

Substituting back into (B13), letting $N \to \infty$ and replacing the momentum sums by integrals yields

$$\langle \bar{\mathcal{H}}^{>}_{4}\rangle_> = 2(n+2)\int^{>}_{g}\frac{1}{(r+eq^2)}\int^{<}_{g}\vec{\sigma}_{g}\cdot\vec{\sigma}_{g}. \quad (B17)$$

Finally, momentum and spin rescaling introduce a factor $\hat{c}^2 b^{-d}$ as discussed earlier. Bearing in mind the factor $-u$ associated with $\langle\bar{\mathcal{H}}^{>}_{4}\rangle_>$ in (B4) and the factor $-1/2$ in the definition of the quadratic part of $\bar{\mathcal{H}}$, we see that to first order in u, the renormalized Hamiltonian has modified values, r' and e', of r and e given by the recursion relations

$$r' = \hat{c}^2 b^{-d}\, [r + 4u(n+2)\int^{>}_{g}\frac{1}{(r+eq^2)} + O(u^2)], \quad (B18)$$

$$e' = \hat{c}^2 b^{-d-2}\, [1 + 0 + O\,(u^2)], \quad (B19)$$

just as stated in (6.40) and (6.41).

The rest of the calculation proceeds in a similar fashion, with intermediate algebraic expressions of the type displayed in (B11) becoming considerably more complex. However, the combinatorial problem of deciding just which terms can contribute is greatly simplified by the use of graphical or diagramatic notation. We define, first, the free inverse "propagator" by

$$[G_0^{\mu\nu}(g)]^{-1} = \delta_{\mu\nu}(r+eq^2) \simeq \overset{\mu}{\longrightarrow}_{g}\overset{\mu}{}. \quad (B20)$$

This carries both a momentum and a spin component index, and serves to represent the quadratic or "free field" part of the total Hamiltonian. In field-theoretic language r represents the "bare mass". To represent the quartic part of the Hamiltonian we introduce the

$$\text{four-point vertex} \simeq \quad {}^{q_1\,\mu}_{q_2\,\mu}\rangle\cdots\cdots\langle^{q_3\,\nu}_{q_4\,\nu} \quad . \tag{B21}$$

This has four incoming lines which carry momentum and spin component indices μ, μ, ν and ν corresponding to the term $\sigma^{\mu}_{q_1}\sigma^{\mu}_{q_2}\sigma^{\nu}_{q_3}\sigma^{\nu}_{q_4}$ in (B1). Momentum is conserved 'through' a vertex in accord with the condition (B2). Each vertex also carries a "coupling constant" factor u ($\equiv u_4$) .

The process of calculating the renormalized (or "dressed") propagator can now be represented graphically as

$$\left(\xrightarrow[q]{\mu\quad\mu}\right)' = \left(\xrightarrow[q]{\mu\quad\mu}\right) + u\,[\,(\,{}^{q\,\mu}_{-q\,\mu}\rangle\cdots\cdots\bigcirc_{q_>\,\nu}) + (\bigcirc\cdots\cdots\langle\,)$$

$$+ (\;{}^{q\,\mu}\rangle\cdots\cdots\langle^{-q\,\mu}_{\;q\,\mu}\;) + (\;\cdots\;) + (\;\cdots\;) + (\;\cdots\;)] + O(u^2). \tag{B22}$$

The first order diagrams, here, have been arranged in the same order as the corresponding terms in (B11) to facilitate comparison. Evidently they correspond simply to all possible ways of joing up two "legs" of the four-point vertex with matching spin components and momenta in order to leave a propagator-like term. The rules that must be adhered to in constructing the integrals associated with the allowed diagrams are as follows: Each vertex line is accompanied by a factor u. Internal lines carry a propagator factor $G_0(q)$ and (in this renormalization group application) imply integration over the outer zone, >. If an "internal line" (i.e., one for which q is to be integrated) forms a closed loop then its spin index is "free" and can be summed over to yield a factor n. In higher orders of the perturbation theory there is a factor of 1/m! arising from the expansion of the exponential as in (B4). Diagrams that decompose into disconnected parts, i.e., "separated" or "unlinked" diagrams factorize and then cancell when $\mathcal{H}'$ is computed. (This is an example of the "linked cluster theorem").

The diagrammatic expansion for the renormalized vertex itself is thus found to be

$$\begin{pmatrix} q_1{}^{\mu} & & q_3{}^{\nu} \\ \succ & \cdots\cdots & \prec \\ q_2{}^{\mu} & & q_4{}^{\nu} \end{pmatrix}' = \begin{pmatrix} q_1{}^{\mu} & & q_3{}^{\nu} \\ \succ & \cdots\cdots & \prec \\ q_2{}^{\mu} & & q_4{}^{\nu} \end{pmatrix} - \frac{u}{2}\left[\begin{pmatrix} q_1{}^{\mu} & & q\,\alpha & & q_3{}^{\nu} \\ \succ & \cdots\cdots & \bigcirc & \cdots\cdots & \prec \\ q_2{}^{\mu} & & q'\alpha' & & q_4{}^{\nu} \end{pmatrix} + (\succ \cdots \prec\!\!\succ \cdots \prec) + \cdots \right.$$

8 (diagrams) × n (free spin components)

$$+ \left(\overset{\mu}{\underset{\mu}{\succ}} \cdots\cdots \overset{\nu}{\prec} \;\; \underset{\mu}{\succ} \cdots\cdots \overset{\nu}{\underset{\nu}{\prec}} \right) + \cdots$$

32 (diagrams) of weight 1

$$\left. + \left(\overset{\mu}{\underset{\mu}{\succ}} \cdots\cdots \overset{\nu}{\underset{\nu}{\prec}} \;\; \underset{\nu}{\succ} \cdots\cdots \overset{\nu}{\prec} \right) + \cdots \right] + O(u^3) \qquad \text{(B23)}$$

32 (diagrams) of weight 1

Note that diagrams such as

$$\overset{q^{>}}{\bigcirc} \cdots\cdots \overset{q'}{\underset{q}{\prec\succ}} \cdots\cdots \prec$$

with an "articulation" line or "cut bond" cannot arise,[51] since momentum conservation would require that q (in $<$) = $-$ q' (in $>$) which is impossible.

The two internal lines in the two-vertex diagrams yield, on integration, a factor

$$\int_q^{>} \int_{q'}^{>} \frac{1}{(r+eq^2)} \; \frac{1}{(r+eq'^2)}, \text{ with } q + q' = q_1 + q_2. \qquad \text{(B24)}$$

A little reflection shows that the renormalized vertex has, in fact, become q-dependent in that it no longer carries only a constant coupling constant factor, u, but rather involves a kernel $u_4(q_1, q_2, q_3)$. However, we may expand this kernel in powers of the q_j and associate the coupling constant u with u_4 (0,0,0). Likewise then, the renormalized coupling constant u' is to be associated with $u'_4(0,0,0)$. Consequently we can put $q_1 = q_2 = 0$ so that $q = -q'$ and the factor for the internal lines thus becomes simply $\int_q^{>}(r+eq^2)^{-2}$. After allowing for spin and spatial rescaling, the recursion relation for u that follows is seen to be

$$u' = \hat{c}^4 \, b^{-3d} \, u \, [1 - \frac{1}{2} u \, (8n + 64) \int_q^{>} \frac{1}{(r+eq^2)^2} + O(u^2)], \qquad \text{(B25)}$$

in agreement with (6.42). Note that the combinatorial factor (8n+24) = 8(n+8) directly represents the breakdown of the diagrams in (B24). With practice one learns how to write down such combinatorial factors by inspection for such simple diagrams as here, and by fairly rapid analysis for more complex diagrams like those that enter in calculation of the ε^2 and ε^3 terms. Reference to the ε-expansions (6.64), etc. shows that it is just these combinatorial factors that determine the n and ε dependence of the various exponents!

The discovery of a q-dependence in the renormalized vertex is typical of how new terms are generated on renormalization. It indicates that, in principle, such q^2, q^4, ... σ^4-terms should have been included in the original Hamiltonian, along with q^0, q^2, q^4, ...σ^6 terms and so on. However, on renormalization each q^2 factor would gain an extra renormalization factor b^{-2}, etc., so that one sees that such terms represent, at least near d=4, successively more irrelevant critical operators. Nonetheless, it is clear that care and thought are required: blind calculation may lead to a correct answer but an awareness of the general structure of the renormalization group process is a necessary guide if pitfalls are to be avoided!

We saw in (B19) that there is no first order contribution to the renormalization of the coefficient e which determines the decay exponent η. It is worthwhile recording that the required leading correction comes from the second order propagator diagrams

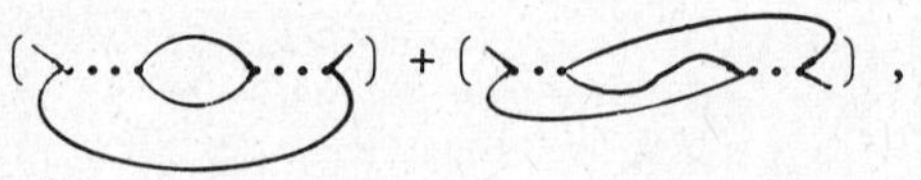

16*n* + 32

which yield the recursion relation (with $e \equiv 1$)

$$e' = b^{-\eta}[1 + 16\,(n+2)\,I\,u^2 \ln b + O(u^3)]\ , \qquad \text{(B26)}$$

where

$$I = \lim_{b\to\infty} \frac{128\pi^4}{\ln b} \int_{\underset{\sim}{q}}^{>} \int_{\underset{\sim}{q}'}^{>} \frac{1}{q^2 q'^2 (q^2+q'^2)} = \frac{1}{2}. \qquad \text{(B27)}$$

This recursion relation then yields η correct to order ε^2 as quoted in (6.71). Thus one need not go to third order in u to find η to $O(\varepsilon^2)$ although this is necessary for the other exponents.

APPENDIX C Dimensionality as a Continuous Variable

In the (ε=4-d)-expansion for critical exponents the spatial dimensionality, d, is treated as a continuously variable parameter. One way of giving definite meaning to this procedure is based on the observation that the only place that dimensionality enters into the calculations is in performing various integrals which are of the form $\int d^d q f(\underset{\sim}{q})$ where, in the simplest case, the integrand, $f(\underset{\sim}{q})$, is actually spherically symmetric and therefore a function only of q^2 rather than of the individual components of q. When d has a standard integral value the simplest way of doing this type of integral is to transform to hyperspherical coordinates. The integrand depends only on the radial component, and so the integration can be performed immediately over the angular coordinates. Thus, as mentioned in the text of Sec. 6.5, one obtains

$$\int d^d q\ f(q^2) = C_d \int_0^\infty f(q^2) q^{d-1} dq, \tag{C1}$$

where the area of a unit d-sphere is given, as in the text, by

$$C_d = 2\pi^{d/2}/\Gamma(\tfrac{1}{2}d). \tag{C2}$$

The gamma function, $\Gamma(\frac{1}{2}d)$, is a well-defined analytic function of its argument so that (C1) is meaningful mathematically even when d is nonintegral. Thus the extension to arbitrary (even complex!) values of d is straightforward for functions which are spherically symmetric.

At the next stage one encounters integrals which also involve scalar products such as $\underset{\sim}{q} \cdot \underset{\sim}{p}$, where $\underset{\sim}{p}$ is some reference momentum. Such integrals can be dealt with by the formula

$$\int d^d q\ f(q^2, \underset{\sim}{q}\cdot\underset{\sim}{p}) = C_{d-1} \int_0^\infty q^{d-1} dq \int_0^\pi (\sin\theta)^{d-2}\, d\theta\, f(q^2, pq\cos\theta). \tag{C3}$$

More generally, following Wilson[52] one only needs the following properties of general d-dimensional integrals:

(a) Linearity: $$\int d^d q\ [f_1(\underset{\sim}{q}) + f_2(\underset{\sim}{q})] = \int d^d q\ f_1(\underset{\sim}{q}) + \int d^d q\ f_2(\underset{\sim}{q}), \tag{C4}$$

(b) Translation Invariance: $$\int d^d q\ f(\underset{\sim}{q} + \underset{\sim}{p}) = \int d^d q\ f(\underset{\sim}{q}), \tag{C5}$$

(c) Scaling: $$\int d^d q\ f(b\underset{\sim}{q}) = b^{-d} \int d^d q\ f(\underset{\sim}{q}), \tag{C6}$$

(d) Normalization: $$\int d^d q\ e^{-q^2} = \pi^{d/2}. \tag{C7}$$

Then, by way of illustration, if one needs an integral such as

$$I_1(\underset{\sim}{p}_1, \underset{\sim}{p}_2) = \int d^d q \frac{(\underset{\sim}{q}\cdot \underset{\sim}{p}_1)^2 (\underset{\sim}{q}\cdot\underset{\sim}{p}_2)^2}{r + q^2} \tag{C8}$$

one first uses the identity

$$\frac{1}{r + q^2} = \int_0^\infty e^{-(r + q^2)s}\, ds \quad , \tag{C9}$$

to reduce the problem to the Gaussian-type integral

$$I_2\,(\underset{\sim}{p}_1,\, \underset{\sim}{p}_2; s) \;=\; \int d^d q\; (\underset{\sim}{q}\cdot\underset{\sim}{p}_1)^2\; (\underset{\sim}{q}\cdot\underset{\sim}{p}_2)^2\, e^{-sq^2}. \tag{C10}$$

But this can be obtained by differentiating the generating function

$$I_0(\alpha; s) \;=\; \int d^d q\; e^{-sq^2 + \underset{\sim}{q}\cdot\sum_i \alpha_i \underset{\sim}{p}_i}, \tag{C11}$$

with respect to α_1 and α_2 twice and setting all the α_i to zero. On the other hand the generating function may be evaluated for general d, using (C4) to (C7), simply as:

$$I_0(\alpha; s) \;=\; s^{-d/2}\, \pi^{d/2}\, \exp[(\sum_i \alpha_i\, \underset{\sim}{p}_i)^2/4s]. \tag{C12}$$

These considerations suffice for field-theoretic applications and hence for the formal developments of ε-expansions. One may, however, be concerned about the use of a lattice cut off such as enters in, say, the exact solution of the spherical model. If one has nearest neighbor couplings on a hypercubic lattice one then encounters d-fold integrals like

$$I^{(d)}(z) = \int_{-\pi}^{\pi} \frac{d\theta_1}{2\pi} \cdots \int_{-\pi}^{\pi} \frac{d\theta_d}{2\pi}\; [z + \sum_{j=1}^{d} \cos\theta_j]^{-1}. \tag{C13}$$

However, by using (C9) and the integral expression for the Bessel function $J_0(x)$ this can be transformed to

$$I^{(d)}(z) = \int_0^\infty [J_0(s)]^d e^{-zs} ds, \tag{C14}$$

which is again well-defined for general d. The d-dependent critical exponents obtained for the spherical model this way agree precisely (to the orders of ε available) with the ε expansion expressions evaluated with $n \to \infty$. (see sec. 4.6).

One may discuss the continuation of dimensionality for lattice models more generally. It is natural to restrict attention to hypercubic lattices which a moment's thought shows have a coordination number 2d. This statement, of course, immediately extends to nonintegral values of d!. To see how to proceed further, consider, to be concrete, the susceptibility of a spin 1/2 Ising model with nearest neighbor coupling of strength J. The susceptibility may be expressed in terms of the spin-spin correlation functions $\langle s_{\underset{\sim}{0}} s_{\underset{\sim}{R}} \rangle$ between sites $\underset{\sim}{0}$ and $\underset{\sim}{R}$ as

$$\bar{\chi}(T) = \sum_{\underset{\sim}{R}} \langle s_{\underset{\sim}{0}} s_{\underset{\sim}{R}} \rangle, \tag{C15}$$

where, as usual, the angular brackets denote the statistical expectation defined here by

$$\langle A \rangle = 2^{-N} \sum_{\{s_i=\pm 1\}} \Big(A(s) \prod_{(ij)} e^{Ks_is_j} \Big) \Big/ 2^{-N} \sum_{\{s_i=\pm 1\}} \prod_{(ij)} e^{Ks_is_j}, \qquad \text{(C16)}$$

the sums running over all spin states of the lattice. In the high temperature limit, $K = J/k_BT \to 0$, one can, as pointed out in Sec. 4.5, expand the exponential factors in powers of K as

$$e^{Ks_is_j} = 1 + Ks_is_j + \frac{1}{2!} K^2 s_i^2 s_j^2 + \dots . \qquad \text{(C17)}$$

Each power of K in the full expansion of (C17) is clearly associated with a nearest neighbor lattice bond. When calculating the susceptibility an extra pair of spins, $S_{\underset{\sim}{0}}$ and $S_{\underset{\sim}{R}}$, will appear in the expression for the numerator of (C16). The resulting expansion in terms of multiple spin products must be summed over all possible spin configurations. The contribution of any given product of spins may then be evaluated by using the identity

$$\frac{1}{2} \sum_{s=\pm 1} s^k = 1, \text{ for k even,} \qquad \text{(C18)}$$
$$= 0, \text{ for k odd.}$$

Finally, by collecting up similar terms one sees that the expansion for the susceptibility can be written for <u>any</u> lattice in the diagrammatic form

$$\bar{\chi} = 1 + a_1 [\,/\,] K + a_2 [\wedge] K^2 + a_3 K^3 [\wedge\vee]$$

$$+ (a_{4,1} [\square] + a_{4,2} [\wedge\wedge]) K^4 + \dots . \qquad \text{(C19)}$$

The coefficients $a_1 = 1$, a_2, a_3, etc. depend on the topology of the associated diagram, representing bonds on the lattice, but are <u>independent</u> of the lattice structure (or dimensionality) which, in turn, is embodied only in the values ascribed to the <u>graph embedding constants</u>, $[\,/\,]$, $[\wedge]$, etc. (The reader should go through the derivation of the first few terms to see how this works (See also Sec. 4.5).

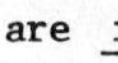

Now, more explicitly, $[\,/\,]$ denotes the number of bonds per lattice site. In a d-dimensional hypercubic lattice, this is evidently $[\,/\,] = \frac{1}{2}(2d) = d$, bearing in mind that each nearest-neighbor bond is shared between two lattice sites. In a similar way, $[\wedge]$ denotes the number of chains of length two bonds (per lattice site) where successive bonds of the chain must not lie on top of one another: since there are $\frac{1}{2}(2d)$ choices per site for the first bond and $(2d-1)$ remaining choices for placing the second bond at one end of the first bond, we obtain $[\wedge] = d(2d-1)$.

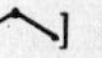

Likewise, the square yields [□] = 2d(d-1), and so on. Evaluation at d=3 then yields the series quoted in (4.19).

For completeness we digress a moment to recall that the Ising series presented in (4.19) is given in terms of the variable

$$v = \tanh K = K - \frac{1}{3}K^3 + \dots , \qquad \text{(C20)}$$

rather than in powers of K directly. The usefulness of this variable, in fact, arises directly from the diagrammatic or graphical expansion technique: thus, for Ising variables, for which $s_i s_j$ can take only the two values +1 or −1, it is simpler to replace the infinite expansion (C17) by the two-term identity

$$e^{Ks_i s_j} = (\cosh K)\,[1 + v s_i s_j], \qquad \text{(C21)}$$

which is easily checked. In making expansions of (C16) in powers of v, each bond now appears only once, with weight v, rather than multiply with weights K, K^2, K^3, ... as entailed in the use of (C17).

To return to the general theme, it should now be clear that even the most complicated diagram entering in a graphical expansion will have embedding constants or weights that are just polynomials in the dimensionality d. It follows that each term in the high temperature expansion of $\overline{\chi}(T)$ can be analytically continued to arbitrary values of d. Thus, at least while the series converges, the susceptibility itself can also be defined for continuous dimensionality.

The same procedure works for all other properties. This lattice definition and the prescription of introducing continuous d through various integrals do not obviously agree in general (and no such proofs have been presented). Wherever they have been tested, however, the different prescriptions appear to coincide and, in particular, it is reasonable to expect that they will all yield the same results in the critical region.

APPENDIX D Hyperscaling and Dangerous Irrelevant Variables

Consider the hyperscaling relation $d\nu = 2-\alpha$. This relation was obtained in Sec. 5.5.4 from the renormalization of the correlation length according to

$$\xi[\overline{\mathcal{H}}] = b\xi[\overline{\mathcal{H}}'], \text{ with } \xi \sim t^{-\nu}, \qquad \text{(D1)}$$

which merely represents the basic rescaling of lengths, and of the free energy according to

$$f[\overline{\mathcal{H}}] = b^{-d} f[\overline{\mathcal{H}}'], \text{ with } f_{\text{sing.}} \sim t^{2-\alpha}. \qquad \text{(D2)}$$

The question is: "How can the arguments go wrong, as the breakdown of hyperscaling in large dimensionalities implies must happen?". There are, in fact, various mechanisms by which hyperscaling can fail. To explain the most likely mechanism, which does not actually violate the basic structure of the renormalization group theory[53] let us recapitulate the argument for hyperscaling.

As seen in Sec. 5.5, near a fixed point the free energy depends on a number of scaling fields, g_1, g_2, ..., in terms of which it should scale asymptotically in the form

$$f(g_1,g_2,\ldots) \approx b^{-d} f(b^{\lambda_1} g_1,\ b^{\lambda_2} g_2,\ \ldots). \tag{D3}$$

If we make the standard choice and identification

$$b^{\lambda_1} = 1/g_1 \approx 1/t, \tag{D4}$$

we obtain

$$f(g_1,g_2,\ \ldots) \approx t^{d/\lambda_1} f(1,\ \frac{g_2}{t^{\lambda_2/\lambda_1}},\ \frac{g_3}{t^{\lambda_3/\lambda_1}},\ \ldots)\ . \tag{D5}$$

The natural example to consider is provided by the simplest continuous spin ferromagnet where $g_2 \approx H$ and $\lambda_2/\lambda_1 = \Delta$, while $g_3 = u$ represents the coefficient of the quartic spin term [see Sec. 6.1] with $\lambda_3/\lambda_1 = \phi_3 \equiv \phi$. More generally, however, we need not specify the nature of u. Then we have

$$f(t,H,u) \approx t^{d/\lambda_1} Y_0(\frac{H}{t^\Delta},\ \frac{u}{t^\phi}), \quad \text{with} \quad Y_0(y,z) \approx f(1,y,z,0,0,\ \ldots), \tag{D6}$$

where, for simplicity, we now ignore all further variables which we thus assume are "harmless" irrelevant variables.

The scaling exponent ϕ may, in principle, be positive, negative or zero. If it is positive then u is actually a relevant variable and its flow under renormalization is <u>away</u> from the fixed point selected. One is then dealing with some sort of <u>multicritical</u> situation which is not pertinant to the present issue. On the other hand, if ϕ is negative u is formally irrelevant and on approach to the critical point, one has

$$\frac{u}{t^\phi} = ut^{|\phi|} \to 0 \quad \text{as} \quad t \to 0. \tag{D7}$$

Therefore however large u was initially, the scaled combination u/t^ϕ becomes arbitrarily small asymptotically close to the critical point, and so, formally, one has

$$f \approx t^{d/\lambda_1} Y_0\left(\frac{H}{t^\Delta}, 0\right) \approx t^{2-\alpha} Y\left(\frac{H}{t^\Delta}\right), \tag{D8}$$

as argued in Sec. 5.5.4. On making the identification

$$2 - \alpha = d/\lambda_1, \tag{D9}$$

and using the general result $\nu = 1/\lambda_1$ [see (5.96)] we arrive at the hyperscaling relation $d\nu = 2 - \alpha$. Evidently the asymptotic scaling function is given by

$$Y(y) \approx Y_0(y,0). \tag{D10}$$

Now this analysis relies implicitly on the <u>assumption</u> that $Y_0(y,0) = f(1,y,0, \ldots)$ has a well-defined value. It may happen, however, that the full function $Y_0(y,z)$ actually <u>diverges</u> when $z \to 0$. Note that the fact that u is an irrelevant variable in no way excludes this possibility! To examine the likely consequences of such a situation let us postulate a simple power law divergence of the form

$$Y_0(y,z) \approx \frac{W(y)}{z^\mu} \text{ as } z \to 0+ \text{ with } \mu > 0. \tag{D11}$$

An irrelevant variable, u, giving rise to this type of behavior is characterized as a <u>dangerous irrelevant variable.</u> Substituting this assumption into (D6) and letting $ut^{|\phi|} \to 0$ as $t \to 0$ now yields

$$f \approx t^{d/\lambda_1} W\left(\frac{H}{t^\Delta}\right)/u^\mu \; t^{\mu|\phi|} \approx t^{(d/\lambda_1)-\mu|\phi|} \tilde{Y}\left(\frac{H}{t^\Delta}\right), \tag{D12}$$

where $\tilde{Y}(y) = W(y)/u^\mu$ evidently represents a new asymptotic scaling function. Interpretation of this new behavior in terms of the standard thermodynamic exponents (still accepting $\nu = 1/\lambda_1$) yields the modified relation

$$2 - \alpha = d\nu - \mu|\phi|. \tag{D13}$$

This clearly represents a breakdown of the original hyperscaling relation! Notice, nevertheless, that the renormalization group framework has been preserved intact: the only flaw in the original argument was a failure to recognize and allow for possible singular behavior of the scaling function.

But how far-fetched is the idea of a scaling function diverging as in (D11)? The answer is "Not at all!". Indeed, when u represents the coefficient of the s^4 term in a continuous spin model, just such a divergence is found when one calculates the form of the free energy scaling function (for nonzero but small u) above four dimensions. Since the us^4 term with $u > 0$ is essential for the convergence of the partition function below the mean field critical temperature a divergence as $u \to 0$ is hardly very surprising. Nevertheless, the actual behavior is, in fact, a little more subtle even than supposed in (D11). What one finds, first, are the renormalization group eigenvalues

$$\lambda_1 = 2, \ \lambda_2 = \tfrac{1}{2}d + 1, \ \text{and} \ \lambda_3 = 4 - d, \tag{D14}$$

which pertain to the Gaussian fixed point studied in Sec. 6.4 which is stable for $d > 4$ since $\lambda_3 \equiv \lambda_u$ and hence,

$$\phi = \lambda_3/\lambda_1 = -\tfrac{1}{2}(d - 4), \tag{D15}$$

are then negative. The standard renormalization group exponent identifications yield $\nu = 1/\lambda_1 = \frac{1}{2}$, which is the expected classical value, but also

$$2 - \alpha = \frac{d}{\lambda_1} = \frac{1}{2}\,d, \ \text{and} \ \Delta = \frac{\lambda_2}{\lambda_1} = \frac{1}{4}\,d + \frac{1}{2}, \tag{D16}$$

neither of which correspond to the classical values, $\alpha = 0$ and $\Delta = 3/2$. However, the scaling function $Y_0(y,z)$ entering (D6) does behave in a singular manner when $z \to 0$: specifically one finds

$$Y_0(y,z) = \frac{1}{z}\,W_0(y\,z^{\frac{1}{2}}), \tag{D17}$$

where $W_0(w)$ is a well-behaved function. This resembles the postulate (D11) with $\mu=1$ and so, via (D13), yields the 'operative' or observed critical exponent

$$2 - \alpha = d\nu - \mu|\phi| = \frac{1}{2}\,d - \frac{1}{2}\,(d-4) = 2. \tag{D18}$$

Thus we obtain $\alpha = 0$ which is now in accord with the classical predictions (and, of course, violates hyperscaling).

Evidently, then, u <u>is</u> a dangerous irrelevant variable at the Gaussian fixed point when $d > 4$. Further, however, u, in the guise of the scaled variable $z = ut^{|\phi|}$, also enters as a factor in the <u>argument</u> of the scaling function W_0. This argument thus becomes

$$w = y\,z^{\frac{1}{2}} = \frac{H}{t^{\lambda_2/\lambda_1}}\,(u^{\frac{1}{2}}\,t^{\frac{1}{2}|\phi|}) = \frac{Hu^{\frac{1}{2}}}{t^{\Delta}}, \tag{D19}$$

where now the operative scaling exponent for the ordering field is seen to be

$$\Delta = \frac{\lambda_2}{\lambda_1} - \frac{1}{2}\,|\phi| = \frac{1}{4}\,d + \frac{1}{2} - \frac{1}{4}\,(d-4) = \frac{3}{2}. \tag{D20}$$

This contrasts sharply with (D15) but agrees with the classical prediction!

The moral of this story is that the standard scaling relations for critical exponents depend, in their derivation, on assumptions, usually left tacit, about the nonsingular or nonvanishing behavior of various scaling functions and their arguments. In many cases these assumptions are valid and may be confirmed by explicit calculation (or other knowledge) but in certain circumstances they may fail, in which case an exponent relation may change its form. Other nontrivial cases of dangerous irrelevant variables are known so that the phenomenon, although not common, is not truly exceptional.

BIBLIOGRAPHY

We list here some reviews and monographs as suggestions for further reading and references concerning renormalization group theory and its applications. Some are at a more introductory and less technical level than the present lectures: these are listed first. Others follow roughly in order of increasing technical sophistication.

1. B. Widom in "Fundamental Problems in Statistical Mechanics" vol. III, Ed. E. V. G. Cohen (North Holland, 1975) pp. 1-45.

2. P. Pfeuty and G. Toulouse, "Introduction to Renormalization Groups and Critical Phenomena" (Wiley, 1977).

3. M. E. Fisher, Rev. Mod. Phys. 46, 597-616 (1974).

4. D. J. Wallace and R. K. P. Zia, Repts. Prog. Phys. 41, 1-85 (1978).

5. A. Z. Patashinskii and V. L. Pokrovskii, "Fluctuation Theory of Phase Transitions" (Pergamon Press, 1979).

6. L. E. Reichl, "A Modern Course in Statistical Physics", (Univ. of Texas Press, 1980), Chap. 4.

7. A. P. Young in "Ordering in Strongly Fluctuating Condensed Matter Systems", Ed. T. Riste (Plenum Press, 1980) pp. 11-31, 271-284.

8. S.-K. Ma, "Modern Theory of Critical Phenomena" (Benjamin, 1976).

9. K. G. Wilson, Rev. Mod. Phys. 47, 773-840 (1975).

10. K. G. Wilson and J. Kogut, Physics Reports 12 C (1974).

11. B. I. Halperin in "Physics of Low-Dimensional Systems", Eds. Y. Nagaoka and S. Hikami (Publ. Office, Prog. Theoret. Phys., Kyoto, 1979).

12. Articles by A. Aharony, by E. Brézin, J. C. Le Guillon and J. Zinn-Justin, by Th. Niemeijer and J. M. J. van Leeuwen, by F. J. Wegner and by others in "Phase Transitions and Critical Phenomena", vol. 6, Eds. M. S. Green and C. Domb (Academic Press, 1976). Other volumes in this series provide a broad coverage of other aspects of critical phenomena.

13. D. J. Amit, "Field Theory, the Renormalization Group and Critical Phenomena" (McGraw-Hill, 1978).

References

1. See First Book of Kings, Chap. 7, verse 23.

2. For example, see the famous pictures reproduced in H. E. Stanley, "Introduction to Phase Transitions and Critical Phenomena" (Oxford Univ. Press, 1971) Fig.1.7.

3. D. Balzarini and K. Ohrn, Phys. Rev. Lett. 29, 840 (1972); D. Balzarini, Can. J. Phys. 52, 499 (1974).

4. M. J. Sienko, J. Chem. Phys. 53, 566 (1970).

5. M. I. Bagatskii, A. V. Voronel' and V. G. Gusak, Sov. Phys. JETP 16, 517 (1963).

6. G. Ahlers, Phys. Rev. Lett. 24, 1333 (1970).

7. B. J. Lipa, unpublished work at Stanford University (1981).

8. See, for example H. E. Stanley loc. cit. Figs. 11.4 and 11.5 or M. E. Fisher, Rep. Prog. Phys. 30, 615 (1967) Fig. 18, etc.

9. See, for example, H. B. Tarko and M. E. Fisher, Phys. Rev. B11, 1217 (1975) which lists data for various universal and nonuniversal parameters of two and three-dimensional Ising models.

10. Based on the general scaling treatment presented in Proc. Nobel Symposium 24, Eds. B. Lundqvist and S. Lundqvist (Academic Press, 1974) pp. 16-37.

11. Frenkel's philosophy was expressed more fully in a 1946 review article devoted to the theory of metals. I came across his formulation two decades ago in I. E. Tamm's obituary of Yakov Il'ich Frenkel which appeared in Soviet Physics Uspekhi 5, No. 2, Sept./Oct. (1962).

12. See e.g. the review "Simple Ising models still thrive": M. E. Fisher, Physica 106A, 28 (1981) and references therein.

13. C. Domb, Adv. Phys. 9, 149 (1960).

14. See e.g. Domb loc. cit.; C. Domb and M. F. Sykes, J. Math. Phys. 2, 63 (1961); M. E. Fisher, Rept. Prog. Phys. 30, 615 (1967) Figs. 13, 14; D. S. Gaunt and A. J. Guttmann, in "Phase Transitions and Critical Phenomena", Vol. 3, Eds. C. Domb and M. S. Green (Academic Press, London, 1974).

15. See e.g. J. Zinn-Justin, J. Physique 42, 783 (1981). J.-H. Chen, M. E. Fisher and B. G. Nickel, Phys. Rev. Lett. 48, 630 (1982).

16. See R. Balian and G. Toulouse, Phys. Rev. Lett. 30, 544 (1973); M. E. Fisher, Phys. Rev. Lett. 30, 679 (1973).

17. We follow D. R. Nelson and M. E. Fisher, Ann. Phys. (N.Y.) 91, 226 (1975).

18. Various concrete definitions of the correlation length and their inter-relations are discussed in an Ising model context by M. E. Fisher and R. J. Burford, Phys. Rev. 15b, 583 (1967) and M. B. Tarko and M. E. Fisher, Phys. Rev. B 11, 1217 (1975). In most simple critical phenomena all definitions agree up to a constant factor as the critical point is approached. In more complex situations, however, there may be different correlation lengths for different degrees of freedom even though (5.62) is maintained. Sometimes, as in dipolar interactions in Ising systems, two differently behaving correlation lengths are associated with orthogonal spatial directions in which case one may also need to generalize (5.62) to allow for different rescaling factors, say $b_\parallel$ and $b_\perp$, in different directions.

19. See R. B. Griffiths, Physica 106A, 59 (1981); R. B. Griffiths and P. A. Pearce, J. Stat. Phys. 20, 499 (1979).

20. F. J. Wegner, J. Phys. C 7, 2098 (1974).

21. L. P. Kadanoff, Physics 2, 263 (1966).

22. M. J. Buckingham in Critical Phenomena, Eds., M. S. Green and J. V. Sengers (National Bureau of Standards Misc. Publ. 273, Washington, D. C., 1966), p. 95.

23. See also P. Seglar and M. E. Fisher, J. Phys. C 13, 6613 (1980).

24. V. L. Ginzburg, Sov. Phys. Solid St. 2, 1824 (1960).

25. L. P. Kadanoff et al., Rev. Mod. Phys. 39, 395 (1967).

26. P. C. Hohenberg in Fluctuations in Superconductors edited by W. S. Goree and F. Chilton (Stanford Research Inst., Menlo Park, 1968) p. 305.

27. J. Als-Nielsen and R. J. Birgeneau, Amer. J. Phys. 45, 554 (1977).

28. Discovered originally by K. G. Wilson and M. E. Fisher, Phys. Rev. Lett. 28, 240 (1972) where the ε-expansion was introduced.

29. E. Brézin, J. C. LeGuillon, J. Zinn-Justin and B. G. Nickel, Phys. Lett. 44A, 227 (1973).

30. A. A. Vladimirov, D. I. Kazakov, and O. Tarasov, Sov. Phys. JETP 50, 521 (1979).

31. M. E. Fisher, S.-K. Ma, and B. G. Nickel, Phys. Rev. Lett. 29, 917 (1972).

32. M. E. Fisher and D. R. Nelson, Phys. Rev. Lett. 32, 1352 (1974).

33. M. E. Fisher and A. Aharony, Phys. Rev. Lett. 30, 559 (1973).

34. A. I. Larkin and D. E. Khmel'nitskii, Sov. Phys. JETP 29, 1123 (1969) discovered some of these features before the general development of renormalization group theory.

35. See also A. Aharony, Phys. Lett. 44A, 313 (1973).

36. E. K. Riedel and F. J. Wegner, Phys. Rev. Lett. 29, 349 (1972).

37. R. Abe, Prog. Theoret. Phys. 48, 1414 (1972).

38. A. M. Polyakov, Phys. Lett. 59B, 79 (1975); see also E. Brézin and J. Zinn-Justin, Phys. Rev. Lett. 36, 691 (1976).

39. J. M. Kosterlitz and D. J. Thouless, J. Phys. C6, 1181 (1973).

40. J. M. Kosterlitz, J. Phys. C7, 1046 (1974).

41. J. V. José, L. P. Kadanoff, S. Kirkpatrick and D. R. Nelson, Phys. Rev. B16, 1217 (1977).

42. B. I. Halperin and D. R. Nelson, Phys. Rev. Lett. 41, 121 (1978); E 41, 519 (1978).

43. A. P. Young, Phys. Rev. B19, 1855 (1979).

44. Th. Niemeijer and J. M. J. van Leeuwen, Physica (Utrecht) 71, 17 (1974).

45. A. A. Migdal, Zh. Eksp.-Teor. Fiz. 69, 810, 1457 (1975).

46. L. P. Kadanoff, Ann. Phys. (N.Y.) 100, 359 (1976).

47. See e.g., B. I. Halperin in "Collective Properties of Physical Systems" Nobel Symp. No. 24., Eds. B. Lundqvist and S. Lundqvist (Academic Press, 1974); P. C. Hohenberg and B. I. Halperin, Rev. Mod. Phys. 49, 435 (1977); J. Tobochnik, S. Sarker, and R. Cordery, Phys. Rev. Lett. 46, 1417 (1981).

48. D. R. Nelson, Phys. Rev. B11, 3504 (1975).

49. J. Rudnick and D. R. Nelson, Phys. Rev. 13, 2208 (1976).

50. K. G. Wilson, Rev. Mod. Phys. 47, 773 (1975).

51. A review of graph theoretical terminology is given by J. W. Essam and M. E. Fisher, Rev. Mod. Phys. 42, 271 (1970).

52. K. G. Wilson, Phys. Rev. D7, 2911 (1973).

53. M. E. Fisher, in "Renormalization Group in Critical Phenomena and Ouantum Field Theory: Proceedings of a Conference", Edited by J. D. Gunton and M. S. Green (Temple University, 1974) pp.65-68.

PHASE TRANSITIONS AND INSTABILITIES

A course of lectures by :

H. Thomas

Institut für Physik der Universität Basel

Manuscript compiled, from lecture notes taken by :

D. Eyre

National Research Institute for Mathematical Sciences

of the CSIR

CONTENTS

1. Historical Introduction

The fact that the same substance may exist in different 'phases' which show *discontinuous* transitions into each other has always been a fascination to scientists. In fact, the difference between different phases of the same substance was felt more significant than the difference between different substances. The 'four elements' certainly characterize phases rather than substances, and their difference was thought to be due to different forms of the 'atoms'. The discovery of the critical point[1] (T Andrews)

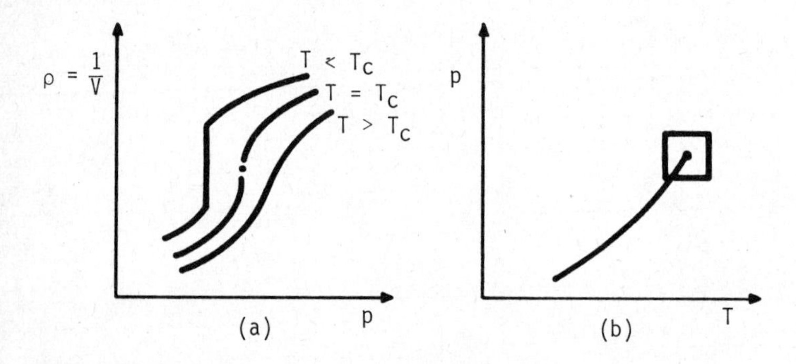

Fig. 1.1 : Projections of the $p\rho T$ surface for a liquid-gas system. (a) Isothermal cross-sections. (b) Vapour pressure curve (coexistence line).

was a great surprise and increased the fascination in phase transitions: The existence of a continuous path for changing one phase into the other was incompatible with the concept of different molecules for different phases, and the appearance of a phase transition was interpreted to arise from a shift of the balance between attractive and repulsive properties of the molecules with temperature.

In the course of time, other critical points were discovered: Critical points in ferromagnets[2] (J Hopkinson), which behaved remarkably similarly to the liquid-gas critical point (apart from the symmetry between H and -H which has no counterpart in the liquid-gas system).

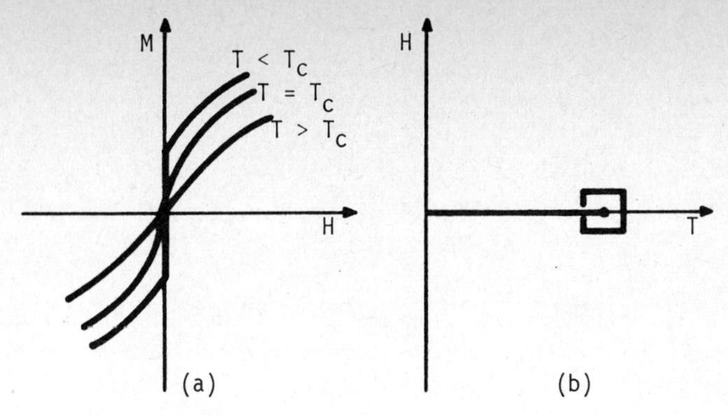

Fig. 1.2: Projections of the HMT surface for a ferromagnetic system.
(a) Isothermal cross-sections. (b) Coexistence line in HT plane.

A wide variety of other physical systems showing critical points include:

Binary liquid mixtures

Binary alloys

Antiferromagnets

Ferroelectrics and crystals with structural phase transitions

Liquid crystals

Superconductors

Superfluids

These systems show a number of interesting phase diagrams, some of which will be discussed during this course.

2. Classical Theory of the Critical Point

2.1 Basic Assumption of Classical Theory

In 1873 J D van der Waals[3] gave an equation of state

$$(p+a/v^2)(v-b) = RT \ , \tag{2.1}$$

which when combined with the Maxwell construction described a liquid-gas system both above and below the critical point. In 1907 P Weiss[4] did the same for the ferromagnet,

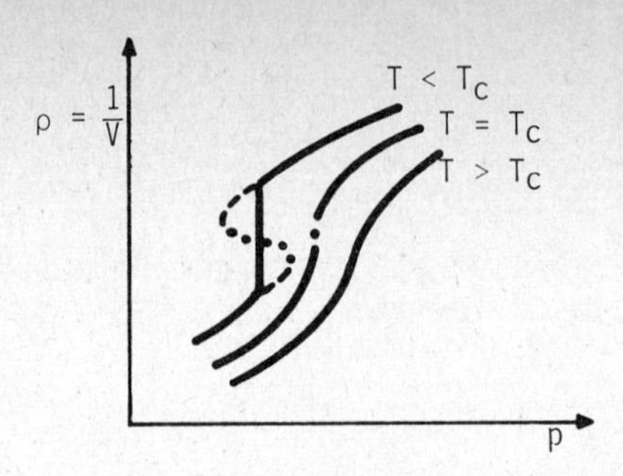

Fig. 2.1 : Classical prediction of isotherms in ρp plane for a liquid-gas system

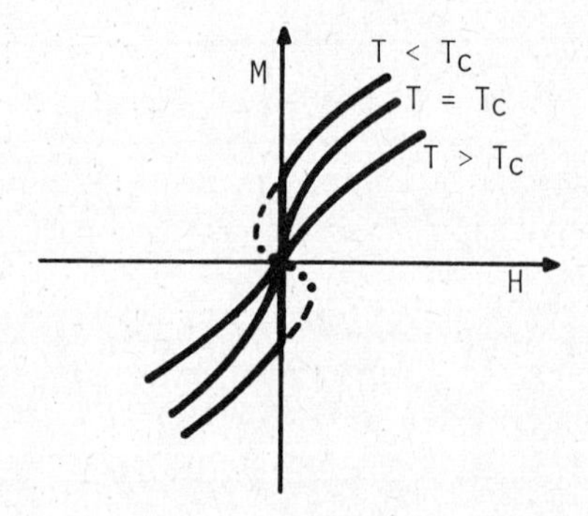

Fig. 2.2 : Classical prediction for isotherms in MH plane for a ferromagnet.

$$M/M_o = L(\mu_o(H+\Lambda M)/RT); \; L(x) = \coth(x) - \frac{1}{x} . \qquad (2.2)$$

These equations of state lead to the prediction of a universal scaling behaviour close to the critical point, known as the 'law of corresponding states'. This behaviour does not depend on the details of the above equations of state, but rests on one basic hypothesis:

The two phases at $T < T_c$ can be connected by a theoretical construction of a continuous set of states, such that the forces (p, H) are *analytic* but *non-monotonous* functions of the system variables (v, M) and temperature T ('van der Waals loops') (dotted lines in Figs. 2.1 and 2.2).

The discontinuity on the coexistence line and the singularities at the critical point arise solely from the *inversion* of these analytic functions and the Maxwell construction.

Thus $p(v,T)$ and $M(H,T)$ are assumed to be analytic functions in both arguments, which satisfy

$$\left.\frac{\partial p}{\partial v}\right|_{p_c T_c} = \left.\frac{\partial^2 p}{\partial v^2}\right|_{p_c T_c} = 0,$$

$$\left.\frac{\partial H}{\partial M}\right|_{H=0,T_c} = \left.\frac{\partial^2 H}{\partial M^2}\right|_{H=0,T_c} = 0 , \tag{2.3}$$

at the critical point. Thus, the leading terms of the Taylor expansion at the critical point are given by

$$\begin{aligned} p-p_c &= a(T-T_c)(v-v_c) + b(v_c-v)^3 + c(T-T_c), \\ H &= a(T-T_c)M + bM^3 . \end{aligned} \tag{2.4}$$

From this expansion one easily obtains the behaviour on the *coexistence line* $p_{liq} = p_{gas}$ and $H = 0$,

$$|\rho-\rho_c| , M \sim (T_c-T)^{\frac{1}{2}}; \tag{2.5}$$

the behaviour of the *isothermal compressibility* $K = -(1/v)(\partial v/\partial p)_T$ and the *isothermal susceptibility* $\chi = (\partial M/\partial H)_T$ along the coexistence line and its continuation

$$K,\chi = C_{\pm}|T_c-T|^{-1} , C_+ = 2C_- , \tag{2.6}$$

where the factor C_+ is twice as large above T_c as the factor C_- below T_c; and the behaviour on the *critical isotherm* $T = T_c$,

$$p-p_c \sim (\rho-\rho_c)^3 , H \sim M^3 , \tag{2.7}$$

where we have used the molar density $\rho = 1/v$ instead of the molar volume v. In modern language, one expresses these results in terms of 'critical exponents' β, γ, δ defined by

$$|\rho-\rho_c| \ , \ M \sim (T_c-T)^{\beta} \ ,$$

$$K,\chi \sim |T_c-T|^{-\gamma} \ , \tag{2.8}$$

$$p-p_c \sim (\rho-\rho_c)^{\delta} \ , \ H \sim M^{\delta} \ ,$$

whence $\beta = 1/2$, $\gamma = 1$, $\delta = 3$ for the classical theory.

2.2 Thermodynamic Description of a System

2.2.1 System Variables and Conjugate Forces

A state of the system is described by a set of system variables Q:

For *lattice models*, an n-component variable Q_ℓ is associated with every lattice site $\ell = 1,2,\ldots N$. For *continuum models*, the states are described by n-component densities $Q(x)$, $x \in R^d$.

With each system variable Q_ℓ or $Q(x)$ we associate a conjugate force F_ℓ or $F(x)$ respectively:

System	Q	F	n
Liquid-gas	ρ	μ	1
Ferromagnet	M	H	3
Ferroelectric	P	E	3
Quadrupolar (e.g. liquid-crystal)	Q (quadrupole moment)	F (field gradient)	5

Examples of Q, F and n for a number of physical systems are given above (μ is the chemical potential).

For the ferromagnetic and ferroelectric systems in anisotropic cases, only the com= ponents along the easy axis ($n = 1$ for the Ising model) or easy planes ($n = 2$ for the XY model) need be considered.

2.2.2 Thermodynamic Potential

We shall consider a system in a heat bath at temperature T. For given values of the system variables and T, the *Helmholtz free energy* $\mathbb{F}(\{Q\}),T)$ yields the thermal and caloric *equations of state*:

$$\frac{\partial \mathbb{F}}{\partial Q_\ell} = F_\ell(\{Q\}),T) \ , \ \frac{\partial \mathbb{F}}{\partial T} = -S(\{Q\},T) \ . \tag{2.9}$$

Note that we have used $\{Q\}$ to denote the variables $Q_1,\ldots,Q_\ell,\ldots Q_N$. The second derivatives of $\mathbb{F}$ yield the *response functions*, in particular the isothermal reci= procal susceptibility matrix

$$\chi^{-1}_{\ell\ell'} = \left(\frac{\partial F_\ell}{\partial Q_{\ell'}}\right)_T = \frac{\partial^2 \mathbb{F}}{\partial Q_\ell \partial Q_{\ell'}} \ , \tag{2.10}$$

and the heat capacity at $Q_\ell = Q = \text{constant}$,

$$C_Q = T\left(\frac{\partial S}{\partial T}\right)_{\{Q\}} = -T\,\frac{\partial^2 \mathbb{F}}{\partial T^2} \ . \tag{2.11}$$

Assuming that the equation of state has a unique inverse, i.e. $Q_\ell = Q_\ell(\{F\},T)$, one obtains the *Gibbs potential* $G(\{F\},T)$ for given values for the forces and temperature by means of a Legendre transformation

$$G(\{F\},T) = \mathbb{F}(\{Q\}(\{F\},T),T) - \sum_\ell F_\ell \cdot Q_\ell(\{F\},T) \ . \tag{2.12}$$

This transformation yields the equations of state in the form

$$\frac{\partial G}{\partial F_\ell} = -Q_\ell(\{F\},T), \ \frac{\partial G}{\partial T} = -S(\{F\},T). \tag{2.13}$$

It is convenient to define a potential

$$\Phi(\{Q\},\{F\},T) = \mathbb{F}(\{Q\},T) - \sum_\ell F_\ell \cdot Q_\ell \tag{2.14}$$

depending both on the system variables and the forces by including the interaction term $-\sum_\ell F_\ell \cdot Q_\ell$ of the system with the external forces. This potential is a minimum at thermodynamic equilibrium, i.e.

$$\Phi(\{Q\},\{F\},T) = \text{minimum} \ , \tag{2.15}$$

at fixed {F} and T, the minimum condition yielding the thermal equation of state. The minimum value of Φ is just the Gibbs potential G({F},T).

One may use this technique of Legendre transformations in order to eliminate irrelevant variables or to re-introduce variables of interest ('deflating or in= flating the thermodynamic potential'). In the case of continuous systems, sums over ℓ must be replaced by integrals over a continuous variable x, and partial derivatives by functional derivatives.

2.2.3 Properties of the Thermodynamic Potential

In order that the thermal equation of state has a *unique inverse* (except on the coexistence line), $\mathbb{F}$ and therefore Φ must be *convex functions* of the configuration. Necessary conditions for convexity are that the susceptibility χ is non-negative definite and that the specific heat $C_Q \geqslant 0$.

It then follows that the Gibbs potential is a continuous function of {F} and T. In the Ehrenfest classification of phase transitions, if at least one first derivative of the appropriate thermodynamic potential is discontinuous for certain values of {F} and T, then it is said that a phase transition of first order occurs at these values: For the liquid-gas system both the molar density $\rho = \partial G/\partial\mu$ and the entropy $S = -\partial G/\partial T$ are discontinuous across the coexistence line ($T\Delta S$ is the latent heat), whereas for the ferromagnet the magnetization $M = \partial G/\partial H$ is discontinuous but $S = -\partial G/\partial T$ is continuous across the coexistence line. If all the first deriva= tives are continuous, and at least one of the second derivatives is discontinuous, then the phase transition is said to be a second order transition: Along the co= existence line for the liquid-gas and ferromagnet, three second derivatives are discontinuous at T_c. This Ehrenfest scheme has not proved very useful, and a modern classification distinguishes first order phase transitions as discontinuous, and all higher order phase transitions as continuous.

Why does temperature play a special role? One needs a coordinate in parameter space which varies along the coexistence line. Temperature is acceptable for all phase transitions for which the coexistence line has a finite slope $(\partial F/\partial T)_{coex}$ at T_c.

2.3 The Landau Potential

We define Q and F such that F = 0 on the coexistence line and Q = 0 in the high temperature phase on the continuation of the coexistence line.

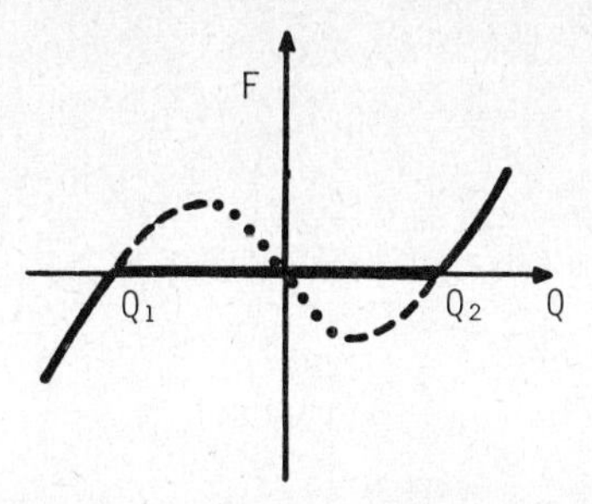

Fig. 2.3 : Equation of state for $T < T_c$. Solid curve is the true equation of state.

The flat portion F = 0 of the true equation of state between the two values $Q_1(T)$ and $Q_2(T)$ characterizing the two coexisting phases gives rise to a flat portion $\mathbb{F}(Q,T) = \mathbb{F}(Q_{1,2},T)$ of the Helmholtz free energy.

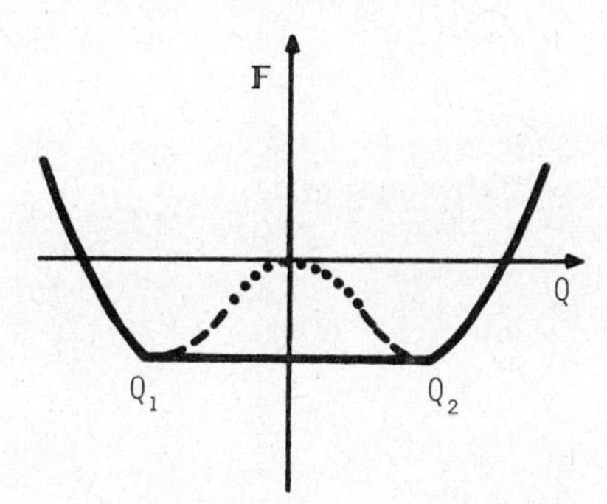

Fig. 2.4 : Helmholtz free energy for $T < T_c$. Solid curve corresponds to the true equation of state.

The assumption of an analytic van der Waals loop connecting the two phases $Q_1(T)$, $Q_2(T)$ is equivalent to the assumption that the two branches of the free energy for $Q < Q_1(T)$ and $Q > Q_2(T)$ are parts of an *analytic* function $\mathbb{F}(Q,T)$. The non-monotonic character of the van der Waals loop is reflected in the non-convexity of this function. We can thus formulate the basic assumption of the Landau Theory:

There exists a thermodynamic potential $\Phi(Q,F,T)$, the Landau potential, which is an analytic function of its arguments, but below T_c is a non-monotonous function of Q, whose convex cover is the true thermodynamic potential.

For the simplest case of a system described by a single variable Q, with n = 1, the leading terms of the equation of state

$$\mathbb{F} = \alpha(T-T_c)Q^2 + b\,Q^4 \;,\quad b > 0 \;, \tag{2.16}$$

yield the leading terms of the Landau potential for *uniform states* ($Q_\ell = Q$ = constant)

$$\Phi = [\tfrac{1}{2}\alpha(T-T_c)Q^2 + \tfrac{1}{4}bQ^4 - FQ]\,V \;. \tag{2.17}$$

We next consider the case of *non-uniform states*: Q(x). The existence of a surface energy between two uniform phases (a surface tension in the liquid-gas system and domain-wall energy in the ferromagnet) shows that there is an energy associated with spatial change. In a continuum model, the free energy density will depend on the gradient ∇Q. Since terms linear in ∇Q vanish by partial integration, the lead= ing term for slow variations will be quadratic in ∇Q. Thus the simplest model Landau potential for a system described by a single continuous variable Q(x) is given by

$$\Phi = \int\{\tfrac{1}{2}aQ^2 + \tfrac{1}{4}bQ^4 + \tfrac{1}{2}c(\nabla Q)^2 - F(x)Q(x)\}\,d^dx,\; a(T) = \alpha(T-T_c),\; b > 0,\; c > 0. \tag{2.18}$$

Introducing the Fourier amplitude

$$Q(x) = \frac{1}{\sqrt{V}}\sum_q Q_q\, e^{iq.x} \tag{2.19}$$

in the linear and quadratic forms yields

$$\Phi = \sum_q\{\tfrac{1}{2}(a+cq^2)\;|Q_q|^2 - F_qQ_{-q}\} + \tfrac{1}{4}b\int Q^4\, d^dx. \tag{2.20}$$

From the expression one may read off immediately the zero-field reciprocal susceptibility above T_c,

$$\chi_q^{-1} = \frac{\partial^2\Phi}{\partial Q_q \partial Q_{-q}} \;, \tag{2.21}$$

whence

$$\chi_q = \frac{1}{a + cq^2} \quad . \tag{2.22}$$

This expression is valid for small q. Similar results are obtained for n > 1 above T_c and for n = 1 below T_c. In the case of a multi-component system, one has to dis= tinquish below T_c the susceptibilities parallel (∥) and perpendicular (⊥) to Q. Only $\chi_{\parallel}$ follows the above behaviour, whereas $\chi_{\perp}$ diverges as $q \to 0$, in the isotropic case.

It must be emphasized again that there is no empirical basis for the assumption of the existence of an analytical Landau potential. In fact, for a finite system, the thermodynamic potential will be analytic, but will be a convex function of Q and will not have any Landau hump. This can be seen by the following simple argument:

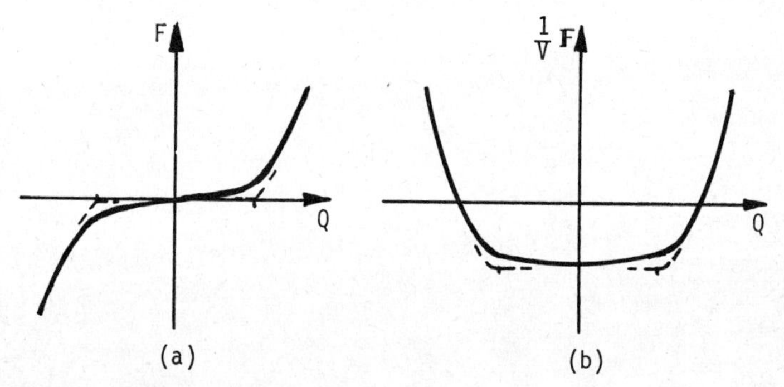

Fig. 2.5 : Analytic behaviour of a finite system. (a) Equation of state. (b) Helmholtz free energy.

A state with an equal distribution over the two phases yielding Q = 0 has a higher entropy, its thermodynamic potential is therefore slightly lower than that of the pure phases. Only in the thermodynamic limit does one obtain a sharp transition and a flat portion of $\mathbb{F}(Q)$. Such finite size effects give rise to the phenomenon of superparamagnetism of small magnetic particles.

We should therefore be prepared for deviations from the Landau theory discussed in other lectures of this course.

2.4 Fluctuations

The thermodynamic variables Q_ℓ we have used in the previous Sections are statistical averages over the fluctuating quantities $\hat{Q}_\ell$. We define the fluctuations δQ_ℓ by

$$\hat{Q}_\ell = Q_\ell + \delta Q_\ell \ , \tag{2.23}$$

where

$$Q_\ell = \langle \hat{Q}_\ell \rangle \ , \ \langle \delta Q_\ell \rangle = 0 \ . \tag{2.24}$$

The $\hat{Q}_\ell$ may be considered either the true microscopic quantities, or one may assume that some 'coarse-graining' (averaging over short-wavelength fluctuations) has already been carried out.

Of experimental interest is the tensor of *equal time correlations*

$$S_{\ell\ell'} = \langle \delta Q_\ell \delta Q_{\ell'} \rangle = \langle \hat{Q}_\ell \hat{Q}_{\ell'} \rangle - Q_\ell Q_{\ell'} \ . \tag{2.25}$$

In a uniform phase, $S_{\ell\ell'}$ depends only on $R_{\ell\ell'} = R_{\ell'} - R_\ell$ on account of the lattice translation symmetry, that is

$$S_{\ell\ell'} = S(R_{\ell\ell'}) \ . \tag{2.26}$$

The correlation between the fluctuations of the Fourier components

$$\delta Q_q = \frac{1}{\sqrt{N}} \sum_\ell \delta Q_\ell \, e^{-iq \cdot R_\ell} \tag{2.27}$$

is given by

$$\langle \delta Q_q \delta Q_{q'} \rangle = \frac{1}{N} \sum_{\ell\ell'} S_{\ell\ell'} \, e^{-i(q.R_\ell + q'.R_{\ell'})} \ . \tag{2.28}$$

For a uniform phase, this reduces to

$$\langle \delta Q_q \delta Q_{q'} \rangle = S_q \delta(q+q') \ , \tag{2.29}$$

where

$$S_q = \sum_{\ell'} S_{\ell\ell'} \, e^{-iq.R_{\ell\ell'}} \tag{2.30}$$

is called the *static structure factor*. It has, apart from a delta function, the significance of a mean square fluctuation of the Fourier components ('fluctuation strength'), and it determines the total cross section for the scattering of light (X-rays) and particles (neutrons) from the fluctuations.

It is important to note that the equal time correlation and the static structure factor can be calculated from thermodynamic quantities by making use of the fluctuation theorem of statistical mechanics. For a classical system

$$S_{\ell\ell'} = kT\chi_{\ell\ell'} \, . \qquad (2.31)$$

Thus, the divergent behaviour of the susceptibility near T_c gives rise to the occurrence of '*critical fluctuations*'. In Landau theory, one obtains from the behaviour of the susceptibility above T_c (for small q) the Ornstein-Zernike behaviour

$$S_q = \frac{kT}{a+cq^2} \, , \quad a = \alpha(T-T_c) \, . \qquad (2.32)$$

Fourier inversion yields the asymptotic behaviour of the correlation function for large $R_{\ell\ell'}$,

$$S_{\ell\ell'} \sim \frac{1}{R_{\ell\ell'}{}^{(d-1)/2}} \cdot e^{-\kappa R_{\ell\ell'}} \, , \quad T > T_c, \qquad (2.33)$$

where

$$\kappa^2 = \frac{a}{c} = \frac{\alpha}{c} (T-T_c) \, , \qquad (2.34)$$

and $\kappa = 1/\xi$ has the significance of a reciprocal *correlation length* ξ. Thus we learn the important fact that the correlation length diverges at the critical point as

$$\xi(T) \sim (T-T_c)^{-\nu} \qquad (2.35)$$

where, in Landau theory, the critical exponent ν has the value $\nu = 1/2$. At the critical point, where the correlation length is infinite, the correlations decay according to the power law

$$S_{\ell\ell'} \sim \frac{1}{R_{\ell\ell'}{}^{d-2}} \; . \tag{2.36}$$

One defines a critical exponent η by

$$S_{\ell\ell'} \sim \frac{1}{R_{\ell\ell'}{}^{d-2+\eta}} \quad : \quad S_q \sim q^{2-\eta} \; , \tag{2.37}$$

which, in Landau theory, has the value $\eta = 0$. The above considerations show that the critical behaviour of the fluctuations is *not* caused by the fluctuations δQ_ℓ themselves becoming large, but rather by the correlation range diverging for $T \to T_c$.

We can make use of the above results in order to demonstrate an internal inconsis= tency of the Landau theory in less than four dimensions: From the analyticity of the thermodynamic potential it follows that the heat capacity, given by

$$C_Q = -T \frac{\partial^2 \mathbb{F}}{\partial T^2} \tag{2.38}$$

is finite at $T = T_c +$, and the heat capacity has a finite jump at T_c. In other words the critical exponent defined by

$$C_Q \sim (T-T_c)^{-\alpha} \tag{2.39}$$

has the value $\alpha = 0$. On the other hand, the heat capacity may also be obtained as the temperature derivative of the internal energy. Now, considering the case of bilinear interactions, i.e. interactions of the form

$$-\tfrac{1}{2} \sum_{\ell\ell'} v_{\ell\ell'} \, Q_\ell Q_{\ell'} \tag{2.40}$$

the interaction contribution to the internal energy may be written

$$E^{int} = -\tfrac{1}{2} \sum_{\ell\ell'} v_{\ell\ell'} \, S_{\ell\ell'} = -\tfrac{1}{2} \sum_q v_q S_{-q} \; . \tag{2.41}$$

Consider now the contribution from small q values ($|q| \leqslant q_o$) such that v_q may be taken as a constant and S_q given by the Ornstein-Zernike form. Then

$$\Delta E = E^{int}(T) - E^{int}(T_c)$$

$$\propto -\sum_{\substack{q \\ |q| \leqslant q_o}} \left[\frac{1}{\kappa^2+q^2} - \frac{1}{q^2} \right] . \tag{2.42}$$

We replace this sum over q by the integral

$$\Delta E \sim \int_o^{q_o} \frac{\kappa^2}{\kappa^2+q^2} q^{d-3} dq \quad , \tag{2.43}$$

and after a change of variable, $\xi = q/\kappa$. we obtain

$$\Delta E \sim \kappa^{d-2} \int_o^{q/\kappa} \frac{\xi^{d-3}}{1+\xi^2} d\xi \quad . \tag{2.44}$$

Consider first the case $d < 4$. As $T \to T_c+$, the integral takes on a finite con= stant value, whence ΔE diverges as

$$\Delta E \sim \kappa^{d-2} \sim (T-T_c)^{d/2-1} \quad , \tag{2.45}$$

and

$$C_Q \sim (T-T_c)^{d/2-2} \quad . \tag{2.46}$$

The exponent $\alpha = 2-d/2$ is less than zero and therefore in disagreement with the Landau theory. For the case $d > 4$ ($T \to T_c+$) the leading term in the integral diverges as $(T-T_c)^{-d/2+2}$, whence

$$\Delta E \sim T-T_c \quad , \tag{2.47}$$

and

$$C_Q \sim \text{constant} \; . \tag{2.48}$$

Thus we obtain a result $\alpha = 0$ which is consistent with Landau theory.

2.5 Validity Condition of Landau Theory

One should distinguish carefully between two potentials used in phase transition theory : One is the thermodynamic Landau potential Φ used in the previous Sections which is a function of the average quantities Q_ℓ. The other, often called the Landau-Ginsburg-Wilson potential, is a distribution potential $\hat{\Phi}$ for the fluctuating quantities $\hat{Q}_\ell$ in the sense that the statistical distribution is given by

$$p(\{Q\},T) = \frac{1}{Z} \exp[-\beta\hat{\Phi}(\{\hat{Q}\},\{F\},T)] \;, \tag{2.49}$$

where

$$\hat{\Phi}(\{\hat{Q}\},\{F\}.T) = \mathbb{F}(\{\hat{Q}\},T) - \sum_\ell F_\ell \hat{Q}_\ell \;. \tag{2.50}$$

If the $\hat{Q}_\ell$ were the true microscopic coordinates then $\hat{\Phi}$ would be the Hamiltonian, and would be independent of T. But we consider the case that some coarse-graining (averaging over short wavelength components) has already been carried out, in which case the distribution potential becomes explicitly T-dependent.

How are the two potentials related? The Gibbs potential $G(\{F\},T)$ is given by

$$\begin{aligned} Z &= \exp[-\beta G(\{F\},T)] \\ &= \int \exp[-\beta\hat{\Phi}(\{\hat{Q}\},\{F\},T)]\, D\hat{Q} \;, \end{aligned} \tag{2.51}$$

and the average quantities Q_ℓ are found from

$$\begin{aligned} Q_\ell &= - \frac{\partial G}{\partial F_\ell} (\{F\},T) \\ &= \frac{1}{Z} \int \hat{Q}_\ell \exp[-\beta\hat{\Phi}(\{\hat{Q}\},\{F\},T)]\, D\hat{Q} \;, \end{aligned} \tag{2.52}$$

which may be inverted to yield $F_\ell(\{Q\},T)$. Legendre transformation yields the Helmholtz free energy

$$\mathbb{F}(\{Q\},T) = G(\{F(\{G\},T)\},T) + \sum_\ell F_\ell(\{Q\},T).Q_\ell \tag{2.53}$$

and thus the Landau potential (without the hump!)

$$\Phi(\{Q\},\{F\},T) = \mathbb{F}(\{Q\},T) - \sum_\ell F_\ell . Q_\ell \;. \tag{2.54}$$

In a compact notation we can write

$$\exp[-\beta\Phi(\{Q\},\{F\},T)] = \int \exp[-\beta\hat{\Phi}(\{Q\},\{F\},T)]\, D\hat{Q}, \tag{2.55}$$

$$Q_\ell \exp[-\beta\Phi(\{Q\},\{F\},T)] = \int \hat{Q}_\ell \exp[-\beta\hat{\Phi}(\{\hat{Q}\},\{F\},T)]\, D\hat{Q} \quad . \tag{2.56}$$

The important fact to realize is that the distribution potential $\hat{\Phi}$ will be an analytic function with a Landau hump at low temperatures. This just expresses the fact that the distribution will peak at the values $\pm$ Q in the low temperature phase. We thus assume a Landau-type form for $\hat{\Phi}$ and ask under what conditions can the above phase-space integral be evaluated by saddle-point integration, i.e. when can the integral be replaced by the maximum value of the integrand[5]. If this condition is satisfied, then the thermodynamic potential at equilibrium is indeed given by the minimum of the analytic function $\hat{\Phi}$, i.e. $\hat{\Phi}$ can be taken as the Landau potential.

We consider specifically an n-component continuum model in zero field

$$\hat{\Phi}(\{\hat{Q}\},T) = \int[\tfrac{1}{2}a\hat{Q}^2 + \frac{1}{4n}\, b(\hat{Q}^2)^2 + \tfrac{1}{2}c(\nabla\hat{Q})^2]\, d^dx \ , \tag{2.57}$$

where $(\nabla\hat{Q})^2$ is defined as

$$(\nabla\hat{Q})^2 = \sum_{\alpha=1}^{n} (\nabla\hat{Q}_\alpha)^2 = \sum_{i=1}^{d} \sum_{\alpha=1}^{n} \left(\frac{\partial\hat{Q}_\alpha}{\partial x_i}\right)^2 \quad , \tag{2.58}$$

and the n- dependence of the quadratic term has been arranged such that a meaningful limit $(n \to \infty)$ exists. We assume $T > T_c$ and introduce new variables

$$x = \sqrt{\frac{c}{|a|}}\, \xi \quad , \quad \hat{Q} = \sqrt{\frac{n|a|}{b}}\ \hat{S} \ , \tag{2.59}$$

such that the integral in the exponent becomes independent of the parameters except for an overall constant Ω. Thus

$$\begin{aligned} \beta\hat{\Phi} &= \Omega\hat{\phi}(\hat{S},T) \ , \\ \hat{\phi}(\hat{S},T) &= \int[\tfrac{1}{2}\hat{S}^2 + \tfrac{1}{4}(\hat{S}^2)^2 + \tfrac{1}{2}(\nabla_\xi\hat{S})^2]\, d^d\xi \quad , \\ \Omega &= \frac{na^{2-d/2}c^{d/2}}{bkT} \quad , \end{aligned} \tag{2.60}$$

such that

$$\exp[-\beta\Phi] = J \int \exp[-\Omega\hat{\phi}(\hat{S})]\, D\hat{S} \quad , \tag{2.61}$$

where J is the Jacobian. The saddle-point integration is justified, i.e. we may write

$$\Phi(Q_{eq}) \approx \frac{1}{\beta}\Omega\,\hat{\phi}(\hat{S}_o) = \hat{\Phi}(\hat{Q}_o) \quad , \tag{2.62}$$

where $\hat{S}_o$ and $\hat{Q}_o$ are the saddle point coordinates, provided

$$\Omega >> 1 \quad . \tag{2.63}$$

It is easy to see that this condition is equivalent to the *Ginsburg condition* $<\delta Q^2> << Q_o{}^2$, where $Q_o(T)$ is the value of Q at a temperature $\sim T - 2(T-T_c)$.

Let us first consider some special cases for which $\Omega >> 1$:

(a) $b \to 0$: Gaussian model. (Undefined at temperatures below T_c).

(b) $c \to \infty$: The interaction range $\xi \propto c^{\frac{1}{2}}$ becomes infinite.

(c) $n \to \infty$: Equivalent to the spherical model.

Now let us consider the case we are really interested in, namely the case of *ordinary phase transitions*. In this case, b, c and n are finite, but $a \propto |T-T_c| \to 0$ as $T \to T_c +$. Thus

$$\Omega \propto a^{2-d/2} \to \begin{cases} 0 \\ \infty \end{cases} \text{for } d \lessgtr 4 \ , \tag{2.64}$$

and we see that Landau theory is correct for d > 4, but for d < 4 Landau theory breaks down close to T_c. In the *displacive limit*, when $T_c = 0$, then the presence of the T in the denominator Ω means that

$$\Omega \propto a^{1-d/2} \to \begin{cases} 0 \\ \infty \end{cases} \text{for } d \lessgtr 2 \quad . \tag{2.65}$$

In this case we see that Landau theory breaks down close to T_c only for d < 2.

It should be noted that the validity condition $\Omega >> 1$ only shows that

$$\Phi(Q_{eq}) = \hat{\Phi}(Q_o) \tag{2.66}$$

but does not guarantee that all exponents are given by the Landau values. In particular, in the $n \to \infty$ case, the exponents require special consideration.

3. Symmetry Aspects

3.1 Symmetry Group of a Phase

Consider state changes occurring on a path in F-T space along the coexistence line, continued beyond the critical point.

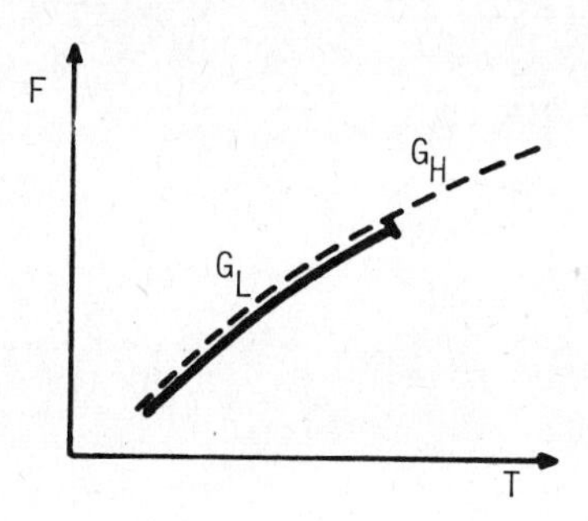

Fig. 3.1 : Continuation of the coexistence line.

All states of one phase have the same symmetry :

Symmetry group G of the phase.

Symmetry group is the set of those transformations permuting like atoms, which are equivalent to combinations of translations, rotations and reflections.

What is the symmetry of states along this path?

In general the symmetry above the critical point is characterized by a symmetry group G_H of the high phase, and below the critical point by a symmetry group G_L of the low phase. We make the following important distinction :

(1) Symmetry-conserving transition : $G_L = G_H$, for example the liquid-gas system.

(2) Symmetry-breaking transition : $G_L \subset G_H$, for example the ferromagnet.

Most of the Landau theory is concerned with symmetry breaking transitions.

3.2 Landau Criteria for Symmetry-Breaking 2nd Order Phase Transition

For symmetry-breaking transitions one may derive a set of criteria for the occurrence of a 2nd order phase transition. Landau theory restricts the number of possibilities: Given a crystal with a certain symmetry Landau theory tells us which transitions can *not* occur by 2nd order phase transition.

For our purpose we will exclude the so called elastic phase transitions, that is the Q's should not be elastic strains because this requires special consideration. Further, for a symmetry-breaking transition, the Q's should not contain a fully symmetric part, i.e. they should transform according to a representation which does not contain the unit representation A_{1g}. For simplicity we will consider only the F = 0 case. In this case the Landau expansion is

$$\Phi(\{Q\},T) = \Phi_o(T) + \sum_{\ell\ell';\lambda\lambda'} a_{\ell\lambda,\ell'\lambda'}(T) Q_{\ell\lambda} Q_{\ell'\lambda'} + \ldots, \qquad (3.1)$$

where Q is a vector and the $Q_{\ell\lambda}$ are components, λ denotes e.g. the x, y or z components, and ℓ stands for the lattice site. Note that there is no linear term in this expansion because Φ is symmetric under G_H. The H-phase is locally stable (stable for small deviations) if the second-order terms are positive definite.

Now, we are interested in the point of 'marginal stability' at T_c. We find the point of marginal stability by diagonalizing the quadratic form in the Landau expansion. This is where the crystal symmetry helps us. The translation symmetry alone can be used to unscramble the spatial part by going over to Fourier components.

$$\Phi(\{Q\},T) = \Phi_o(T) + \tfrac{1}{2} \sum_{q,\lambda\lambda'} a_{q,\lambda\lambda'}(T) Q_{q\lambda} Q_{-q\lambda'} + \ldots \; . \qquad (3.2)$$

We can diagonalize further by making use of the irreducible representation Γ_q of the 'small group'

$$\Phi(\{Q\},T) = \Phi_o(T) + \tfrac{1}{2} \sum_{q,\Gamma_q} a_{q,\Gamma_q}(T) \sum_{\gamma} | Q_{q,\Gamma_q,\gamma}|^2 + \ldots \; , \qquad (3.3)$$

where $Q_{q,\Gamma,\gamma}$ transforms according to the irreducible space group representation $\Gamma = \{q^*,\Gamma_q\}$, having dimension n. We have now diagonalized the quadratic form.

We next want to test the stability of the H-phase. The H-phase is stable as long as all the $a_{q\Gamma}$ are positive. Thus a candidate for T_c is that value of T for which

$$\min a_{q\Gamma}(T_c) = 0 \; . \qquad (3.4)$$

We assume that at T_c only one of the coefficients becomes zero. If more coefficients simultaneously become zero then we have a multicritical point (see Aharony's lecture, this volume).

The 1st Landau criterion states that at T_c the H-phase is marginally stable against a mode which transforms as an irreducible representation $\neq$ the unit representation A_{1g} of G_H. This mode is called the 'order parameter'. The dimension of Γ is n (the number of components of the order parameter). We note the following special cases:

(1) $q = o$: 'ferro' $\left\{\begin{array}{l}\text{electric}\\ \text{magnetic}\\ \text{distortive}\end{array}\right.$ transition.

(2) $q = \frac{1}{2}K$: 'antiferro' transition.
(K is the reciprocal lattice site)

If the value of q is general then there may be an incommensurate phase.

A *second criterion*, due to Lifshitz[6] (1941), is stability against slow spatial changes to the order parameter. The order parameter is allowed to vary slowly in space, $Q_{q\Gamma}(x)$. This caused some confusion in the literature, e.g. Dimmock[7] and Kaplan[8] (1964), and was eventually cleared up by Dzyaloshinsky[9] (1964) and by Goshen, Mukamel and Shtrikman[10] (1974). Let us consider the q-dependence of $a_{q\Gamma}$, viz.

$$a_{q\Gamma} = a_{q_o\Gamma} + b(q-q_o) + c(q-q_o)^2 + \ldots . \tag{3.5}$$

The linear and quadratic terms give rise to Lifshitz invariants of 1st and 2nd order in Φ respectively. If $b \neq 0$ then a_{q_o} cannot be a minimum of a_q.

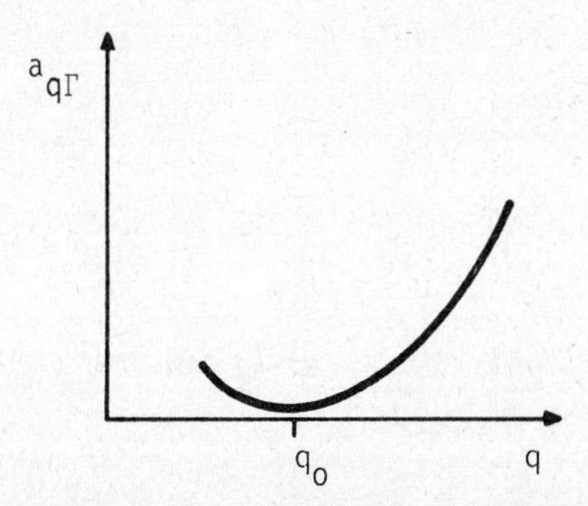

Fig. 3.2 : The curve of $a_{q\Gamma}$ at $T = T_c$ showing a minimum at q_o.

The Lifshitz condition says that b = 0, i.e. no 1st order Lifshitz invariant exists, on account of symmetry:

$$\left[\Gamma^2_{op}\right]_{\text{antisymmetric}} \not\ni \Gamma_{\text{vector}} \quad . \tag{3.6}$$

Note that the coefficient b may not only vanish because of symmetry, but could also vanish accidientially at T_c. Then the system may have a type of phase transition such that $q = q(T)$, $q(T_c) = q_o$. This is a non-Landau transition and required special consideration.

The Lifshitz criterion yields important restrictions for phase transitions into phases with constant q.

If the 2nd criterion is satisfied we must look at the sign of the quadratic term: If $c > 0$ this is a candidate for a second order phase transition.

The third criterion, due to Landau, is obtained by keeping only the order parameter coordinate, and eliminating all other $Q_{q\Gamma}$ by minimization,

$$\Phi(Q) = \Phi_o + \tfrac{1}{2} a_2 \sum_{\gamma} |Q_\gamma|^2 + \frac{1}{3} \sum_{\gamma\gamma'\gamma''\alpha} a^{(\alpha)}_{3\,\gamma\gamma'\gamma''} Q_\gamma Q_{\gamma'} Q_{\gamma''}$$

$$+ \tfrac{1}{4} \Sigma\, a_4\, QQQQ + \ldots \quad . \tag{3.7}$$

For a 2nd order phase transition $a_3 = 0$, i.e. no 3rd Landau invariant exists because of symmetry:

$$\left[\Gamma^3_{op}\right]_{\text{symmetric}} \not\ni A_{1g} \tag{3.8}$$

Necessary conditions for Landau transitions are thus

(1) Min $a_{q\Gamma}(T_c) \equiv a_\Gamma(T_c) = 0$; $Q_{q\Gamma}$ transforms as $\Gamma \neq A_{1g}$.

(2) No 1st order Lifshitz invariants; coefficient of the 2nd order Lifshitz invariant > 0.

(3) No 3rd order Landau invariants; coefficient of the 4th order Landau invariant > 0.

If $a_4 = 0$ this leads to a tricritical phase transition (see Aharony's lecture).

Note: The 3rd criterion is always satisfied for a phase transition from a para= magnetic to magnetically ordered state on account of time reversal symmetry.

Symmetry of the L-phase ≡ symmetry of equilibrium value of Q in the n-dimensional order parameter space, determined (in the case of $n > 1$) by the form of the 4th order invariants and the sign of the a_4's. The lowest possible symmetry is given by the 'kernel' of the representation $\Gamma = \{q,\Gamma_q\}$ = subgroup which leaves each vector Q of the order parameter space invariant. If the equilibrium value of Q is along one of the symmetry directions, then the symmetry is higher. For example:

G_H = cubic group, $q = 0$, Γ_q vector representation: Q_x, Q_y, Q_z. Two 4th order invariants:

$$\Phi_4 = \tfrac{1}{4}[a_4^{(1)} (Q_x^2 + Q_y^2 + Q_z^2)^2 + a_4^{(2)} (Q_x^4 + Q_y^4 + Q_z^4)] \quad . \tag{3.9}$$

$a_4^{(1)} > 0$, $a_4^{(2)} < 0$: $Q \parallel$ cubic axis, tetragonal phase (e.g. Fe)

$a_4^{(1)} > 0$, $a_4^{(2)} < 0$: $Q \parallel$ body diagonal, trigonal phase (e.g. Ni)

3.3 Critical Points of Symmetry-Conserving Transition

We all know of a symmetry-conserving transition from the liquid-gas model. However, we now argue that this model is a special case. Consider the case of a fully symmetric order parameter Q (every element of the G group leaves Q invariant). An example is the ferroelectric KH_2PO_4 (KDP) in an applied electric field.

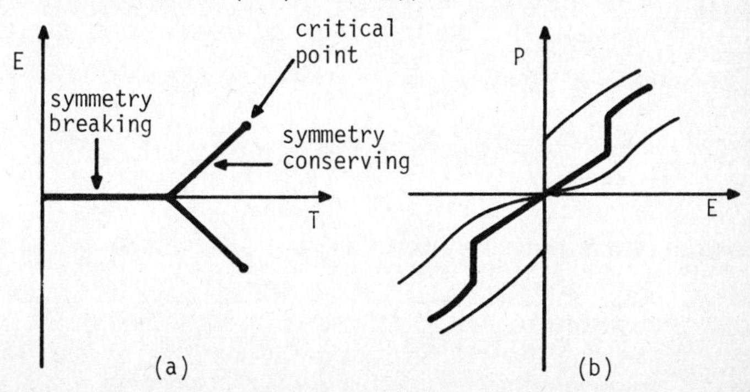

Fig. 3.3 : (a) Coexistence line. (b) Polarization curve for the ferroelectric KH_2PO_4 . P is a fully symmetric coordinate.

This substance has been studied by Courtens and Gammon[11]. Whenever there is a fully symmetric coordinate Q (in this case the polarization P) there is a linear coupling to the symmetric part of the strain tensor, i.e. the polarization P gives rise to a fully symmetric uniform strain,

$$\varepsilon = gP \quad , \tag{3.10}$$

and the fluctuations P_q give rise to acoustic mode coordinates $Q_q = \tilde{g}P_q$. For KDP four of the six components of the strain tensor ε_{xx}, ε_{yy}, ε_{zz} and ε_{xy} are fully symmetric.

Now let us eliminate the polarization P and express the potential in terms of the strain tensor and the acoustic mode coordinates Q_q. We write the Landau potential

$$\Phi = \tfrac{1}{2} \Sigma\, C_{\nu\nu'}(T)\varepsilon_\nu \varepsilon_{\nu'} + \tfrac{1}{2} \sum_{q\neq o,\lambda} c_\lambda^2(T) q^2 |Q_{q\lambda}|^2 \quad + \ldots. \quad , \tag{3.11}$$

where $C_{\nu\nu'}$ is the tensor of the isothermal elastic constants and $c_\lambda(T)$ is the isothermal sound velocity. We now diagonalize the elastic energy to obtain

$$\Phi_{elastic} = \tfrac{1}{2} \sum_n C_n(T)\varepsilon_n^2 \quad , \tag{3.12}$$

where C_n are the eigenvalues of $C_{\lambda\lambda'}$. At $T = T_c$ the smallest eigenvalue of the fully symmetric part is zero,

$$C_1(T_c) = 0 \; . \tag{3.13}$$

The critical point of symmetry-conserving transition is always an elastic instability.

The important point to realize is that *none* of the isothermal sound velocities c_λ becomes zero at T_c, i.e. only the *uniform part becomes unstable*. This is due to the fact that no acoustic wave can be formed from a fully symmetric strain. A strain tensor from which an acoustic wave may be formed is called a 'wave strain'. A wave strain has the form $\varepsilon = \frac{1}{2}(ab+ba)$, $a \perp b$. It can be shown that the symmetry-breaking eigenspace of the elastic tensor always contains a wave strain, but a fully symmetric eigenspace does not. Therefore, a symmetry-breaking elastic transition is always associated with a zero sound velocity c given by

$$\rho c^2 = C^{\text{symm-br}}(T) , \tag{3.14}$$

but a symmetry-conserving transition is not: In the case of a symmetry-conserving transition, contributions always occur from some non-vanishing eigenvalues

$$\rho c_\lambda^2 = a_1 C_1(T) + \sum_{n=2}^{6} a_n C_n(T) , \tag{3.15}$$

not all $a_n = 0$. This behaviour is related to the fact that there are six components of uniform strain but only three acoustic modes at $q \neq 0$. Thus, not every elastic instability can be associated with an acoustic mode. Thus, near the critical point only one term of the thermodynamic potential becomes zero and all the $q \neq 0$ terms stay at a finite positive value. This means that there are no critical fluctuations, and Landau theory is exact.

Why does one have critical behaviour for the liquid-gas transition? There is no shear elasticity, and only one elastic constant, namely the bulk modulus, is non-zero. In this special case the isothermal sound velocity is given by

$$\rho c^2 = C_T^{\text{bulk}} \tag{3.16}$$

where C_T^{bulk} is the isothermal bulk modulus. Thus, the isothermal sound velocity goes to zero as $C_T^{\text{bulk}} \to 0$, and this is the reason for the occurrence of critical fluctuations in this system.

4. Mean-Field Approximation

4.1 Variational Formulation of Equilibrium Statistical Mechanics

The task of statistical mechanics is to derive the thermodynamic potential from a microscopic model Hamiltonian. Let

$$H(\{\hat{Q},\hat{P}\}) = H_0 - \sum_{\ell} \hat{Q}_\ell \cdot F_\ell , \tag{4.1}$$

where $\hat{Q}_\ell$ are the microscopic coordinates and $\hat{P}_\ell$ are the canonical momenta. The statistical distribution in phase space Γ is

$$\rho(\{\hat{Q},\hat{P}\}) \; : \; \int \rho d\Gamma = 1 \; , \; d\Gamma = D\hat{Q}D\hat{P} \; . \tag{4.2}$$

This distribution determines the average value

$$<\psi> = \int \psi \rho \, d\Gamma \tag{4.3}$$

of any phase function $\psi(\{\hat{Q},\hat{P}\})$. We define the free energy functional

$$\Phi[\rho] = <H> + kT<\ell n\rho>$$

$$= \int (H\rho + kT\rho \ell n\rho) d\Gamma \; , \tag{4.4}$$

with the property that for a system in equilibrium at temperature T

$$\Phi[\rho] = \text{minimum under } \int \rho d\Gamma = 1 \; . \tag{4.5}$$

Explicitly

$$\delta(\Phi - \lambda\rho) = \int (H + kT\ell n\rho + kT - \lambda)\delta\rho d\Gamma = 0 \; , \tag{4.6}$$

where λ is a Lagrange parameter.

This condition yields the equilibrium distribution ρ_{eq},

$$\rho_{eq} = \frac{1}{Z} e^{-\beta H} \; , \; Z = \int e^{-\beta H} d\Gamma \; , \; \beta = 1/kT \; , \tag{4.7}$$

and the Gibbs potential

$$G = \Phi[\rho_{eq}] = -kT\ell nZ \; . \tag{4.8}$$

For a *classical system* one can separate the kinetic and configuration part of the Hamiltonian

$$H = H_{kinetic}(\{\hat{P}\}) + H_{config}(\{\hat{Q}\}) \; , \tag{4.9}$$

and as a consequence ρ_{eq} factorizes

$$\rho_{eq} = \rho_{kinetic}(\{\hat{P}\})\rho_{config}(\{\hat{Q}\}) \; . \tag{4.10}$$

The kinetic energy part is trivial to deal with, so in practice one is left only with the configuration part. In this case $d\Gamma \rightarrow D\hat{Q}$.

For a *classical spin system* there is no kinetic energy part. The complete Hamiltonian consists only of the configurational part : $H = H_{config}$.

In a *quantum mechanical system* then H, ρ are operators. The description above remains valid if we replace the phase space integral by the trace

$$\int d\Gamma \rightarrow \text{trace} \; . \tag{4.11}$$

4.2 Mean Field Approximation by Variational Ansatz

Consider a model system for which H consists of a single-particle part and a bilinear interaction part

$$H = \sum_{\ell} H_{\ell}(\hat{Q}_{\ell}) - \tfrac{1}{2} \sum_{\ell\ell'} v_{\ell\ell'} \hat{Q}_{\ell}\hat{Q}_{\ell'} \; , \tag{4.12}$$

where

$$H_{\ell} = H_{\ell o} - \hat{Q}_{\ell}F \; . \tag{4.13}$$

The problem is to evaluate the partition function. In the absence of interactions we have

$$\rho = \prod_{\ell} \rho_{\ell} \quad , \quad \rho_{\ell} = \frac{1}{Z_{\ell}} e^{-\beta H_{\ell}} \; . \tag{4.14}$$

The idea of the mean field approximation is to minimize the functional $\Phi[\rho]$, not for arbitrary statistical densities, but for the product distribution

$$\rho = \prod_{\ell} \rho_{\ell}(\hat{Q}_{\ell}) \quad . \tag{4.15}$$

That is we disregard correlations between different sites.

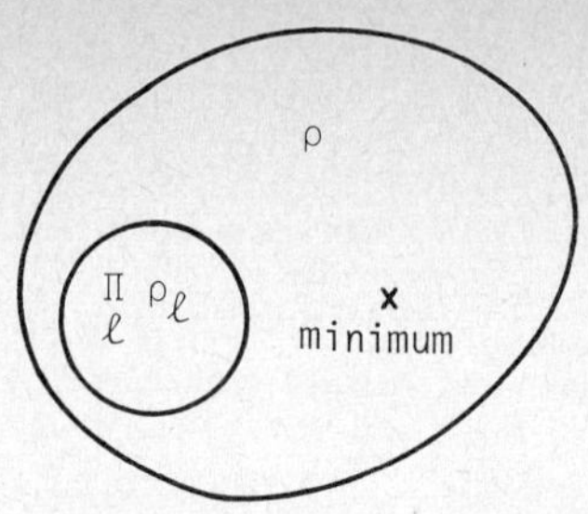

Fig. 4.1 : Schematic diagram showing the set of all possible ρ. The true minimum of Φ will be outside the subset $\prod_\ell \rho_\ell$.

Let us put this Ansatz into the functional

$$\Phi\left[\prod_\ell \rho_\ell\right] = \sum_\ell \int (H_\ell \rho_\ell + kT\rho_\ell \ell n\rho_\ell)\, d\hat{Q}_\ell - \tfrac{1}{2}\sum_{\ell\ell'} v_{\ell\ell'} Q_\ell Q_{\ell'} \ , \tag{4.16}$$

and minimize under the constraints

$$\int \rho_\ell d\hat{Q}_\ell = 1 \ , \ \int \hat{Q}_\ell \rho_\ell d\hat{Q}_\ell = Q_\ell \ . \tag{4.17}$$

The hope is that the minimum of this functional will not lie too far away from the true minimum given by all possible ρ. We write

$$\delta(\Phi - \sum_\ell \lambda_\ell \rho_\ell - \sum_\ell \phi_\ell \hat{Q}_\ell \rho_\ell)$$

$$= \sum_\ell \int (H_\ell + kT\ell n\rho_\ell + kT - \lambda_\ell - \phi_\ell \hat{Q}_\ell)\delta\rho_\ell d\hat{Q}_\ell = 0 \ , \tag{4.18}$$

where λ_ℓ and ϕ_ℓ are Lagrange parameters. This yields the distribution

$$\rho_\ell = \frac{1}{Z_\ell} \exp[-\beta(H_\ell - \phi_\ell \hat{Q}_\ell)] \ , \ Z_\ell = \int \exp[-\beta(H_\ell - \phi_\ell \hat{Q}_\ell)]\, d\hat{Q}_\ell \ , \tag{4.19}$$

and the average coordinates

$$Q_\ell = \langle \hat{Q}_\ell \rangle = \frac{d\ell n Z_\ell}{d(\beta\phi_\ell)} = B_\ell(\phi_\ell + F_\ell) \ . \tag{4.20}$$

This last equation is the *single particle equation of state*, with the external force F_ℓ replaced by $\phi_\ell + F_\ell$.

Now we assume we can calculate all the single particle properties. Thus we can calculate the Lagrange parameter ϕ_ℓ from the inverse

$$\phi = B_\ell^{-1} (Q_\ell) - F_\ell \; . \qquad (4.21)$$

We also want to know what is the value of the functional Φ at the minimum. We find

$$\Phi = \sum_\ell \{<H_\ell> + kT[-\beta(<H_\ell> - \phi_\ell Q_\ell) - \ell n \, Z_\ell]\}$$

$$- \tfrac{1}{2} \Sigma \, v_{\ell\ell'} \, Q_\ell Q_{\ell'} \; . \qquad (4.22)$$

Thus

$$\Phi(\{Q\},T) = \sum_\ell \Phi_\ell(Q_\ell,T) - \tfrac{1}{2} \sum_{\ell\ell'} v_{\ell\ell'} \, Q_\ell Q_{\ell'} \; , \qquad (4.23)$$

where the *single particle potential* for a given $Q_\ell = <\hat{Q}_\ell>$ is

$$\Phi_\ell(Q_\ell,T) = - kT \, \ell n \, Z_\ell + \phi_\ell Q_\ell \; . \qquad (4.24)$$

We can write this in the form

$$\exp[-\beta\Phi_\ell] = \int \exp\{-\beta[H_\ell - \phi_\ell \, (\hat{Q}_\ell - Q_\ell)]\} d\hat{Q}_\ell \; . \qquad (4.25)$$

The complete thermodynamic equilibrium is found by minimizing this potential Φ with respect to the Q's. Taking the derivative

$$\frac{\partial\Phi_\ell}{\partial Q_\ell} = - kT \frac{\partial}{\partial Q_\ell} \ell n \, Z_\ell + \phi_\ell + Q_\ell \frac{\partial\phi_\ell}{\partial Q_\ell} \; , \qquad (4.26)$$

and using the fact that

$$kT \frac{\partial}{\partial Q_\ell} \ell n \, Z_\ell = \frac{\partial \, \ell n \, Z_\ell}{\partial(\beta\phi_\ell)} \cdot \frac{\partial\phi_\ell}{\partial Q_\ell} = Q_\ell \frac{\partial\phi_\ell}{\partial Q_\ell} \; , \qquad (4.27)$$

we obtain

$$\frac{\partial\Phi_\ell}{\partial Q_\ell} = \phi_\ell \; . \qquad (4.28)$$

Thus, at *thermodynamic equilibrium* the Lagrange parameters have the simple interpretation

$$\phi_\ell = \sum_{\ell'} v_{\ell\ell'} \, Q_{\ell'} \equiv F_\ell^{mol} \; , \qquad (4.29)$$

where F_ℓ^{mol} is the mean molecular field. This is the reason for the name 'molecular field theory'. After replacing ϕ_ℓ by F_ℓ^{mol}, we can write the equation of state

$$Q_\ell = B(F_\ell^{mol} + F_\ell) \quad . \tag{4.30}$$

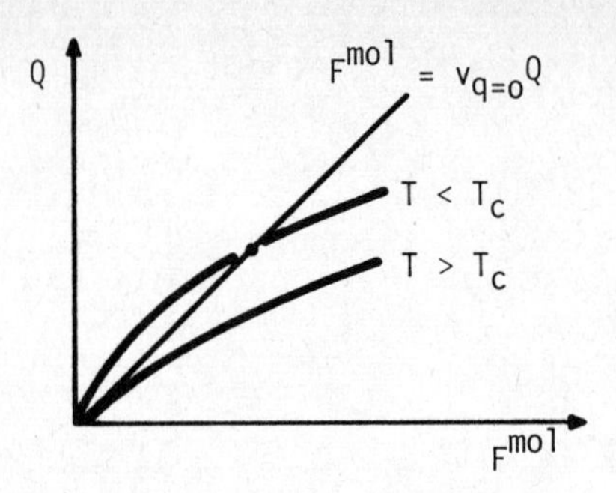

Fig. 4.2 : Solution of the self-consistency equation.

The mechanism to get self-consistent solutions to this equation is to look for the point where the line of $F^{mol} = vQ$ intersects the equation of state.

The *reciprocal susceptibility* is the second derivative of the thermodynamic potential

$$\chi_{\ell\ell'}^{-1} = \frac{\partial^2\Phi}{Q_\ell \partial Q_{\ell'}} \quad . \tag{4.31}$$

We obtain

$$\chi_{\ell\ell'}^{-1} = \chi_{s,\ell}^{-1}(Q_\ell)\delta_{\ell\ell'} - v_{\ell\ell'} \quad , \tag{4.32}$$

where $\chi_{s,\ell}$ is the single particle susceptibility coming from the second derivative of the single particle potential Φ_ℓ. Thus we obtain *Dysons equation* for the susceptibility

$$\chi_{\ell\ell'} = \chi_{s,\ell}\,\delta_{\ell\ell'} + \chi_{s,\ell} \sum_{\ell''} v_{\ell\ell''}\,\chi_{\ell''\ell'} \quad . \tag{4.33}$$

Although the correlations between different sites are disregarded, one obtains an approximation for the equal-time correlation functions by using the fluctuation theorem

$$S_{\ell\ell'} = kT\chi_{\ell\ell'} \quad . \tag{4.34}$$

4.3 Mean Field Approximation from Linearization of Fluctuations

An alternative derivation of the molecular-field equations is obtained in the following way:

Each particle responds with a *single-particle equation of state* B to the sum of an external field F_ℓ and the *true* interaction field F_ℓ^{mol}.

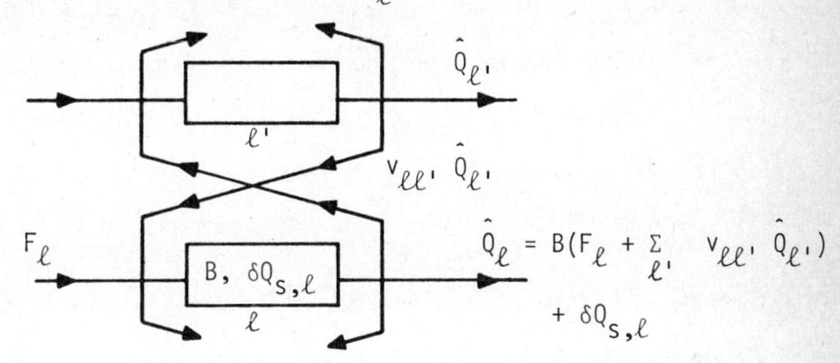

Fig. 4.3 : System response to the external field and the interaction field

We assume that the response B is not affected by fluctuations of F_ℓ^{mol}, i.e. that the system responds linearly to its own fluctuations, $\hat{Q}_\ell = Q_\ell + \delta Q_\ell$:

$$Q_\ell = B(F_\ell + \sum_{\ell'} v_{\ell\ell'}\, Q_{\ell'}) \ , \tag{4.35}$$

$$\delta Q_\ell = \chi_{s,\ell} \sum_{\ell'} v_{\ell\ell'}\, \delta Q_{\ell'} + \delta Q_{s,\ell} \ . \tag{4.36}$$

One may use this picture to derive the mean field equations and equations for the fluctuations. The single particle fluctuation theorem

$$S_{s,\ell} = \langle \delta Q_{s,\ell} \rangle = kT\chi_{s,\ell} \ , \tag{4.37}$$

implies the fluctuation theorem for the coupled system

$$S_{\ell\ell'} = \langle \delta Q_\ell \delta Q_{\ell'} \rangle = kT\chi_{\ell\ell'} \ . \tag{4.38}$$

In this model the mean field approximation becomes consistent with the fluctuation theorem.

4.4 Results of Mean Field Approximation for a Uniform System

Consider the H-phase, all atoms equal so that we can replace all $\chi_{s,\ell} = \chi_s$. Then

$$\chi_{\ell\ell'} = \chi_s \delta_{\ell\ell'} + \chi_s \sum_{\ell''} v_{\ell\ell''} \chi_{\ell''\ell'} \ . \tag{4.39}$$

Take the Fourier transform and solve for

$$\chi_q = \frac{\chi_s(T)}{1 - v_q \chi_s(T)} \ . \tag{4.40}$$

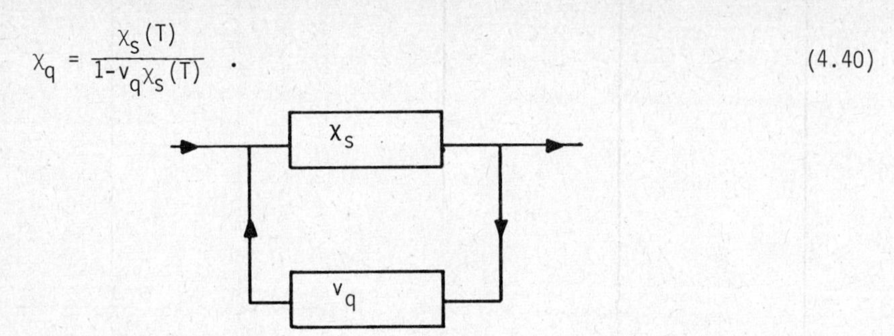

Fig. 4.4 : Feedback system.

Because of feedback there is a critical temperature at which the denominator becomes zero. The type of instability is determined by max $v_q = v_{q_o}$

$$v_{q_o} \chi_s(T_c) = 1 \ . \tag{4.41}$$

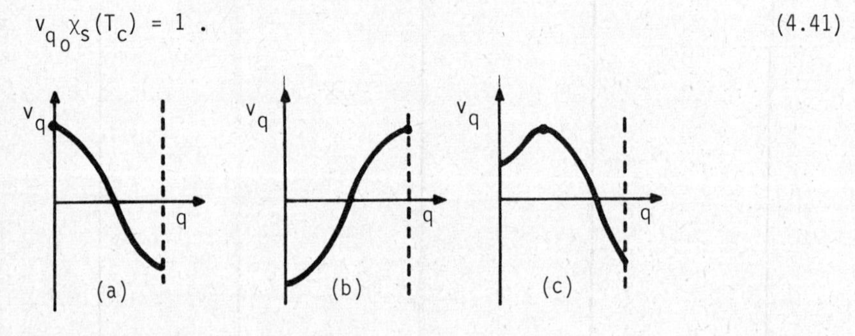

Fig. 4.5 : Three cases where v_q is a maximum.

(a) At $q_o = 0$: ferro. (b) At $q_o = \frac{1}{2}K$: antiferro. (c) At q_o inside Brillouin zone (BZ) : incommensurate phase.

Three cases are considered in Fig. 4.5. Let us consider the ferro-case in a little more detail. For small q we expand

$$v_q = v_o - \alpha q^2 \ , \quad v_o \chi_s(T_c) = 1 \ . \tag{4.42}$$

The single particle susceptibility becomes

$$\chi_q = \frac{1}{[\chi_s^{-1}(T) - v_o] + \alpha q^2} \ . \tag{4.43}$$

This is the Ornstein-Zernike form. This yields the thermodynamic potential

$$\Phi = \sum_q \tfrac{1}{2}\{[\chi_s^{-1}(T) - v_o] + \alpha q^2\}|Q_q|^2 + \tfrac{1}{4}b \sum_{\ell} Q_{\ell}^4 , \qquad (4.44)$$

which for small values of Q_q is just the Landau potential (2.20).

5. Introduction to Driven Systems

What do we mean by driven systems? A system is coupled to several external baths.

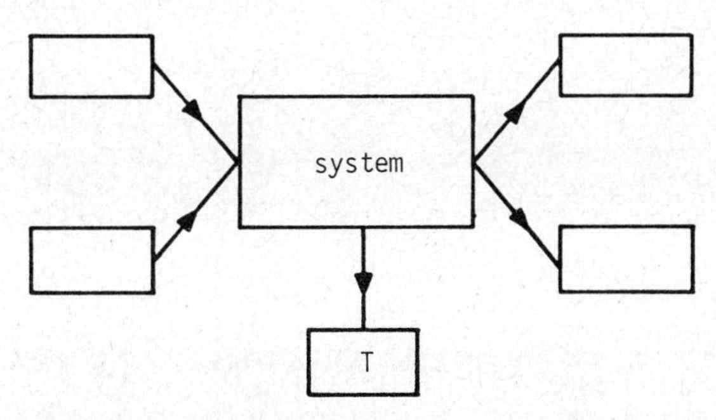

Fig. 5.1 : System coupled to external baths.

The feature that distinguishes a driven system from an equilibrium system is that even in the steady state there are fluxes going through the system. We will assume only stationary driven systems, i.e. the baths are time independent. In such a stationary driven system there is always dissipation, and in order to dump heat generated in the system one of the boxes must be a temperature bath.

5.1 Representative Examples

Let us consider a few representative examples. The first example is the *Rayleigh-Bénard instability*.

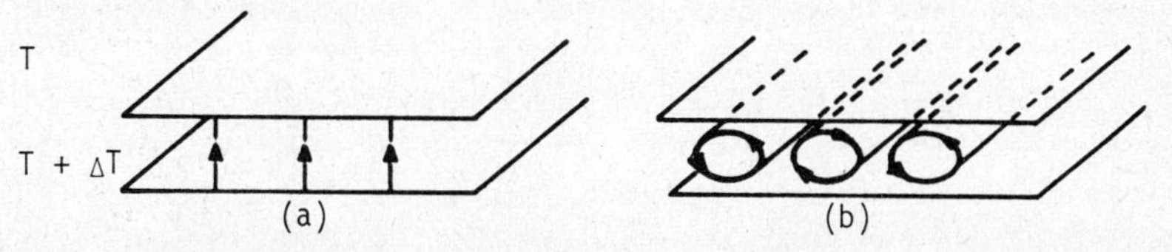

Fig. 5.2 : Rayleigh-Bénard instability.

The instability arises whenever a horizontal layer of fluid with positive coefficient of volume expansion is heated from below. If the fluid is heated only a little then the heat is transported by conduction (Fig. 5.2(a)). The reason the hot fluid does not rise is due to competition between viscosity and buoyancy in the fluid. But, if the system is driven hard enough then at some ΔT it becomes unstable and the hot fluid indeed rises: the driving force overcomes the frictional forces in the fluid. The heat is now transported by convection rolls (Fig. 5.2(b)). By driving the system harder these rolls begin to oscillate. This is a time-dependent state. If the system is driven even harder then the motion finally becomes turbulent.

A second example is the *Taylor instability*.

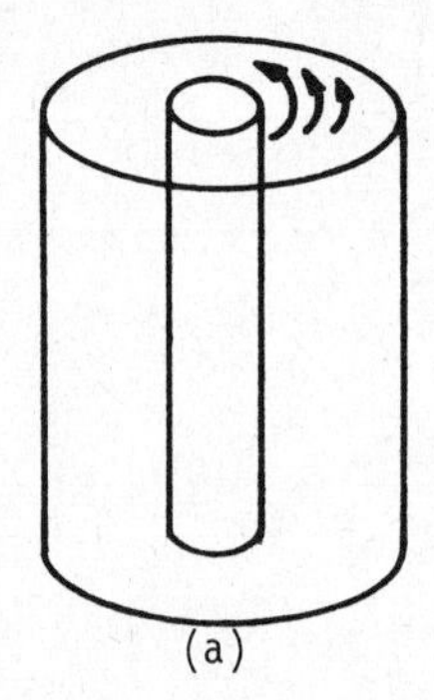

(a)

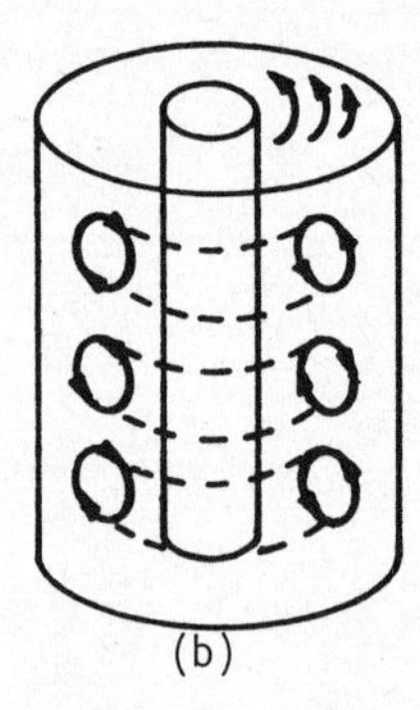

(b)

Fig. 5.3 : Taylor instability.

Rotating the inner cylinder gives a transverse momentum to the fluid. The fast moving fluid on the inside of the cylinder wants to move outward because of centri= fugal forces, and replace the slow layer at the outside. If the inner cylinder is rotated slowly then there is ordinary Couette flow, where angular momentum fed into the fluid from the inner cylinder is transported outside by viscosity (Fig. 5.3(a)). At a critical rotation speed the system becomes unstable and stationary annular convection cells transport this momentum (Fig. 5.3(b)). For faster rotation an instability occurs against oscillations, and finally a transition to turbulent motion.

Other examples are the *laser*: pump the system and at the laser threshold there is a transition to an oscillating state, and at higher pumping rate another transition to a pulsing state; also *autocatalytic chemical reactions*: fix the chemical potential and the system goes into a state of oscillatory motion.

As a final example we consider the current instability that occurs in certain semiconductors:

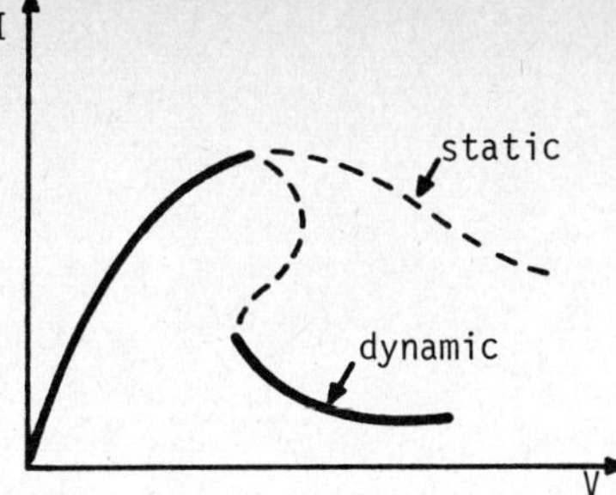

Fig. 5.4 : Current-voltage characteristic of a semiconductor showing bulk negative differential conductivity.

The differential conductivity becomes negative at high electric fields. A well known example is the *Gunn instability* in GaAs. A uniform current state with negative differential conductivity is unstable against the build-up of charge density fluctuation:

δE

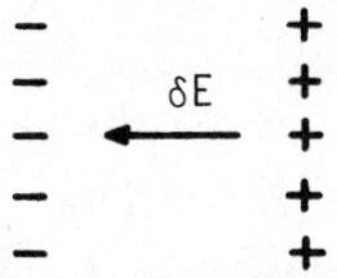

Fig. 5.5 : Stability of charge density fluctuation.

The Coulomb force induces an electric field δE which drives an additional current

$$\delta j = \sigma \delta E \ . \tag{5.1}$$

This current enhances the original charge fluctuations when $\sigma < 0$ until the process is counter-balanced by carrier diffusion. It was found that in a non-uniform state a high-field domain travels through the sample with the drift velocity of the carriers.

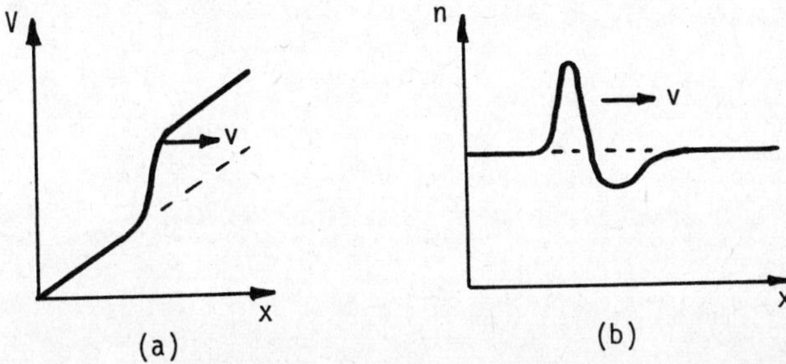

Fig. 5.6 : Travelling dipole domain in a semiconductor. (a) Potential distribution. (b) Carrier density.

Only when the dipole domain arrives at the anode is a new dipole domain created at the cathode.

5.2 Symmetry Breaking

We recall that symmetry is an important feature for phase transitions in equilibrium systems. The symmetries we have been considering so far have been spatial symmetries: symmetries under translation, rotation and reflection, and in the case of magnetic systems also time-reversal symmetry. In driven non-equilibrium systems, not only these symmetries but also symmetry under *time translations* may be spontaneously broken at an instability, leading to time-dependent (non-stationary) states under stationary driving conditions. (We have seen in the above examples that the driven system may develop such a time dependence.) From the symmetry point of view this is the new feature of driven systems. Because of this feature the structures that appear in driven systems are sometimes called 'dynamic structures', although it should be borne in mind that some of these structures are not dynamic.

For the case of one space dimension the following types of situations can arise:

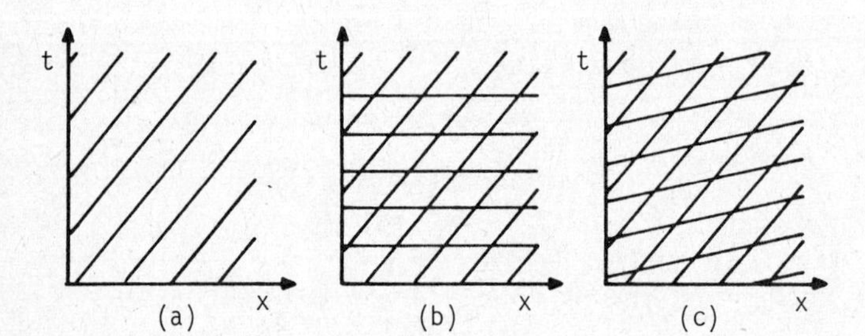

Fig. 5.7 : Examples of time-dependent structures. (a) Travelling waves. (b) Standing waves on a flow. (c) Doubly periodic waves.

Travelling waves (Fig. 5.7(a))

$$\phi = f[q(x-ut)] \quad , f(\xi+2\pi) = f(\xi) \quad , \tag{5.2}$$

standing waves on a moving fluid (Fig 5.7(b))

$$\phi = F[q(x-ut), \omega t] \quad ,$$

$$F(\xi+2\pi,\eta) = F(\xi,\eta+2\pi) = F(\xi,\eta) \quad , \tag{5.3}$$

and *doubly periodic waves* (Fig. 5.7(c))

$$\phi = F[q_1(x-u_1t)\ ,\ q_2(x-u_2t)]\ . \tag{5.4}$$

We will concentrate on the first case, where one finds as a function of time at a fixed point in space a periodic pattern, and at a fixed time a periodic structure in space, but one still has translation symmetry along the direction x-ut = const. This is the simplest case of a 'dynamical structure'.

NOTE The term 'process' is often used to describe the whole function $\phi(t)$ while the term 'state' denotes the field $\phi(t)$ at a fixed time. We shall continue to use the term 'state' to describe both.

6. Description of Driven Systems

6.1 The Time Evolution Equation and Main Problem

The *state* at a given time t is described by a set of fields

$$\phi(t) = \{\phi_n(r,t)\}\ . \tag{6.1}$$

We are given a set of parameters $\alpha = \{\alpha_n\}$ which can be externally controlled, and we want to know how the system moves. We assume that the change in time is given by the values of the fields and values of the control parameters at that given time. Such a locality in time is based on the existence of different time scales for the macrovariables and eliminated microvariables. In order to eliminate memory effects the state space is chosen to be sufficiently large that all *slow* variables are included. The *time evolution equation* of the fields is of the form

$$\phi(t) = B[\phi(t),\alpha]\ , \tag{6.2}$$

together with appropriate boundary conditions, where B is a *non-linear* functional. We assume that B is contracting, namely that in the course of time a volume in phase space will become smaller. We want to consider a homogeneous (uniform) system where B is invariant against translations in space and time. Usually B is a partial differential operator acting on the fields $\phi_n(r,t)$.

The main problem is to find what would correspond in the equilibrium system to the *stable* 'states': such states are called attractors. What is an attractor? If the system is already on an attractor it will stay on the attractor for all later times. If the system is close to an attractor it will approach the attractor.

We recall that in equilibrium systems, for a given set of control parameters, the thermodynamic equilibrium state is usually uniquely defined, except at special points α_c in control space, namely at the coexistence lines, where different phases coexist. This is not true for the bifurcation theory of driven systems. For certain types of the functional B there exists a whole range of α-values where different states can coexist and are simultaneously stable.

The task is:

(1) Find solutions of the non-linear equation (6.2).

(2) Test the stability of these solutions.

If both these problems are solved, then we have found the attractors. The problem is that it is not possible, in general, to solve the non-linear equation explicitly. Nevertheless, it is still possible to say quite a lot about the first instability.

6.1.1 Examples

In order to illustrate the structure of the time evolution equation we now consider a few examples. The first example is taken from *hydrodynamics*. The *state*

$$\phi(t) = \{\rho(r,t),\ v(r,t),\ T(r,t)\} \tag{6.3}$$

is described by the density field $\rho(r,t)$, velocity field $v(r,t)$, and temperature field $T(r,t)$. The *time evolution* of these fields is described by a continuity equation

$$\dot{\rho} = -\nabla.(\rho v) \ , \tag{6.4}$$

the Navier-Stokes equation

$$\dot{v} = -v.\nabla v - \frac{1}{\rho}\nabla p(\rho,T) + \nu : \nabla\nabla v + g \quad , \tag{6.5}$$

and the heat conduction equation

$$\dot{T} = -v.\nabla T + \kappa\nabla^2 T - \frac{1}{\rho c_v} \tag{6.6}$$

Here the pressure $p(\rho,T)$ is given by the equation of state, ν is the kinematic viscosity tensor , c_v is the specific heat, and κ is the heat conductivity. The choice of control variable depends on the problem under consideration. For the Rayleigh-Bénard problem a suitable control parameter is the *Rayleigh number*

$$Ra = (\frac{g\alpha\ell^3}{\kappa\nu})\Delta T_0 \quad , \tag{6.7}$$

which is a dimensionless quantity describing the temperature difference ΔT_0 across the fluid. Here α is the coefficient of volume expansion and ℓ is the thickness of the layer. In the Taylor problem a suitable choice is the *Reynolds number*

$$Re = v_0 \frac{\ell}{\nu} \quad , \tag{6.8}$$

where v_0 is the velocity of the fluid at the inner boundary. For both examples the type of behaviour also depends strongly on the *Prandtl number*

$$Pr = \frac{\nu}{\kappa} \; . \tag{6.9}$$

The next example is the *laser*. In the rotating wave approximation the *state*

$$\phi(t) = \{\sigma(r,t), \; P(r,t), \; E(r,t)\} \tag{6.10}$$

is described by the inversion $\sigma(r,t)$ of the medium, and complex amplitudes for the polarization $P(r,t)$ and electric field $E(r,t)$. The *time evolution equations* are

$$\dot{\sigma} = -\gamma_{\parallel}\,(\sigma-\sigma_0) + \frac{2i}{\hbar}\,(EP^*-E^*P) \quad , \tag{6.11}$$

$$\dot{P} = -\gamma_{\perp}P - \frac{i}{3}\frac{|M|^2}{\hbar}\,E\sigma \quad , \tag{6.12}$$

$$\dot{E} = -\kappa E - c\,\frac{\partial E}{\partial x} + i\,\frac{\omega_0}{2\varepsilon}\,P \; . \tag{6.13}$$

Here $\gamma_{\parallel}$ and $\gamma_{\perp}$ are relaxation rates for the inversion and polarization respectively, κ describes the cavity losses, M is a dipole matrix element of the laser transition, ε is the dielectric constant, and c the velocity of light in the medium. The *control variable* is the pump parameter σ_o.

A further example is that of *chemical reactions* where the set of fields has as many components as there are concentrations c_α in the system. These concentrations represent the *state* variables

$$\phi(t) = \{c_\alpha(r,t)\} \ . \tag{6.14}$$

The *time evolution equations* are

$$\dot{c}_\alpha = R_\alpha(\{c\},\{\mu\}) + D_\alpha \nabla^2 c_\alpha \ , \tag{6.15}$$

where R_α are the reaction rates and D_α the diffusion constants. In this case the *control parameters* are those chemical potentials μ_α that can be fixed externally.

The last example is *electronic conduction.* The *state*

$$\phi(t) = \{n(r,t),\ p(r,t),\ w(r,t),\ E(r,t),\ B(r,t)\} \tag{6.16}$$

is described by the carrier density n(r,t), mean carrier momentum p(r,t), mean carrier energy w(r,t), and the Maxwell fields E(r,t) and B(r,t). *Time evolution* is described by the equation of continuity

$$\dot{n} + \nabla.(nv) = 0 \ , \tag{6.17}$$

the equation of momentum balance

$$\dot{p} + v.\nabla p + \frac{1}{n}\nabla.\Pi = eb.(E\text{-}vxB) - \gamma_p p \ , \tag{6.18}$$

and energy balance

$$\dot{w} + v.\nabla w + \frac{1}{n}\nabla.\Sigma = ev.E - \gamma_w(w\text{-}w_o) \ , \tag{6.19}$$

and Maxwells equations

$$\dot{B} + \nabla \times E = 0 \quad , \quad \nabla . B = 0$$

$$\varepsilon_L \dot{E} - \nabla \times B = - env \quad , \varepsilon_L \nabla . E = e(n - n_D) \ . \qquad (6.20)$$

Here the velocity v, the electronic stress tensor Π, effective acceleration tensor b, the energy flow vector Σ, and relaxation rates γ_p and γ_w for momentum and energy must be given for each model as functionals of the state ϕ. The equilibrium energy w_0, the lattice dielectric constant ε_L and donor density n_D are given constants. The *control parameter* is the externally applied field E_0.

6.1.2 Comparison with Equilibrium System

We consider the motion of the fields $\phi(t)$ as a flow $B[\phi(t)]$ in state space. In the case of systems close to equilibrium this flow is a gradient flow of the Landau potential $\mathbb{F}$,

$$B[\phi] = - \Lambda \nabla_\phi \mathbb{F}[\phi] \quad , \qquad (6.21)$$

where Λ is positive definite. We can make the following comparison between equilibrium systems and driven systems:

	Equilibrium	Non-equilibrium
statics :	$\mathbb{F}[\phi] = \min$	$B[\phi] = 0$
dynamics :	$\dot{\phi} = -\Lambda \nabla_\phi \mathbb{F}[\phi]$	$\dot{\phi} = B[\phi]$

We may eliminate fast variables by the procedure of *adiabatic elimination*. This procedure has been termed by Haken[12] as 'slaving': the slow variables *slave* the fast variables. Of course the system will have fast motion due to fluctuations, but the coupled motion follows that of the slow variables.

6.1.3 Model Systems

By the adiabatic elimination principle it is always possible to eliminate a sufficient number of components of the field so that, as in Landau theory, one can start a systematic investigation of various model systems. Here we present a few examples classified under the name of *time-dependent Ginsburg-Landau equations*. The simplest case is the equation for a one-component ($n = 1$) field

$$\phi = \alpha\phi \; - \phi^2 - \gamma\phi^3 - v\phi_x + \phi_{xx} \, , \tag{6.22}$$

where ϕ_x denotes a spatial derivative of the field . This equation may be applied to the study of current instabilities. The above equation has no internal symmetry, i.e. there is no transformation which can be applied to the field ϕ that will leave the equation unchanged. Other examples are an equation with reflection symmetry $\phi \to -\phi$ ($n = 1$) ,

$$\dot{\phi} = \alpha\phi - \phi^3 - v\phi_x + \phi_{xx} \, , \tag{6.23}$$

and an equation for a complex field ($n = 2$) with continuous symmetry $\phi \to e^{i\varphi}\phi$,

$$\dot{\phi} = \alpha\phi - |\phi|^2\phi - v\phi_x + \phi_{xx} \, , \tag{6.24}$$

which may be used to discuss hydrodynamic instabilities and the laser transition. The particular applications we shall discuss will be concerned with these three types of equations.

6.2 Linear Response of Driven Systems

One may study the response of the system to *dynamical test fields* $f^{ext}(t)$. The time evolution of the perturbed system is

$$\dot{\phi}(t) = B[\phi(t),\alpha] + f^{ext}(t) \, . \tag{6.25}$$

We are interested in that part of the response which is linear in the test forces. For the linear response $\phi(t) = \phi_o(t) + \delta\phi(t)$ one finds

$$\delta\dot{\phi}(t) = -L[\phi_o(t),\alpha]\delta\phi(t) + f^{ext}(t) \, , \tag{6.26}$$

where

$$L = -\left.\frac{\delta B}{\delta\phi}\right|_{\phi_o} \, . \tag{6.27}$$

This response to dynamic perturbations gives valuable information on the dynamics of the system.

The linear response of the system to f^{ext} has the form

$$\delta\phi(t) = \int_{-\infty}^{t} \chi(t,t') f^{ext}(t')dt' , \tag{6.28}$$

where the dynamic *susceptibility* $\chi(t,t')$ is determined by the operator equation

$$\frac{\partial}{\partial t} \chi(t,t') + L[\phi_0(t),\alpha].\chi(t,t') = \delta(t-t').1 . \tag{6.29}$$

In the *steady state*, ϕ_0 is independent of time, and $\chi(\tau)$ depends only on the time difference $\tau = t-t'$. The equation can be analysed by a Fourier transformation

$$\chi(\tau) = \frac{1}{2\pi} \int \chi(\omega) e^{-i\omega t} d\omega , \tag{6.30}$$

and in Fourier space we obtain the result

$$\chi(\omega) = \{-i\omega.1 + L[\phi_0,\alpha]\}^{-1} . \tag{6.31}$$

6.3 Fluctuations in Driven Systems

In any system there occur fluctuations about the deterministic motion. So far we have disregarded the effects of such fluctuations: The theory we have described is on the same level as Landau theory for equilibrium systems. We know, from equilibrium systems, that Landau theory breaks down close to a transition (if the dimensionality is too low). We must expect the same thing to happen for driven systems.

The inclusion of stochastic Langevin forces should be a good starting point for the investigation of the fluctuations, in the same way as the thermodynamic Landau potential is a good starting point for the renormalization group. We need at least an approximate description of the *stochastic* nature of the process $\phi(t)$.

6.3.1 Stochastic Description

Fluctuations about the deterministic state can be described by adding *stochastic fields* $f(t)$ (Langevin forces), to the time evolution equation,

$$\dot{\phi}(t) = B[\phi(t),\alpha] + f(t) . \tag{6.32}$$

By definition the stochastic field has a zero average

$$<f(x,t)> = 0 \ . \tag{6.33}$$

We assume that the stochastic properties of the fields $f(t)$ are determined by the correlation matrix

$$<f(x,t)f(x',t')> = C(x,x')\delta(t-t') \ . \tag{6.34}$$

The problem is to calculate the stochastic properties of the field fluctuations

$$\delta\phi(x,t) = \phi(x,t) - <\phi(x,t)> \ , \tag{6.35}$$

and in particular the time-dependent correlation function

$$S(xt,x't') = <\delta\phi(x,t)\delta\phi(x',t')> \ . \tag{6.36}$$

For a detailed investigation of the stochastic behaviour it may be convenient to introduce a probability distribution $P[\phi,t]$ in state space. Its time evolution is described by a *functional Fokker-Planck equation*

$$\frac{\partial P}{\partial t}[\phi,t] = -\frac{\delta}{\delta\phi}\{B[\phi,\alpha] - \hat{D}\cdot\frac{\delta}{\delta\phi}\}P[\phi,t] \ . \tag{6.37}$$

6.3.2 Linear Theory of Fluctuations

In an equilibrium system the fluctuation-dissipation theorem provides a relationship between the correlation function and the linear response: Given the linear response function one can calculate the correlation function. This result is not applicable to driven systems. In general it is not possible to predict the correlation function from a linear response of the system.

We have seen in equilibrium systems that a way to derive the *mean field approximation* is to linearize the response of the system to its own fluctuations. The same approxi= mation can also be applied to driven systems. We shall assume that the fluctuations are small enough that the response of the system can be treated in linear response theory. This is known as the *random phase approximation* (RPA). Although the quantitative validity of this approximation breaks down in the vicinity of an instability, it is still expected to give a qualitatively correct picture of the fluctuation phenomena.

The linear response $\delta\phi(t)$ to the stochastic forces $f(t)$ is

$$\delta\phi(t) = \int\chi(t,t')f(t')dt' , \tag{6.38}$$

where $\chi(t,t')$ is the dynamic susceptibility. The time-dependent correlation function can be written

$$S(t,t') = \int\chi(t,t'').\hat{c}.\chi(t'',t')dt'' . \tag{6.39}$$

For the case of a steady state this operator depends only on the time difference $\tau = t-t'$. In Fourier space

$$S(\omega) = \chi(\omega).\hat{c}.\chi(\omega) \tag{6.40}$$

Thus, in the RPA the linear-response operator $\chi(\omega)$ also determines the behaviour of the fluctuation spectrum.

7. Bifurcation from the Steady State

We now consider *instabilities* in the system. When does a system become unstable? What other state does the system go into after it goes through an instability? This situation is described by *bifurcation theory*.

7.1 Discrete Systems

In this section we consider a system with a set of *discrete* variables

$$\phi = \{\phi_1,\dots,\phi_n\} . \tag{7.1}$$

An example of a bifurcating system is the *van der Pol oscillator*. This system has two degrees of freedom: The state $\phi = \{x,p\}$ where the position x and momentum p satisfy the equations

$$\dot{x} = p ,$$

$$\dot{p} = (\alpha-bx^2)p - \tfrac{1}{2}\omega_0^2x . \tag{7.2}$$

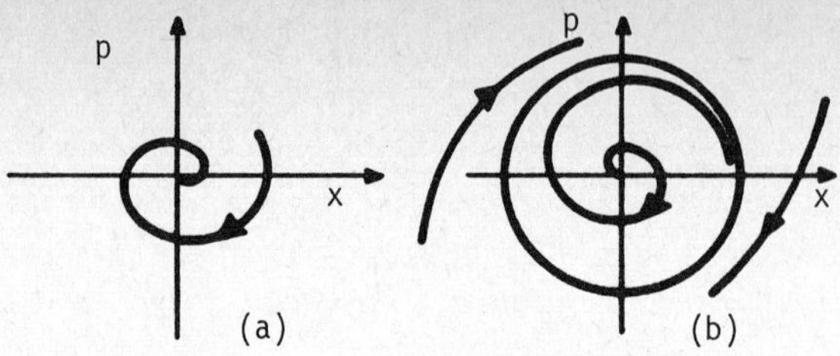

Fig. 7.1 : (a) $\alpha < 0$: the point $x = p = 0$ is an attractor.
(b) $\alpha > 0$: the point $x = p = 0$ is a repeller.

For $\alpha < 0$ this system has ordinary damping and the point $x = p = 0$ is a *stable point*. There is a critical point at $\alpha_c = 0$, and for $\alpha > 0$ the system has negative damping for $x^2 < \alpha/b$ and positive damping for $x^2 > \alpha/b$. We therefore expect a stable limit cycle at $x = \sqrt{\alpha/b}$.

7.1.1 Stability Limit

We assume that $B[0,\alpha] = 0$ such that

$$\phi = 0 \tag{7.3}$$

is a steady state. In order to determine the linear dynamic stability of this solution, we study the equation of motion of a small perturbation $\delta\phi e^{-i\omega t}$,

$$[L(\alpha) - i\omega 1]\delta\phi = 0 \ , \quad L = -\nabla_\phi B \ . \tag{7.4}$$

This represents a linear eigenvalue problem which determines the normal modes $\phi_n e^{-i\omega_n t}$ of the natural motion about the stationary state with eigenfrequencies $\omega_n(\alpha)$. The eigenvalue ω_1 with the largest $\mathrm{Im}\,\omega_n$ corresponds to the least stable mode. If $\mathrm{Im}\,\omega_1 < 0$ then all modes will decay in time, and $\phi = 0$ is an attractor. If $\mathrm{Im}\,\omega_1 > 0$ then the mode ϕ_1 grows in time, and $\phi = 0$ is not an attractor. Thus, the stability limit of the phase $\phi = 0$ occurs at a critical value α_c of the control parameter determined by the condition

$$\mathrm{Im}\,\omega_1(\alpha_c) = 0 \ . \tag{7.5}$$

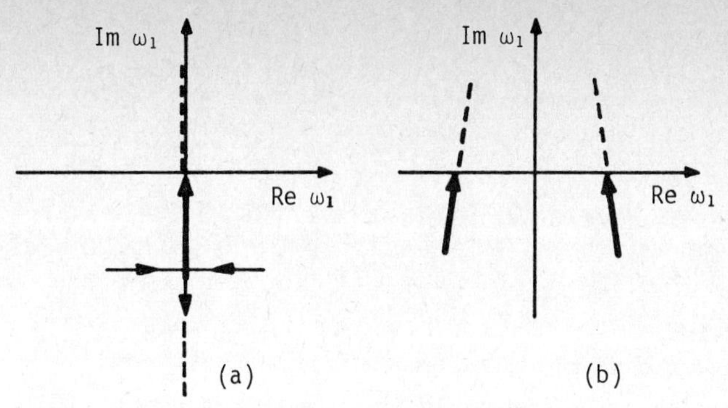

Fig. 7.2 : Undamping of normal modes. (a) Soft mode. (b) Hard mode.

Such undamping of a normal mode can occur either for $\omega_1(\alpha)$ along the imaginary axis (Fig. 7.2(a)),

$$\text{Re}\ \omega_1(\alpha_c) = 0 \ : \text{Soft mode instability,} \tag{7.6}$$

or for a pair of oscillating modes with non-vanishing real part (Fig. 7.2(b)),

$$\text{Re}\ \omega_1(\alpha_c) \neq 0 \ : \text{Hard mode instability.} \tag{7.7}$$

Since we can write the susceptibility

$$\chi(\omega,\alpha) = [L(\alpha) - i\omega.1\,]^{-1} \ , \tag{7.8}$$

it follows that the undamping of a normal mode leads to a divergence of the linear response function at $\alpha = \alpha_c$. In the RPA the dynamic structure factor

$$S(\omega,\alpha) = \chi^*(\omega,\alpha).c(\alpha).\chi(\omega,\alpha) \tag{7.9}$$

gives rise to a divergence of the fluctuation spectrum.

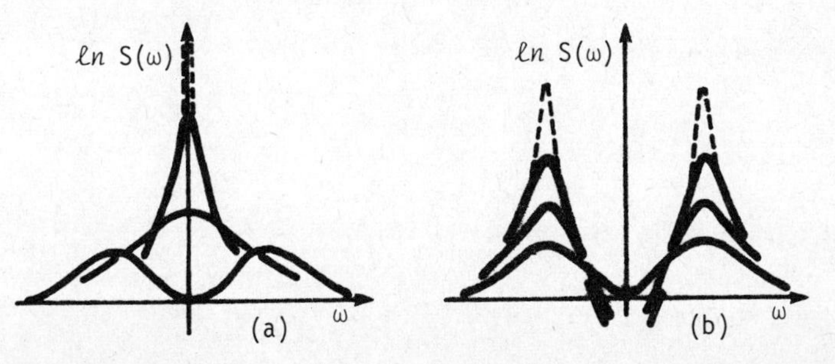

Fig. 7.3 : Critical fluctuations. (a) Soft-mode instability. (b) Hard-mode instability.

Because of the reality of ϕ the eigenvalues $\lambda_n = i\omega_n$ are either *real*, or they occur in *complex conjugate pairs*.

7.1.2 Bifurcation from the Steady State

We now describe what happens on going through the instability. We always assume that the steady state is stable for $\alpha < \alpha_c$ and unstable for $\alpha > \alpha_c$.

(1) $\lambda_1 = i\omega_1$ is real

It is assumed that ϕ can be expanded in a complete set of eigenvectors ψ_n,

$$\phi = \sum_n \phi_n \psi_n . \tag{7.10}$$

At the transition point it is the amplitude that goes with the eigenvector ψ_1 (the unstable mode) which does not decay. We assume that $\rho = |\phi_1|$ can be used as an expansion parameter. Expanding to higher order in ρ gives

$$\alpha = \alpha_c + \sum_k \alpha^{(k)}\rho^k \quad , \; \phi_n = \sum_k \phi_n^{(k)}\rho^k \; ,$$

$$B = B^{(o)} + \sum_k B^{(k)}\rho^k \; , \; L = L^{(o)} + \sum_k L^{(k)}\rho^k \; . \tag{7.11}$$

It is the basis of any bifurcation theory that one must find the right expansion parameter, and in every case this expansion parameter is the amplitude of the norm of the *unstable mode*. (One does *not* use $\alpha - \alpha_c$ as the expansion parameter.)

The result of the analysis is as follows:

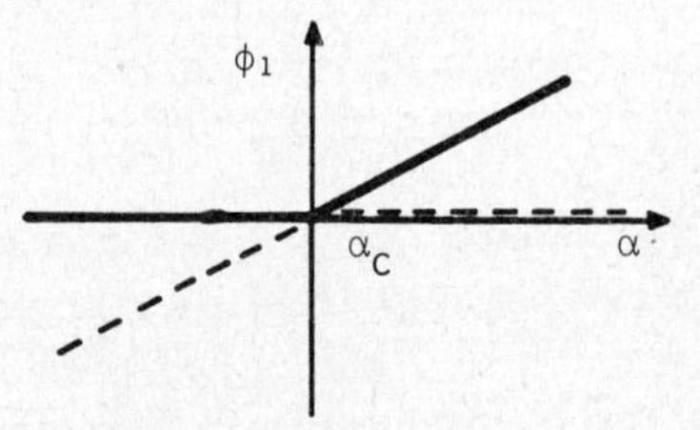

Fig. 7.4 : Exchange of stability at $\alpha = \alpha_c$.

The attractor $\phi = 0$ is stable up to the transition $\alpha = \alpha_c$ and then becomes unstable above the transition. From the expansion one finds another stationary state that is unstable below the transition and becomes stable above the transition ('exchange of stability').

(2) <u>$\lambda_{1,2}$ are complex conjugates</u>

We write

$$\lambda_{1,2} = \lambda(\alpha) \pm i\omega_o(\alpha) \ . \tag{7.12}$$

In this case the two eigenvectors are ψ_1 and $\psi_2 = \psi_1^*$. In state space, the undamped mode at $\alpha = \alpha_c$ has the form

$$\phi(t) = A\psi_1 e^{-i\omega_o^o t} + A^*\psi_2 e^{i\omega_o^o t} \ . \tag{7.13}$$

We assume that the bifurcating solution is also a periodic function of time and expand this function in a Fourier series

$$\phi(t) = \sum_n \phi_n e^{-in\omega t} \quad , \quad \phi_{-n} = \phi_n^* \ . \tag{7.14}$$

Again the amplitude of the first Fourier component

$$\phi_{1,1} = \rho e^{i\psi} \tag{7.15}$$

can be used as an expansion parameter. Expanding to higher order in ρ,

$$\alpha = \alpha_c + \sum_k \alpha^{(k)}\rho^k \quad , \quad \omega_o = \omega_o^o + \sum_k \omega_o^{(k)}\rho^k \ ,$$

$$B = \sum_k B^{(k)}\rho^k \quad , \quad \omega = \omega_o + \sum_k \omega^{(k)}\rho^k \ ,$$

$$\phi_n = \sum_k \phi_n^{(k)}\rho^k \ . \tag{7.16}$$

The results of this analysis are known as the 'Hopf bifurcation'. In the case $\omega_o^o \neq 0$ ('hard-mode instability') the bifurcating solution is for small $\alpha - \alpha_c$ a circle in state space known as the *limit cycle*, which is traversed with frequency ω_o. There are only two possibilities:

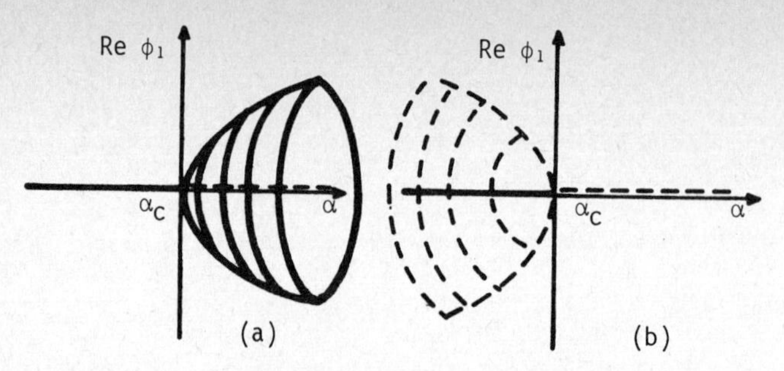

Fig. 7.5 : Hopf bifurcation at $\alpha = \alpha_c$. (a) Stable limit cycle for $\alpha > \alpha_c$. (b) Unstable limit cycle for $\alpha < \alpha_c$.

Either there is a stable limit cycle for $\alpha > \alpha_c$ (normal bifurcation, Fig. 7.5(a)), or there is an unstable limit cycle for $\alpha < \alpha_c$ (inverted bifurcation, Fig. 7.5(b)). For the case of an inverted bifurcation there must occur at α_c a discontinuous transition to another state.

In the case $\omega_0^0 = 0$ ('soft-mode instability') there is either a normal or an inverted bifurcation into a new steady state.

7.2 Continuous Systems

Following Ref. 13 we now consider a system described by a set of (macroscopic) fields

$$\phi(t) = \{\phi_1(x,t),\ldots.,\phi_n(x,t)\} \quad . \tag{7.17}$$

in one space dimension.

7.2.1 Stability of the Uniform Steady State

We consider a continuous system. The uniform steady state ϕ_s is a solution of the non-linear equation

$$B[\phi_s,\alpha] = 0 \ . \tag{7.18}$$

The dynamic stability of the uniform state ϕ_s is determined by the behaviour of small perturbations $\delta\phi(t)$. Because of spatial and temporal translational invariance, these perturbations are plane waves

$$\delta\phi(x,t) = \phi_{q,\omega}\, e^{-i(\omega t - qx)} \quad . \tag{7.19}$$

This leads to the linear eigenvalue problem

$$[L_q(\alpha) - i\omega.1].\delta\phi_{q,\omega} = 0 \ . \tag{7.20}$$

The solution of this equation gives the normal modes with eigenvalues $\omega_\nu(q)$ belonging to the real wave vector q.

The eigenvalue $\omega_1(q,\alpha)$ with the largest Im $\omega_\nu(q,\alpha)$ determines the most weakly damped mode. The stability limit is given by

$$\mathrm{Im}\ \omega_1(q_c,\alpha_c) = 0 \ , \tag{7.21}$$

and the type of instability is determined by Re $\omega_1(q_c,\alpha_c)$:

$$\mathrm{Re}\ \omega_1(q_c,\alpha_c) = 0 : \text{Soft-mode instability,} \tag{7.22}$$

$$\mathrm{Re}\ \omega_1(q_c,\alpha_c) \neq 0 : \text{Hard-mode instability.} \tag{7.23}$$

Here q_c is the wave vector of the critical mode.

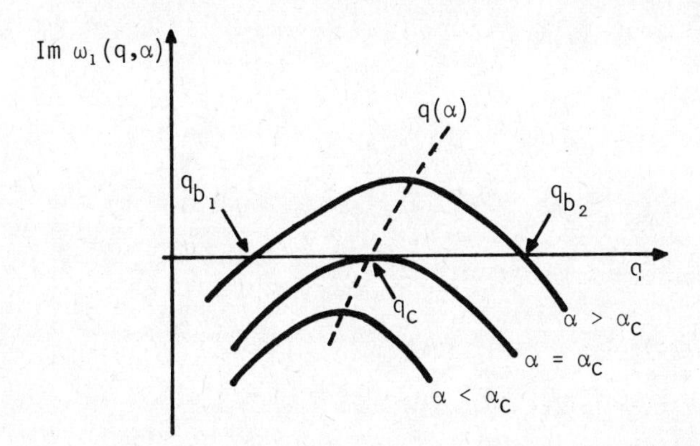

Fig. 7.6 : Behaviour $\omega_1(q,\alpha)$ close to the stability limit $\alpha = \alpha_c$. For $\alpha > \alpha_c$ all modes with wave vectors between q_{b_1} and q_{b_2} have become undamped.

After passing the stability limit, there occurs *continuous undamping* of more and more modes. What is different from the discrete case is that these modes form a continuous spectrum as a function of the wave vector q.

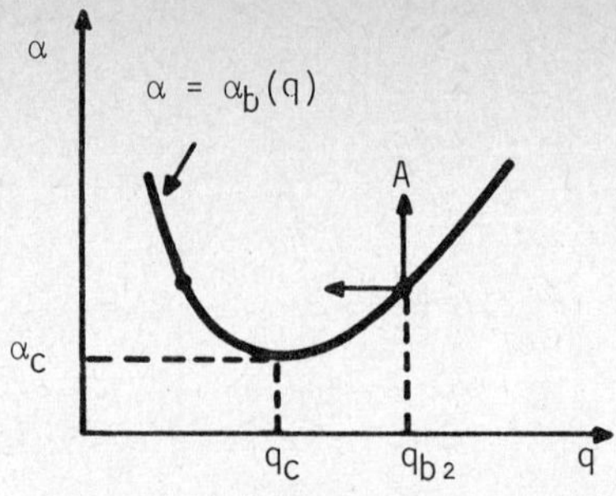

Fig. 7.7 : Bifurcation line $\alpha = \alpha_b(q)$ giving the control parameter α_b at which the mode q becomes undamped.

Setting $\text{Im}\ \omega_1(q,\alpha) = 0$ one obtains the *bifurcation line* giving the values of $\alpha = \alpha_b(q)$ of the control parameter where the mode with wave vector q becomes undamped. All undamped modes are candidates for the new steady state.

7.2.2 Primary Bifurcations of Travelling Wave States from the Steady State

We now consider the problem of determining which states the system may assume after passing through the stability limit. We shall consider the special case of *travelling wave* (TW) states $\phi_T(\xi)$ that depend on space and time in the combination $\xi = x-ut$. For $\alpha > \alpha_c$, more and more modes become unstable along the bifurcation line $\alpha_b(q_b)$,

$$\text{Im}\ \omega_1(q_b,\alpha_b) = 0 , \tag{7.24}$$

with frequency

$$\text{Re}\ \omega_1(q_b,\alpha_b) \equiv \omega_b . \tag{7.25}$$

Each undamped mode gives rise to the bifurcation of a *periodic travelling wave*

$$\phi_T(\xi),\ \phi(\xi+\Lambda_b) = \phi_T(\xi),\ \Lambda_b = 2\pi/q_b . \tag{7.26}$$

Assuming that we have a solution ϕ_T to the equation of motion

$$u\frac{d\phi}{d\xi}T + B[\phi_T,\alpha] = 0 , \tag{7.27}$$

what can we say about its stability properties? We first expand such a solution into its Fourier series

$$\phi_T(\xi) = \phi_s + \sum_n \phi_n\, e^{inq\xi} . \tag{7.28}$$

For real fields $\phi_n = \phi^*_{-n}$.

We want to follow how such a state bifurcates out of the uniform state ϕ_s. We use the amplitude

$$A = |\phi_1| \tag{7.29}$$

to expand

$$\phi_n = \sum_k \phi_n^{(k)} A^k ,$$

$$\alpha = \alpha_b + \sum_k \alpha^{(k)} A^k , \Lambda = \Lambda_b ,$$

$$u = u_b + \sum_k u^{(k)} A^k . \tag{7.30}$$

To the lowest order in the expansion parameter A, ϕ_1 satisfies the linear eigenvalue equation

$$[L_{q_b} (\phi_s, \alpha) - iuq_b . 1] . \phi_1 = 0 . \tag{7.31}$$

Thus the period $\Lambda_b = 2\pi/q_b$ and the pulse velocity u_b at the bifurcation are given by

$$\mathrm{Im}\ \omega(q_b, \alpha_b) = 0 , \tag{7.32}$$

$$\mathrm{Re}\ \omega(q_b, \alpha_b) \equiv \omega_b(q_b) = u_b q_b . \tag{7.33}$$

Expansion to higher powers of A at constant $\Lambda = \Lambda_b$ yields a power series for $\alpha - \alpha_b$, $u - u_b$ and higher Fourier coefficients ϕ_n of the TW state. The type of bifurcation is determined by the leading term of $\alpha - \alpha_b$,

$$\alpha - \alpha_b = \psi(\alpha_b) A^{1/\beta} \tag{7.34}$$

which defines a *bifurcation exponent* $1/\beta$. The sign of $\psi(\alpha_b)$ determines the type of bifurcation

$$\psi(\alpha_b) > 0 : \text{normal bifurcation} , \tag{7.25}$$

$$\psi(\alpha_b) < 0 : \text{inverted bifurcation} . \tag{7.36}$$

Note that $\psi(\alpha_b)$ may change sign at a particular α_b .

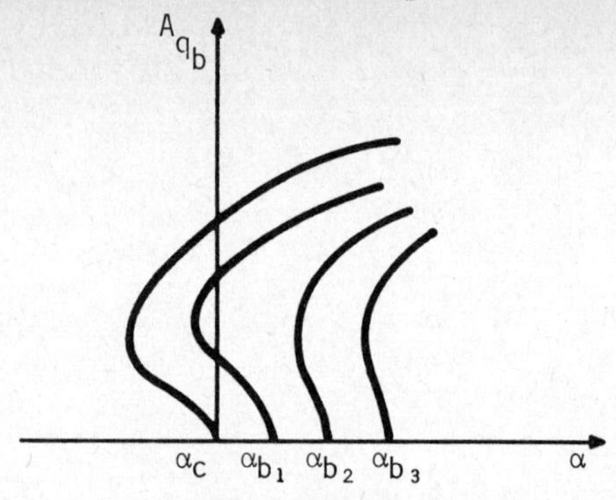

Fig. 7.8 : Amplitude A_{q_b} along the path A (see Fig. 7.7). An example is shown in which the bifurcation of periodic TW states changes from inverted to normal as α_b increases

7.2.3 Stability of TW States

We now test the linear stability of the TW state $\phi_T(\xi)$ against small perturbations $\delta\phi(\xi,t)$. We seek solutions of the form

$$\delta\phi(\xi,t) = \delta\phi_\lambda(\xi)e^{-\lambda t} , \tag{7.37}$$

leading to the eigenvalue problem

$$\{L[\phi_T(\xi),\alpha] - u\frac{d}{d\xi}\}.\delta\phi_\lambda(\xi) = \lambda\delta\phi_\lambda(\xi) . \tag{7.38}$$

The stability of the TW state is determined by the spectrum of

$$\{L[\phi_T(\xi)] - u\frac{d}{d\xi}\} . \tag{7.39}$$

The state $\phi_T(\xi)$ is stable if Re $\lambda > 0$ for all modes except the Goldstone mode.

We now give some methods for investigating the spectrum of $\{L[\phi_T(\xi)] - u\frac{d}{d\xi}\}$.

(1) Breaking of Translational Symmetry

The non-uniform states $\phi_T(\xi)$ are states of broken translational symmetry. A set of equivalent states $\phi_T^{(\alpha)}(\xi)$ is generated by translations. The infinitesimal translation represents a *Goldstone Mode* (GM)

$$\delta\phi_{GM} = \frac{\partial\phi_T(\xi)}{\partial\xi} . \tag{7.40}$$

with eigenvalue $\lambda = 0$.

(2) Periodicity of TW State

The operator $L[\phi_T(\xi)]$ has at least the symmetry of the TW state ϕ_T . The operator L is invariant under the discrete group $\{T_\Lambda\}$, and possibly even under the higher group $\{T_{\Lambda/m}\}$ of translations $n\Lambda/m$. The eigenvalues $\lambda_n(q)$ are multivalued functions of the reduced wave vector over the Brillouin zone $-m\pi/\Lambda \leqslant q \leqslant m\pi/\Lambda$.

(3) Zero-Amplitude Spectrum

At the bifurcation, the spectrum $\lambda_b(q)$ is determined by the spectrum of $\omega_b(q)$ of the uniform state at $\alpha = \alpha_b$ by

$$\lambda_b(q) \to -i\ [\omega_b(\tilde{q}) - u_b\tilde{q}] \quad \text{as } A \to 0 , \tag{7.41}$$

where $\tilde{q} = q + \kappa$ and κ is a reciprocal lattice vector. All small-amplitude solutions with $\alpha_b > \alpha_c$ are unstable at the bifurcation.

(4) Noncrossing Rules

No crossing of eigenvalues can occur in the BZ for modes of the same symmetry.

(5) Perturbation Theory

The curvature of the eigenvalue $\lambda(q)$ near a symmetry point may be obtained from k.p perturbation theory.

7.2.4 Secondary Bifurcation from TW States

Consider a stable TW state $\phi_T(\xi)$. Bifurcation from this state is called *secondary bifurcation*. Whenever Re $\lambda_n(q) = 0$ for some $q = q_{2b}$ there will occur secondary bifurcation of a doubly periodic state

$$\phi_2(x-u_1t,\ x-u_2t) , \tag{7.42}$$

where

$$\phi_2(\xi_1+\Lambda_1,\ \xi_2+\Lambda_2) = \phi_2(\xi_1,\ \xi_2) . \tag{7.43}$$

Here Λ_2 and u_2 at the bifurcation are determined by

$$\Lambda_{2b} = 2\pi/q_{2b} , \tag{7,44}$$

$$u_{2b} = u_1 + \text{Im}\ \lambda_n(q_b)/q_{2b} . \tag{7.45}$$

The same type of argument may be repeated for doubly periodic states leading to tertiary bifurcation of triply periodic states, etc. Under what conditions a bifur= cation of a strongly non-periodic state may occur is discussed in a later section on the onset of turbulence.

7.2.5 Examples

We now illustrate the general considerations given above by three specific examples taken from Section 6.1.3.

(1) Field Equation Without Internal Symmetry

We first study the field equation

$$\dot{\phi} = \alpha\phi - \phi^2\ (-\gamma\phi^3) - v\phi_x + \phi_{xx} . \tag{7.46}$$

Such an equation has an application to current instabilities in semiconductors where $\phi \propto$ excess field, $\alpha \propto$ total current through the sample, and v is the drift velocity of the carriers.

For simplicity, we consider the case $\gamma = 0$. This will leave the field equation invariant under the transformation $\phi \to \alpha - \phi$, but this symmetry will have no con= sequences. The equation has a uniform stationary solution $\phi_s = 0$, with normal mode frequencies

$$\omega_q = vq + i\alpha - iq^2 , \tag{7.47}$$

and a bifurcation line $\alpha = q^2$.

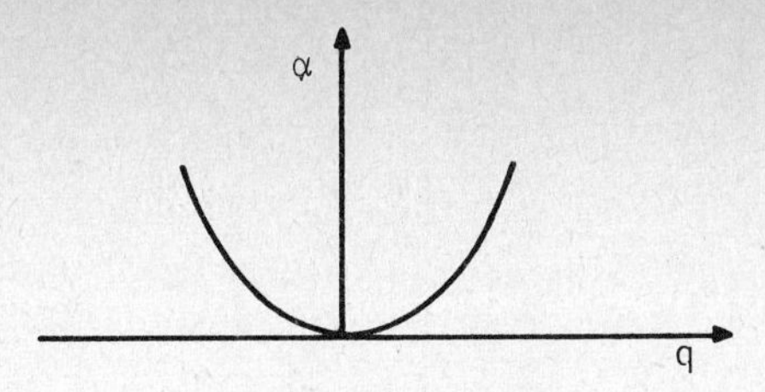

Fig. 7.9 : Bifurcation line for time-dependent Ginsburg-Landau equation.

We find that for $q_b \neq 0$ there is a normal bifurcation ($\beta = \frac{1}{2}$) of a periodic TW state travelling with velocity $u = v$, while for $q_b = 0$ there is a normal bifurcation ($\beta = 1$) of a uniform state $\phi_s = \alpha$ *and* of a solitary wave with $u = v$ (plus unstable family with $u \neq v$). Now let us look at the stability spectrum. For the periodic TW states, with period $\Lambda = 2\pi/q_b$, the spectrum $\lambda = -\alpha_b + q^2$ of the zero-amplitude

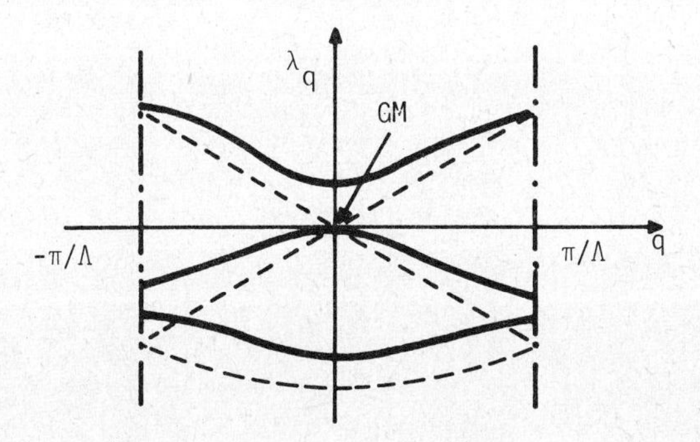

Fig. 7.10 : Spectrum of unstable state (broken line) and of a periodic TW with period Λ. Brillouin zone (BZ) has extension $2\pi/\Lambda$, and Goldstone mode (GM) lies at the centre of the BZ in the second-lowest band.

solution $\alpha = \alpha_b$ contains two unstable bands (broken lines in Fig. 7.10). The Goldstone mode occurs at $q = 0$, and thus the lowest band cannot cross to positive values.

All periodic TW states are unstable, at least against part of the lowest band.

We next consider the stability of the solitary wave ($\Lambda = \infty$). The solitary state has a discrete spectrum.

The solitary state is unstable, but only against one discrete mode.

In the current-instability case, this instability can be removed by coupling the sample to an external circuit with low impedance.

(2) Field equation with reflection symmetry

As a second example we consider the field equation

$$\dot{\phi} = \alpha\phi - \phi^3 - v\phi_x + \phi_{xx} , \tag{4.48}$$

with reflection symmetry $\phi \to -\phi$. For $q_b \neq 0$ there is a normal bifurcation ($\beta = \frac{1}{2}$) of periodic TW states with $u = v$, while for $q = 0$ there is a normal bifurcation ($\beta = \frac{1}{2}$) of two uniform states $\phi_s = \pm\alpha$, *and* two solitary waves with $u = v$.

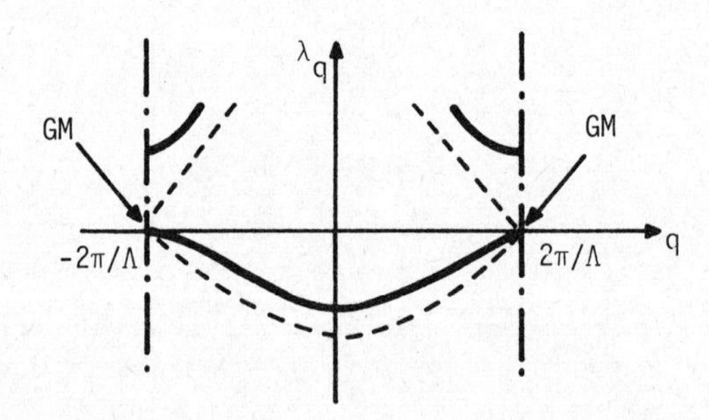

Fig. 7.11 : Spectrum of unstable uniform state (broken line) and TW with period Λ. The Brillouin zone (BZ) has extension $4\pi/\Lambda$, and the Goldstone mode (GM) lies at the boundary of the BZ and belongs to the lowest band.

The periodic TW states have the symmetry

$$\phi_T(\xi+\Lambda) = -\phi(\xi) . \tag{7.49}$$

The BZ is $(-2\pi/\Lambda, 2\pi/\Lambda)$, and the GM lies on the BZ boundary. The spectrum of the zero-amplitude solution contains one unstable band. k.p-perturbation calculation shows that:

All periodic TW states are unstable, at least against part of the lowest band.

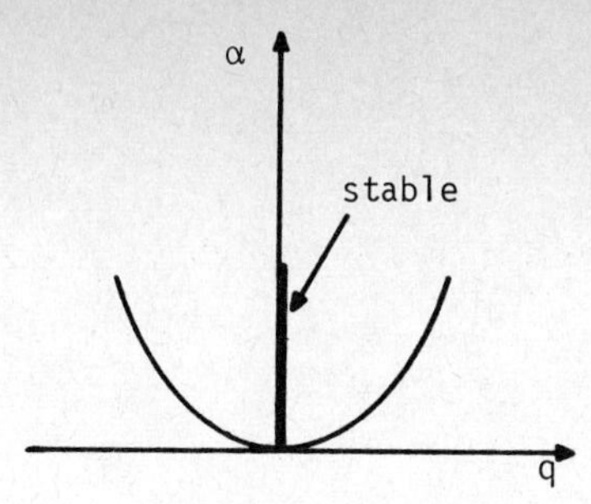

Fig. 7.12 : Bifurcation line.

The two solitary waves, on the other hand, are stable.

(3) Field Equation with Continuous Symmetry

The third example is the field equation

$$\dot{\phi} = \alpha\phi - |\phi|^2\phi - v\phi_x + \phi_{xx} , \tag{7.50}$$

which is invariant under the continuous group of rotations and reflections in the complex ϕ-plane. This equation describes with $v = 0$ the onset of convection in the Rayleigh-Bénard instability, and with v = group velocity of electromagnetic waves the onset of coherent laser action at the laser threshold.

Non-uniform TW states are expressed in terms of an amplitude $R(\xi)$ and phase $\theta(\xi)$,

$$\phi_T(\xi) = R(\xi)e^{i\theta(\xi)} , \tag{7.51}$$

where R and θ are real. We find

$$R\theta_{\xi\xi} + (u-v)R\theta_\xi + 2R_\xi\theta_\xi = 0 , \tag{7.52}$$

$$R_{\xi\xi} + (u-v)R_\xi - R\theta_\xi^2 + \alpha R - R^3 = 0 . \tag{7.53}$$

For $u = v$ the first equation yields

$$J \equiv R^2\theta_\xi = \text{constant.} \tag{7.54}$$

There are two types of non-linear solutions, namely *amplitude waves* ($J = 0$): waves of constant phase, and *phase waves* ($J \neq 0$): waves of constant amplitude, and the phase varies linearly with ξ,

$$R(\xi) = (\alpha - k^2)^{\frac{1}{2}} \quad , \quad \theta(\xi) = k\xi + \theta_0 \tag{7.55}$$

For a given value of J,

$$J = k(\alpha - k^2) = R^2(\alpha - R^2)^{\frac{1}{2}} \quad , \tag{7.56}$$

one finds two solutions, one with a small R and one with a large R.

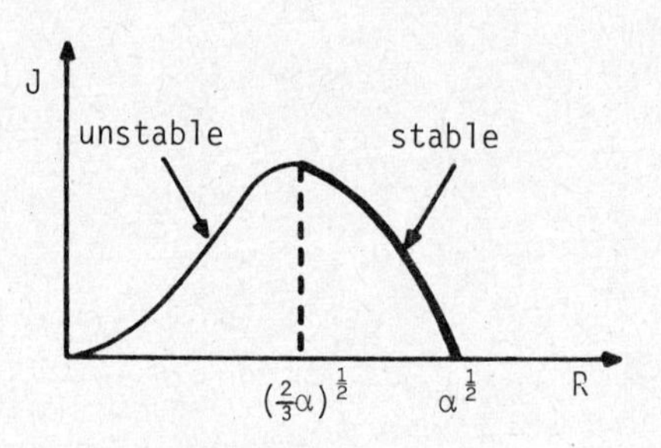

Fig. 7.13 : Phase portrait of the TW's .

These states describe rolls in the Bénard problem and coherent waves in the laser. The stability analysis shows that for given J the solution with small R is unstable and the solution for large R is stable (Fig. 7.13).

8. Onset of Turbulence

8.1 Hopf-Landau Picture[14].

We consider a system which as a function of the control variable α undergoes a sequence of Hopf-type bifurcations (see Fig. 8.1), each bifurcation introducing a new frequency ω_n (n=1,2,...) incommensurate with all previous ones.

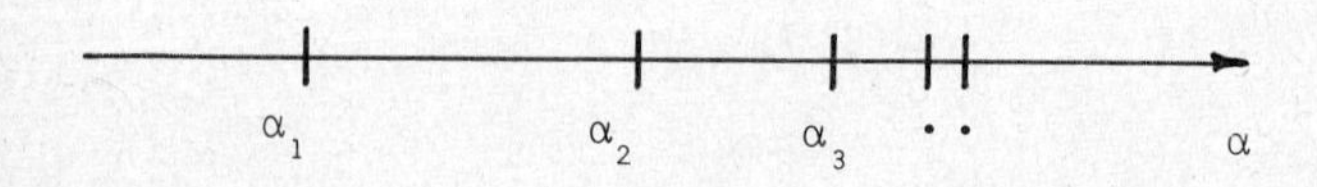

Fig. 8.1 : Sequences of Hopf bifurcations.

Then, after k bifurcations one obtains a state

$$\phi(t) = \phi_k\ (\omega_1 t,\ \omega_2 t,\ \ldots,\ \omega_k t) \quad , \tag{8.1}$$

which is a *multiply periodic* or *almost periodic* function of time. The function ϕ_k is periodic (with period 2π) in each of its arguments,

$$\phi_k(\omega_1 t+2\pi n_1,\ldots,\omega_k t+2\pi n_k) = \phi_k(\omega_1 t,\ldots,\omega_k t) \quad . \tag{8.2}$$

By expanding ϕ_k into a k-fold Fourier series, one obtains

$$\phi(t) = \sum_{\nu_1 \ldots \nu_k} a_k(\nu_1,\ldots,\nu_k)\exp(-i\sum_j \nu_j\omega_j t) \quad , \tag{8.3}$$

which shows that the state $\phi(t)$ contains the higher harmonics as well as the sum and difference frequencies. As more frequencies enter, the motion looks more and more irregular. For the transition to turbulence, one obtains in this picture the sub= jective criterion: when is it no longer possible to resolve the motion into its frequency components?

The flow described by $\phi(t)$ takes place on a k-dimensional torus in phase space. It is *ergodic* (the flow covers the torus densely) but not *mixing* (an initial surface element "stays together"). The coefficients a_k contain phase factors depending on the initial conditions. A quantity independent of the initial conditions is the *autocorrelation function*

$$C(\tau) = \lim_{T\to\infty} \frac{1}{T} \int_0^T \phi^*(t)\phi(t+\tau)\ dt$$

$$= \sum_{\nu_1 \ldots \nu_k} |\ a_k(\nu_1,\ldots,\nu_k)\ |^2 \exp\ (-i\sum_j \nu_j\omega_j\tau) \quad , \tag{8.4}$$

which is an almost periodic function of τ. Its Fourier transform, the *power spectrum*

$$S(\omega) = \int C\ (\tau)\ \exp\ (i\omega\tau)\ d\tau$$

$$= \sum_{\nu_1 \ldots \nu_k} |\ a_k(\nu_1,\ldots,\nu_k)\ |^2\ \delta(\omega-\sum_j \nu_j\omega_j) \quad , \tag{8.5}$$

consists of a dense but countable set of spectral lines at the harmonics and sum and difference frequencies. Both properties - the almost periodic nature of $C(\tau)$

and the dense line spectrum of $S(\omega)$ - are characteristic for an ergodic non-mixing flow. For turbulent motion, however, one expects mixing behaviour, i.e. decay of correlations and a continuous power spectrum, which cannot be obtained from the Hopf-Landau picture.

8.2 The Lorenz Model

Lorenz[15] studied a simplified model of the Bénard instability containing only three variables, one velocity Fourier component x and two temperature Fourier components y, z. For these three variables, the truncated Navier-Stokes equations take the form

$$
\begin{aligned}
x &= \sigma(y-x) \ , \\
y &= -xz+ry-y \ , \\
z &= xy-bz \ . \qquad (8.6)
\end{aligned}
$$

Here r is the Rayleigh number, σ is the Prandtl number, and $b=8/3$. An analysis of this simple system of equations yields as a function of the control parameter r for $\sigma>b+1$ (see Fig. 8.2)

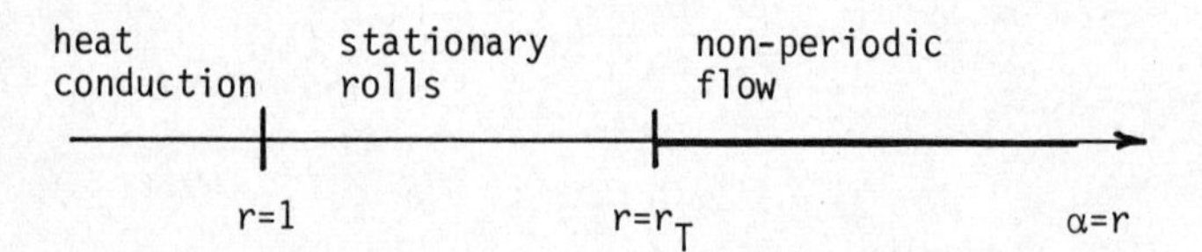

Fig. 8.2 : Bifurcation in the Lorenz model.

$r<1$: Heat conduction state, $x=y=z=0$ stable,

$r=1$: normal bifurcation,

$1<r<r_T$: stationary convection rolls,

$$x = y = \pm[b(r-1)]^{\frac{1}{2}} \ , \quad z=r-1 \quad \text{stable} \ ,$$

$$r_T = \sigma(\sigma+b+3)/(\sigma-b-1) \ , \qquad (8.7)$$

$r=r_T$: inverted bifurcation ,

$r>r_T$: nonperiodic behaviour .

From analogy with phase transitions in equilibrium systems one would have expected that the initially unstable state bifurcating at r_T "bends over" and becomes stable above a certain amplitude, giving rise to a first-order transition. But Lorenz found by integrating the above equations on a computer a transition to a *non-periodic* (*chaotic*) state. The region occupied by the non-periodic orbit has the form of a surface consisting of an uncountable set (a "Cantor set") of sheets. In contra=distinction to the known types of attractors: point attractor (stationary state), limit cycle ≙ 1-torus (periodic state), k-torus (multiply periodic state), this has been called a *strange attractor*. The various attractors may be characterised by the *Lyapunov exponent* λ which describes the asymptotic behaviour of the distance $d \sim \exp(\lambda t)$ between two initially close phase points for large t. One has

- for point attractors $\lambda<0$: The distance from the stationary point decreases exponentially
- for k-tori $\lambda=0$: The distance between two phase points becomes asymptotically con=stant or grows at most algebraically
- for strange attractors $\lambda>0$: All trajectories on the attractor diverge exponentially. Therefore, there is not enough space on a single sheet or even on a countable number of sheets; the attractor must be "folded over" an uncountable number of times.

The exponential divergence of the trajectories leads to mixing behaviour on the attractor, with decay of correlations and a continuous power spectrum. It there=fore seems reasonable that the transition discovered by Lorenz marks the onset of turbulence, although the chaotic behaviour occurs only with respect to time (and not, as in true turbulence, also with respect to the spatial co-ordinates).

8.3 The Ruelle-Takens Picture

Ruelle and Takens[16] showed that under fairly general conditions the Hopf-Landau picture is structurally unstable : They proved that in every neighbourhood of a triply periodic state (a 3-torus) there is a strange attractor. Thus, a triply periodic state will in general not exist over a finite interval of the control parameter.

In fact, already a doubly periodic state is not structurally stable : In every neighbourhood of a doubly periodic state there is a periodic state in which the two frequencies are "locked" in a rational ratio. But it should be noted that neither of these statements says anything about the relative *measure* of multiply periodic states, chaotic states or locked states on the α-axis, i.e. about the chance to have one or the other state for a given value of α.

8.4 Period Doubling Sequences

Another transition to turbulence occurs through an infinite sequence of successive period-doubling (i.e. subharmonic) bifurcations. This has been studied mostly for finite-difference time-evolution equations of the form

$$x_{t+n} = f(x_t \; ; \alpha) \tag{8.8}$$

("maps of the interval"). Particularly well studied is the equation

$$x_{t+1} = \alpha \, x_t \, (1-x_t) \quad . \tag{8.9}$$

It was found independently by Grossmann and Thomae[17], Feigenbaum[18], and Coullet and Tresser[19] that this sequence of bifurcations shows some remarkable regularities.

As function of the control parameter α, there occurs a sequence of bifurcation points $\alpha_1 < \alpha_2 < \ldots$ accumulating at α_∞ (see Fig. 8.3). For the above example, $\alpha_\infty = 3.57$.

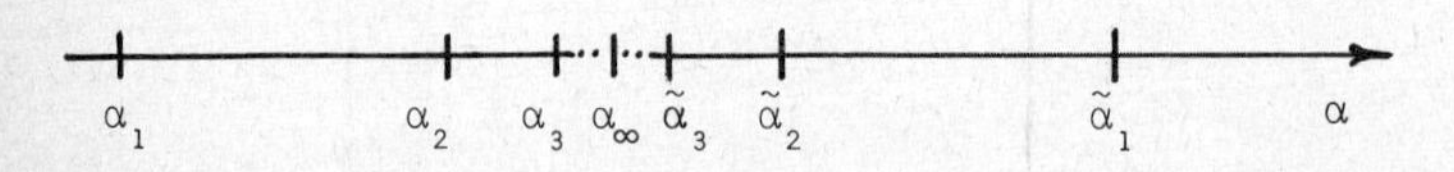

Fig. 8.3: Bifurcation points accumulating at α_∞.

At $\alpha=\alpha_n$, the 2^{n-1}-cycle becomes unstable, which gives rise to the bifurcation of a 2^n-cycle. One finds that asympototically for $n \to \infty$ the distances $\alpha_\infty-\alpha_n$ form a geometric sequence,

$$\alpha_\infty-\alpha_n \sim \delta^{-n} \quad , \tag{8.10}$$

with

$$\delta = 4.669\ldots, \tag{8.11}$$

a universal number, independent of the details of the smooth function f.

For $\alpha>\alpha_\infty$ one finds aperiodic ("chaotic") behaviour characterised by a Lyapumov exponent $\lambda>0$ which varies for $\alpha \to \alpha_\infty$ as[20]

$$\lambda(\alpha) \sim (\alpha-\alpha_\infty)^\tau \quad , \tag{8.12}$$

with

$$\tau = \ln 2/\ln \delta = 0.4\,498 \quad \ldots \tag{8.13}$$

The chaotic region $\alpha>\alpha_\infty$ is interrupted by narrow windows with periodic behaviour and other period-doubling sequences, (see Ref. 21).

For a given value $\alpha>\alpha_\infty$, the chaotic orbits x_t $(t=1,2,\ldots)$ form a structure consisting of 2^n bands in which the x_t are dense, separated by 2^{n-1} gaps. The motion may be viewed as a regular motion on a 2^n cycle, superimposed by an irregular motion ("periodic chaos")[17]. At critical values $\tilde{\alpha}_n$ the bands merge pairwise, such that one obtains a chaotic period-doubling sequence $\tilde{\alpha}_1>\tilde{\alpha}_2>\ldots$ accumulating also at α_∞ (see Fig. 8.3) and one finds asymptotically

$$\tilde{\alpha}_n-\alpha_\infty \sim \delta^{-n} \tag{8.14}$$

with the same value of δ as above.

This behaviour as well as a number of scaling properties may be understood in terms of a renormalisation-group transformation introduced by Feigenbaum[18] and Collet and Eckmann[22]. Of particular interest is the scaling behaviour of a system under the influence of an external noise source[23,24]. It turns out that the noise intensity plays for the transition to turbulence a role which is quite analogous to that of a symmetry—breaking field at an ordinary second-order phase transition.

Such period-doubling sequences have been observed in actual systems showing a trans= ition to turbulence [25,26].

8.5 Transition to Turbulence through Intermittency

We mention briefly a further type of transition to turbulence which has been described by Pomeau and Manneville[27,28]. Here, the motion consists of long periods of regular flow interrupted by intermittent turbulent bursts of random duration. This type is expected to occur when as a function of the control parameter a stable and unstable fixed point collide.

References

1. T. Andrews, *Phil. Trans. R. Soc.*, 159, 575 (1869).
2. J. Hopkinson, *Proc. R. Soc.*, 48, 1 (1890).
3. J.D. van der Waals, Ph.D. Thesis, Univ. of Leiden (1893).
4. P. Weiss, *J. Phys. Radium*, Paris 5, 153 (1907).
5. I owe this argument to J. Lajzerowicz (private communication).
6. E.M. Lifshitz, *J. Phys. Moscow* 6. 61 (1942).
7. J.O. Dimmock, *Phys. Rev.* 130, 1337 (1963).
8. T.A. Kaplan, *Bull. Acad. Sci.* USSR 28, 328 (1964).
9. I.E. Dzyaloshinsky, *Soviet Phys.* -JETP 19, 960 (1964).
10. S. Goshen, D. Mukamel and S. Shtrikman, *Int. J. Magnetism*, 6, 221 (1974)
11. E. Courtens and R.W. Gammon, *Phys. Rev.* B 24, 3890 (1981).
12. H. Haken : Synergetics. *An Introduction.* Springer, Berlin-Heidelberg-New York (1977).
13. M. Büttiker, H. Thomas, *Phys. Rev.* A24, 2635 (1981).
14. See for instance : L.D. Landau and E.M. Lifshitz, Hydrodynamics, Akademie Verlag, Berlin (1966).
15. E.N. Lorenz, *J. Atmos. Sci.* 20, 130 (1963).
16. D. Ruelle, F. Takens, *Comm. Math. Phys.* 20, 167 (1971).
17. S. Grossmann, S. Thomae, Z. *Naturforsch,* 32a, 1353 (1977).
18. M.J. Feigenbaum, *J. Stat. Phys.* 19, 25 (1978), 21, 669 (1979)
19. P. Coullet, C. Tresser, *J. de Phys.*, Colloque 39, C5-25 (1978)
20. B.A. Huberman, J. Rudnick, *Phys. Rev. Letters* 45, 154 (1980).
21. T. Geisel, J. Nierwetberg, *Phys. Rev. Letters* 47, 975 (1981).
22. P. Collet, J.P. Eckmann, Iterated Maps on the Interval as Dynamical Systems, Birkhauser, Basel (1980).
23. B. Shraiman, C.E. Wayne, P.C. Martin, *Phys. Rev. Letters* 46, 933 (1981).
24. J.P. Crutchfield, N. Nauenberg, J. Rudnick, *Phys. Rev. Letters* 46, 935 (1981).
25. A. Libchaber, J. Maurer, *J. de Phys.*, Colloque 41, C3-51 (1980).
26. M. Giglio, S. Musazzi, U. Perini, *Phys. Rev. Letters* 47, 243 (1981).
27. P. Manneville, Y. Pomeau, *Phys. Letters* 75A 1 (1979).
28. Y. Pomeau, P. Manneville, *Comm. Math. Phys.* 74, 189 (1980).

Recent Reviews on Phase Transitions and Critical Phenomena

R.A. Cowley and A.D. Brûce, *Advances in Physcis* 29 No.1 (1980).

C. Domb, M.S. Green (Eds.): Phase Transitions and Critical Phenomena, Vols.1-6. Academic Press, New York 1972-1976.

W. Gebhardt, U. Krey: Phasenübergänge und kritische Phänomene. Vieweg, Braunschweig 1980

F. Fisher: The States of Matter - A Theoretical Perspective. Proceedings of the Robert A. Welch Conferences on Chemical Research XXIII. Modern Structural Methods. Houston, Texas, Nos. 12-14, 1979.

L. Kadanoff et al., *Rev. Mod. Phys.* 39, 395 (1967).

S.K. Ma: Modern Theory of Critical Phenomena. Benjamin, Reading 1976.

K.A. Müller, A Rigamanti (Eds.): Local Properties at Phase Transitions. Proceedings of the International School of Physics Enrico Fermi, Course LIX, Societa Italiana di Fisica, Bologna 1976.

K.A. Müller, H. Thomas (Eds.): Structural Phase Transitions I. Springer, Berlin, Heidelberg, New York 1981.

T. Riste (Ed.): Anharmonic Lattices, Structural Transitions and Melting. Noordhoff, Leiden 1974.

T. Riste (Ed.): Electron-Phonon Interactions and Phase Transitions. Plenum, New York 1977.

T. Riste (Ed.): Ordering in Strongly Fluctuating Condensed Matter Systems. Plenum, New York 1980.

E.T. Samuelsen, E. Anderson, J. Feder (Eds.): Structural Phase Transitions and Soft Modes. Universitetsforlaget, Oslo 1971.

H.E. Stanley: Introduction to Phase Transitions and Critical Phenomena. Clarendon, Oxford 1971.

Recent Reviews on Stabilities in Driven Systems and Onset of Turbulence

J.-P. Eckmann, *Rev. Mod. Phys.* 53, 643 (1981).

H. Haken (Ed.): Chaos and Order in Nature. Springer, Berlin, Heidelberg, New York 1981.

H. Haken (Ed.): Synergetics. Springer, Berlin, Heidelberg, New York 1977.

R.H.G. Helleman, in: Fundamental Problems in Statistical Mechanics 5, ed. by E.G.D. Cohen. North-Holland, Amsterdam 1980, p.165.

O.E. Lanford: Strange Attractors and Turbulence. Springer, Berlin 1981.

R.M. May, *Nature* 261, 459 (1976).

C. Normand, Y. Pomeau, M.G. Velarde, *Rev. Mod. Phys.* 49, 581 (1977).

E. Ott, *Rev. Mod. Phys.* 53, 655 (1981).

T. Riste (Ed.): Fluctuations, Instabilities, and Phase Transitions. Plenum, New York 1975.

T. Riste (Ed.): Nonlinear Phenomena at Phase Transitions and Instabilities. Plenum, New York 1982.

D. Ruelle, *The Mathematical Intelligencer* 2, 126 (1980).

R. Shaw, *Z. Naturforsch.* A36, 80 (1981).

H.L. Swinney, J.P. Gollub (Eds.): Hydrodynamic Instabilities and the Transitions to Turbulence. Springer, Berlin, Heidelberg, New York 1981.

H. Thomas, in: Noise in Physical Systems, ed. by D. Wolf. Springer, Berlin, Heidelberg, New York 1978, p.278

MULTICRITICAL POINTS

Lectures presented at Advanced Course in
Theoretical Physics: Critical Phenomena

January 1982, Stellenbosch, S.A.

Amnon Aharony
Department of Physics and Astronomy
Tel Aviv University, Tel Aviv 69978
ISRAEL

TABLE OF CONTENTS

MULTICRITICAL POINTS

Amnon Aharony
Department of Physics and Astronomy, Tel Aviv University
Tel Aviv 69978, ISRAEL

I. GENERAL REVIEW: TRICRITICAL POINTS

Under simple circumstances, a phase transition is reached when the temperature, T, approaches a special value, T_c, keeping the ordering field H (e.g. the magnetic field for a ferromagnet) equal to zero. In many cases, some additional parameters (e.g. magnetic and electric fields, stresses, chemical potentials, etc.) can be varied experimentally. This enlarges the dimensionality of the phase diagram (equal to two when only T and H are available) and turns the transition point into a line (or a surface) of such points. As one moves on this transition line (surface), one may reach special points (lines) at which some properties of the transition change abruptly. Such points, which are called multicritical points, are the subject of this series of lectures. We start with a qualitative review of several multicritical points, and this lecture is devoted to tricritical points.

I.1. Dilute Annealed Magnets

Consider a lattice of sites, {i}, which are either occupied by magnetic Ising spins, $\{S_i\}$, with $S_i = \pm 1$, or empty. Assigning an "occupation" variable to each site, t_i, so that $t_i = 1$ if the site is occupied and $t_i = 0$ if the site is empty, the total number of spins is given by $N = \sum_i t_i$ (the sum is over all sites). The actual spin at site i is thus $t_i S_i$, and the nearest neighbor exchange interaction may be written as $J\, t_i t_j\, S_i S_j$. Working in the grand canonical ensemble, the partition function is

$$Z(T, h, \mu) = \underset{\{S_i=\pm 1\}}{\mathrm{Tr}} \quad \underset{\{t_i=0,1\}}{\mathrm{Tr}} \quad \exp(-\beta\mathcal{H}) \quad , \tag{1}$$

with

$$-\beta\mathcal{H} = K \sum_{<ij>} t_i t_j\, S_i S_j + h \sum_i t_i S_i + \beta\mu \sum_i t_i \quad , \tag{2}$$

where μ is the chemical potential, ij denotes nearest neighbors,

$$\beta = 1/k_B T, \quad K = \beta J, \quad h = \beta\mu_o H \tag{3}$$

(μ_o is the magnetic moment per spin). Note that the trace in Eq.(1) is also taken over the occupation variables, $\{t_i\}$. This reflects the assumption that the mixture (of magnetic and non-magnetic ions) is annealed, i.e. the ions are moving around and reaching thermal equilibrium within the time scale of the relevant experiments.

We can now change the (dummy) variables S_i and t_i, and define[1]

$$\sigma_i = t_i S_i \quad . \tag{4}$$

Clearly, σ_i has the values +1, -1 or 0. The value zero is found for two values of the original parameters, i.e. $S_i = 1$, $t_i = 0$ and $S_i = -1$ and $t_i = 0$. Replacing the trace over S_i and t_i by a trace over σ_i thus requires an additional factor of 2 whenever $\sigma_i = 0$. Note also that $t_i = \sigma_i^2$. Eq.(2) thus becomes

$$-\beta H = K \sum_{\langle ij\rangle} \sigma_i \sigma_j + h \sum_i \sigma_i + \beta\mu \sum \sigma_i^2 \quad , \tag{5}$$

and Eq.(1) is replaced by

$$Z = \mathop{\mathrm{Tr}}_{\{\sigma_i = \pm 1, 0\}} 2^{\sum_i (1-\sigma_i^2)} e^{-\beta H} \quad . \tag{6}$$

The first factor, whose role is to preserve the total number of original states per site, can simply be absorbed in the chemical potential,

$$\beta\mu \longrightarrow \beta\tilde{\mu} = \beta\mu - \ln 2 \quad . \tag{7}$$

Eq.(5) now represents a new Hamiltonian, for an Ising model with spin one (instead of spin 1/2). This model was first discussed by Blume and Capel,[2,3] to describe the mixture of ^{4}He and ^{3}He. Although the order parameter of superfluid ^{4}He is a complex number, represented by two components ($n = 2$, XY model) rather than by one ($n = 1$, Ising model), the qualitative results agree with many of the experimental observations.

We now set $h = 0$. For $\tilde{\mu} \longrightarrow -\infty$ it is clear that Eq.(5) will always prefer the values $\sigma_i = \pm 1$. The value $\sigma_i = 0$ will never occur, and the problem reduces to that of the spin-1/2 Ising model, exhibiting a second order continuous phase transition. When $\mu \longrightarrow \infty$, all sites will have $\sigma_i = 0$, and no magnetic properties will be observed at any temperature. Consider now the limit of zero temperature, $T = 0$. In that limit, there exists a competition between two ground states, one with $\sigma_i \equiv 0$, i.e. energy $E = 0$, and the other with $\sigma_i \equiv 1$ (or $\sigma_i \equiv -1$), i.e. energy per site $E = -(\frac{1}{2} J z + \tilde{\mu})$ (z is the coordination number). Clearly, a first order transition (from $|\sigma_i| = 1$ to $\sigma_i = 0$) occurs at $\tilde{\mu} = -\frac{1}{2} J z$. By continuity, it is reasonable to expect a line of first order transitions, beginning at $T = 0$, $\tilde{\mu} = -\frac{1}{2} J z$, and going to finite temperatures and lower values of $\tilde{\mu}$. On the other hand, a line of second order transitions (at which $\langle\sigma_i\rangle$ decreases continuously towards zero) is expected for $\tilde{\mu} \longrightarrow -\infty$. A simple way by which these two lines might meet is illustrated in Fig.1. The lines meet at a special multicritical point, called the tricritical point.[4]

Fig.2 illustrates the origin of the name "tricritical": when the ordering field H is non-zero and its magnitude is very large, there is no distinction between the "ordered" and the "disordered" phases. However, at small values of H, the first order line (shown in Fig.1) becomes a first order surface, which ends at a critical

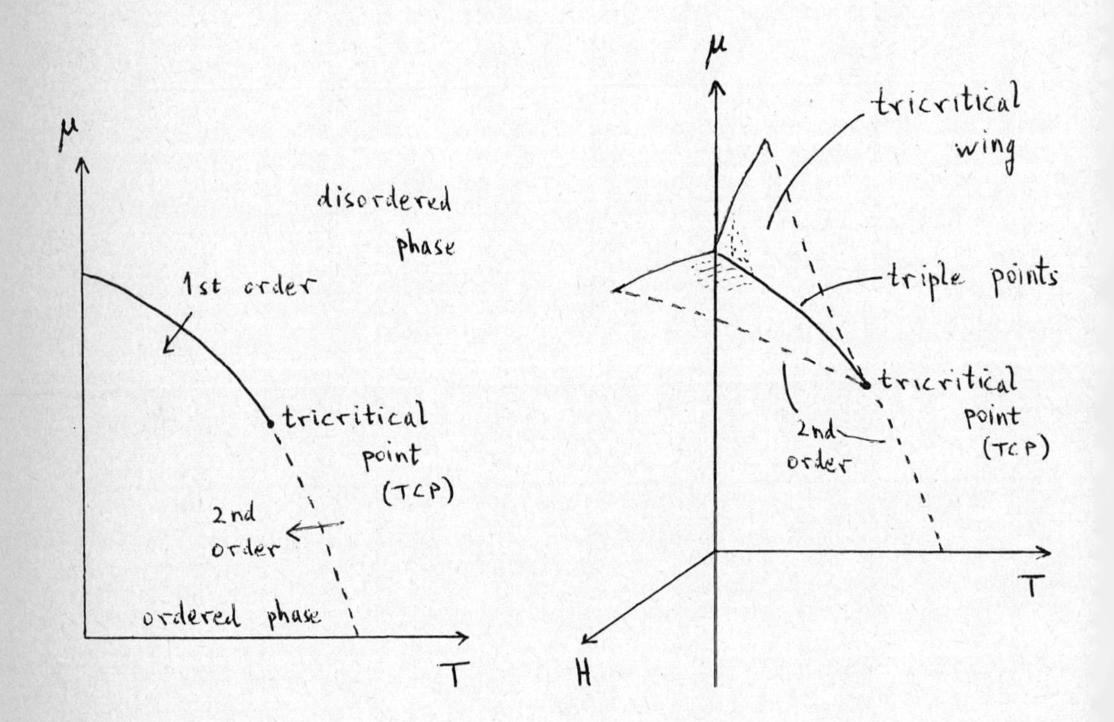

Fig.1. Tricritical point in a dilute magnet.

Fig.2. Tricritical "wings".

(second order) line. These two surfaces, for $H > 0$ and for $H < 0$ (called "tricritical wings") meet the first order $H = 0$ surface (separating $\langle\sigma_i\rangle > 0$ from $\langle\sigma_i\rangle < 0$) at a line of triple points, which ends at the tricritical point.

It should be noted that Fig.1 is not the only possible way by which the first order and second order lines might meet. Fig.3 shows an alternative: the first order line may end at a critical point, within the ordered phase, and the second

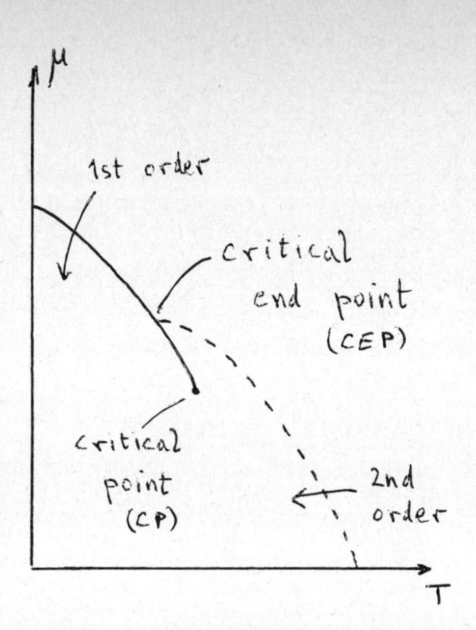

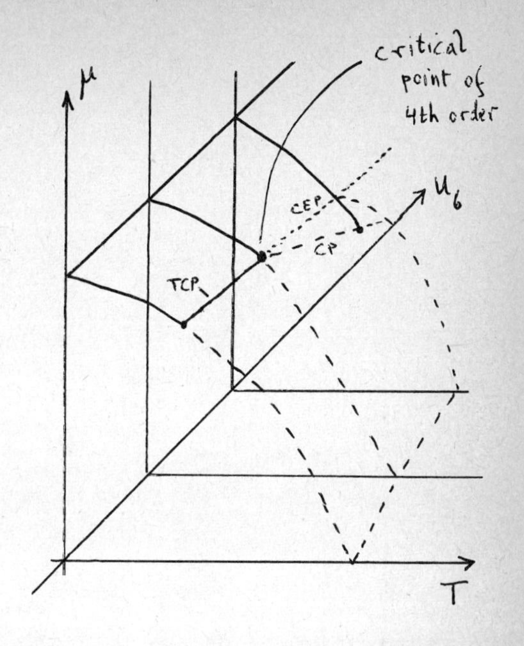

Fig.3. Critical End Point. Fig.4. Critical Point of Fourth Order.

order line then ends where it meets the first order line, at a "critical end point". In fact, there exists an additional parameter, u_6, which may be varied so that Fig.1 and Fig.3 become part of the same phase diagram: as function of u_6, the line of critical points and that of the critical end points meet at another multicritical point, called the "critical point of fourth order" (Fig.4). Beyond that point there appears a line of tricritical points. We shall return to these possibilities below.

I.2. Metamagnets

A simple antiferromagnet is described by a negative exchange interaction. For a spin-1/2 nearest neighbor Ising model, the Hamiltonian may be written as

$$H = - J \sum_{\langle ij \rangle} S_i S_j - \mu_0 H \sum_i S_i \quad , \tag{8}$$

with $J < 0$, $S_i = \pm 1$. On simple hypercubic lattices (e.g. the square lattice, see Fig.5, or the simple cubic lattice), the exchange term in Eq.(8) prefers an antiferromagnetic ground state, in which the spins alternate in sign, as illustrated in Fig.5. A simple way to describe this state is

$$\langle S_i \rangle = M^{\dagger} e^{i \vec{q}_o \cdot \vec{r}_i} \quad , \tag{9}$$

with (for two dimensions)

$$\vec{q}_0 = \left(\frac{\pi}{a}, \frac{\pi}{a}\right) , \tag{10}$$

where a is the lattice constant. The spins are now divided into two interpenetrating *sublattices*, one with $\langle S_i \rangle = M^{\dagger}$ and the other with $\langle S_i \rangle = - M^{\dagger}$. $M^{\dagger}$ is called the "*staggered magnetization*", and serves as the *antiferromagnetic order parameter*.

↓ ↑ ↓ ↑ ↓ ↑
↑ ↓ ↑ ↓ ↑ ↓
↓ ↑ ↓ ↑ ↓ ↑
↑ ↓ ↑ ↓ ↑ ↓
↓ ↑ ↓ ↑ ↓ ↑
↑ ↓ ↑ ↓ ↑ ↓

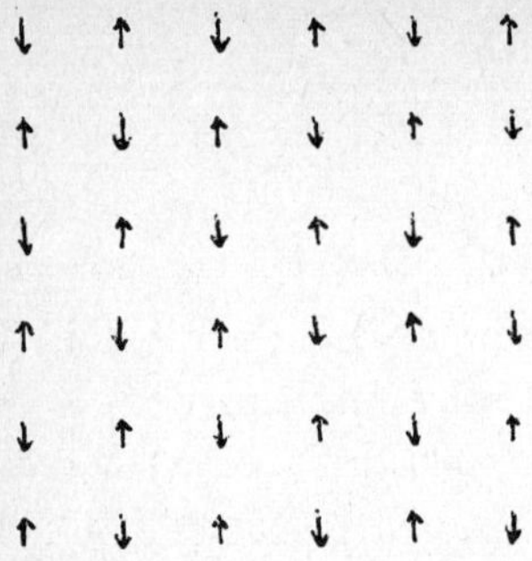

Fig.5. Simple antiferromagnetic ground state.

Since $\{S_i\}$ are "dummy" variables in the partition function, we may change variables into

$$\sigma_i = S_i \; e^{i \vec{q}_0 \cdot \vec{r}_i} , \tag{11}$$

and find that

$$Z = \operatorname*{Tr}_{\{\sigma_i = \pm 1\}} \exp(-\beta \tilde{\mathcal{H}}) , \tag{12}$$

with

$$\tilde{\mathcal{H}} = - |J| \sum_{\langle ij \rangle} \sigma_i \sigma_j - \mu_0 H \sum_i e^{-i\vec{q}_0 \cdot \vec{r}_i} \sigma_i . \tag{13}$$

The first term now represents a *ferromagnetic* coupling between the σ_i's. However, the magnetic field acting on σ_i now oscillates in sign, and $H_i = H \exp(i\vec{q}_0 \cdot \vec{r}_i)$ is called a "*staggered field*". When $H = 0$, the thermodynamics of Eq.(8) is thus equivalent to that of Eq.(13), and one expects a regular continuous transition from the disordered (paramagnetic) phase into the ordered (ferromagnetic or antiferromagnetic) one. Such a continuous transition may also be expected for small non-zero values of H.

The second term in Eq.(8) favors a paramagnetic ordering, in which all S_i are equal to +1. This will certainly be the situation for sufficiently large values of H. There is thus a *competition* between the two terms in Eq.(8). This competition is also reflected in the two possible *ground states* at zero temperature: either $S_i \equiv e^{i \vec{q}_0 \cdot \vec{r}_i}$, and $E = + \frac{1}{2} J z$, or $S_i \equiv 1$ and $E = - \frac{1}{2} J z - \mu_0 H$. Clearly, a first order transition occurs at $\mu_0 H = |J| z$. As described before, the first order line (which begins at $T = 0$, $\mu_0 H = |J| z$) will usually meet the second order line (which begins at $H = 0$) at a *tricritical point*, as illustrated in Fig.6. Phase diagrams like that of Fig.6 have been observed in systems like $Fe\,C\ell_2$,[5] which

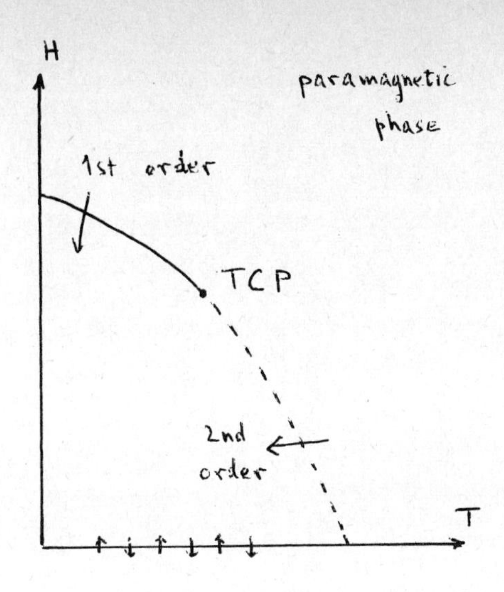

Fig.6. Tricritical in an antiferromagnet.

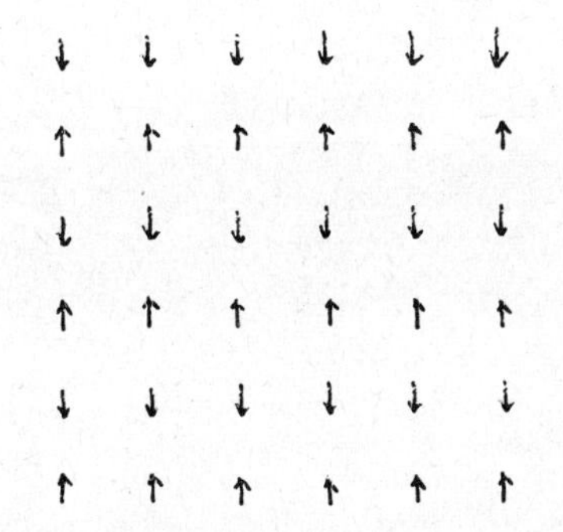

Fig.7. Metamagnetic ground state.

are called "metamagnets". These are somewhat different than the "simple" antiferromagnet described above, in that they are spatially anisotropic: the coupling between spins within two dimensional layers is ferromagnetic, while that between nearest neighbor layers is antiferromagnetic. The resulting ground state, illustrated in Fig.7, may still be described by Eq.(9), with (in three dimensions)

$$\vec{q}_0 = (0, 0, \frac{\pi}{a}) \quad . \tag{14}$$

The remainder of the above discussion, which led to Fig.6, remains unchanged.

Note that although Fig.6 is similar to Fig.1, it is usually very difficult to realize Fig.2, since it is not easy to create the ordering field for the antiferromagnet, i.e. a staggered field.[6]

II. GENERAL REVIEW: LIFSHITZ AND BICRITICAL POINTS

An important element which led to the occurrence of the tricritical points described in the first lecture is that of competition. As we shall see below, this is a very general situation. We now describe two additional examples of multicritical points.

II.1. The Lifshitz Point

Consider a one dimensional array of spins, with ferromagnetic interactions between the nearest neighbors and with antiferromagnetic interactions between the next nearest neighbors,

$$H = - J_1 \sum_i S_i S_{i+1} - J_2 \sum_i S_i S_{i+2} \quad , \tag{15}$$

with $J_1 > 0$, $J_2 < 0$. The ground state will be ferromagnetic, $S_i \equiv 1$, if $J_1 > 2|J_2|$, and will have the structure $+ + - - + + - - + +$ if $J_1 < 2\,|J_2|$. At zero temperature, one expects a first order transition (as function of the ratio $\kappa = |J_2|/J_1$, which may be varied e.g. by pressure) at $\kappa = 1/2$. The antiferromagnetic structure $+ + - - + + - -$ is more complicated than that discussed in Sec. I.2, but may still be described by

$$\langle S_i \rangle = M_1 \sin(\vec{q}_o \cdot \vec{r}_i) + M_2 \cos(\vec{q}_o \cdot \vec{r}_i) \quad , \tag{16}$$

with $q_o = \pi/(4a)$. A similar structure will result for three dimensional layered systems, in which the spins within a layer are coupled ferromagnetically and those in different layers are described by Eq.(15).[7] [In that case $\vec{q}_o = (\frac{\pi}{4a}, 0, 0)$].

In general, one may consider an ordering of the form (16) with a general "wave vector" $\vec{q}_o$. The ordering is commensurate with the underlying lattice when the components of $\vec{q}_o$ are rational products of (π/a), as in the example given above, and incommensurate otherwise. The number of components of the order parameter which describes the phase transition is determined by the number of independent values of $\vec{q}_o$ which represent degenerate different ordered phases. For an incommensurate structure, this number may be as large as 48.[8]

In the example of the Hamiltonian (15), for $J_1 < 2|J_2|$, the ground state is unique and commensurate with the lattice. However, it turns out that at finite temperatures the vector q_o prefers to vary as function of temperature and of κ. For small values of κ, the transition from the paramagnetic phase goes into a ferromagnetic phase, with $\vec{q}_o = 0$. However, beyond a special value of κ the vector $\vec{q}_o$ (at the transition) begins to vary, continuously, gradually increasing in magnitude. This special point has been called the Lifshitz point.[9] The ferromagnetic phase is separated from the ("sinusoidal") phase by a first order line, connecting the point $T = 0$, $\kappa = 1/2$ and the Lifshitz point (Fig.8).

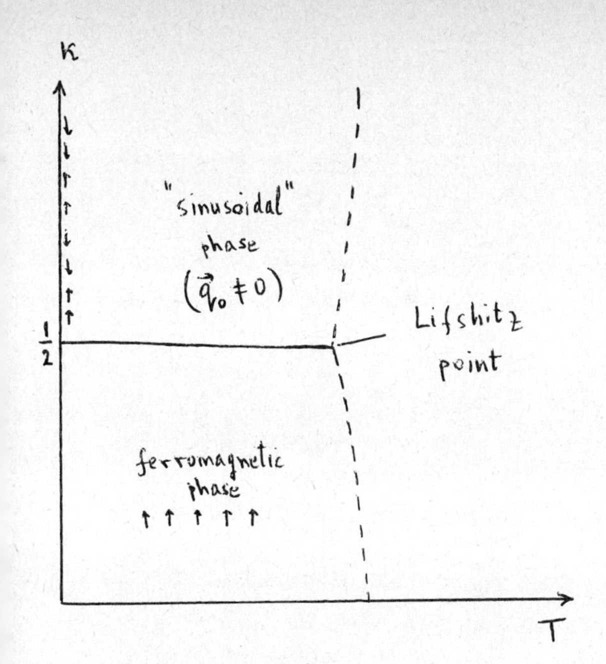

Fig.8. Lifshitz Point.

Rather than separating a second order line from a first order one, as done by the tricritical point, the Lifshitz point separates two second order lines, into two distinct ordered phases. Note that the order parameters which characterize these two phases are different [e.g. M_1 and M_2, Eq.(16), for $\vec{q}_0 \neq 0$, and the usual magnetization for $\vec{q}_0 = 0$]. Both order parameters become degenerate at the Lifshitz point.

II.2. Bicritical Points

Up to this point we have discussed only Ising models, i.e. spins which are limited to one direction in (spin-) space. In reality, all spins are three dimensional. When the interactions are isotropic, the Hamiltonian of ferromagnets may be described by the Heisenberg model,

$$H = - J \sum_{<ij>} (\vec{S}_i \cdot \vec{S}_j) \quad . \tag{17}$$

More generally, we may be interested in n-component spins.

Real crystals never obey to the idealized Hamiltonian (17). The coupling to the lattice degrees of freedom usually breaks the rotational symmetry, and generates easy axes along which the spins align. For example, a spatial uniaxial anisotropy of the lattice may generate, via the spin-orbit coupling, a uniaxial anisotropy in the spins, via single ion terms like

$$\beta H_a = \frac{1}{2} g \sum_i \{(S_i^z)^2 - \frac{1}{2}[(S_i^x)^2 + (S_i^y)^2] \quad . \tag{18}$$

Similar terms may be generated experimentally by the application of uniaxial stress (proportional to g). For $g < 0$, the Hamiltonian (18) prefers ordering of the spins along the z-axis. For $g > 0$, it prefers ordering in the XY plane. At low temperatures, this competition yields a first order "spin flop" transition at $g = 0$, at which the magnetization rotates discontinuously by 90°. The complete T-g phase diagram is thus expected to have the qualitative shape shown in Fig.9: the ordering is

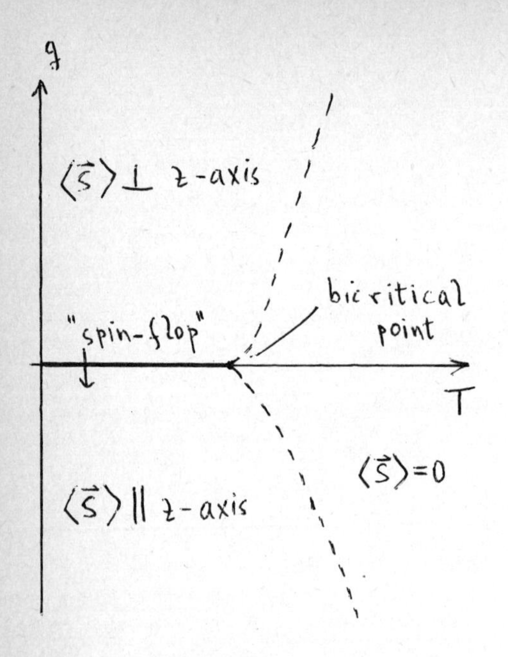

Fig.9. Bicritical point.

along the z-axis for $g < 0$, and in the XY plane for $g > 0$. For very large $|g|$, the fluctuations in the transverse XY plane (when $g < 0$) are expected to be negligible. One may trace over the x and y components of the order parameter in the partition function, and end up with an Ising model for S_i^z. The transition for $g < 0$ is thus expected to exhibit critical properties characteristic for the Ising ($n = 1$) model. Similarly, the transition for $g > 0$ is expected to exhibit XY ($n = 2$) critical behavior. The point at which these two lines meet, at $g = 0$, is called the bicritical point.[10,11] It is only at this point that one expects to observe the critical behavior of the "true" Heisenberg ($n = 3$) model, Eq.(17).

In addition to the (quadratic) uniaxial anisotropy (18), real systems usually also have higher order symmetry breaking interactions. In cubic systems, one expects the single ion cubic Hamiltonian

$$\beta H_C = v \sum_i \sum_{\alpha=1}^{n} (S_i^{\alpha})^4 \quad . \tag{19}$$

Such a term prefers ordering of the spins along cubic axes (e.g. [100]) if $v < 0$, and along cubic diagonals (e.g. [111]) if $v > 0$. One of our aims later in this course will be to find out if such terms affect the asymptotic critical properties, i.e. to see if isotropic Heisenberg systems differ from ones with cubic symmetry.

When both the uniaxial anisotropy, Eq.(18), and the cubic one, Eq.(19), arise simultaneously, a competition between them may arise. Indeed, when $v > 0$ then Eq.(19) prefers ordering along diagonals while Eq.(18) prefers ordering along axes. The resulting phase diagram is shown in Fig.10: the flop line is now replaced by two second order lines, and the multicritical point is now called a "tetracritical point."[12]

In fact, one may show that the tetracritical and the bicritical point are the same point, when viewed in a larger parameter space.[13] An example of this will be shown below.

Some beautiful realizations of the phase diagrams involving bicritical points

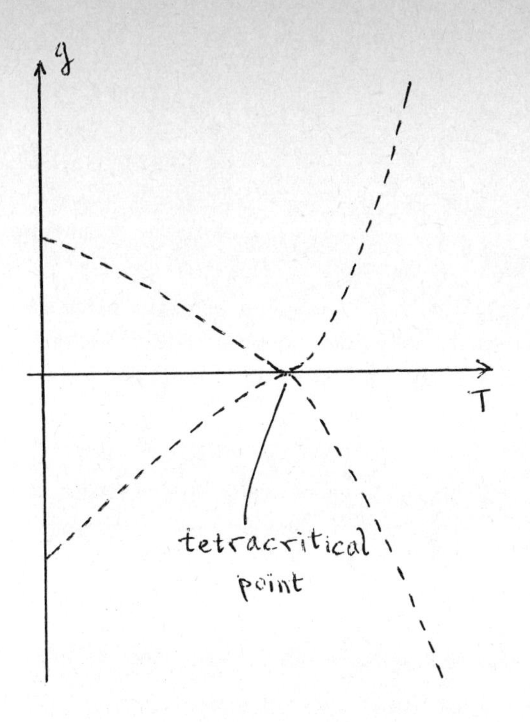

Fig.10. Tetracritical point.

have been observed in structural phase transitions, which occur e.g. in $SrTiO_3$ and $LaA\ell O_3$.[14,15] Another realization, in which these were in fact first discussed,[11,16] concerns anisotropic antiferromagnets. Similar to Eq.(8), consider now an anisotropic Heisenberg antiferromagnet,

$$H = -J\sum_{<ij>} \vec{S}_i\cdot\vec{S}_j + \frac{1}{2}g\sum_i\{(S_i^z)^2 - \frac{1}{2}[(S_i^x)^2 + (S_i^y)^2]\} - \mu_0 H\sum_i S_i^z, \quad J < 0 . \qquad (20)$$

As before, the last term competes with the antiferromagnetic coupling. Consider first the case $g = 0$. For any non-zero value of H it is clear that the system gains energy by having the antiferromagnetic ordering perpendicular

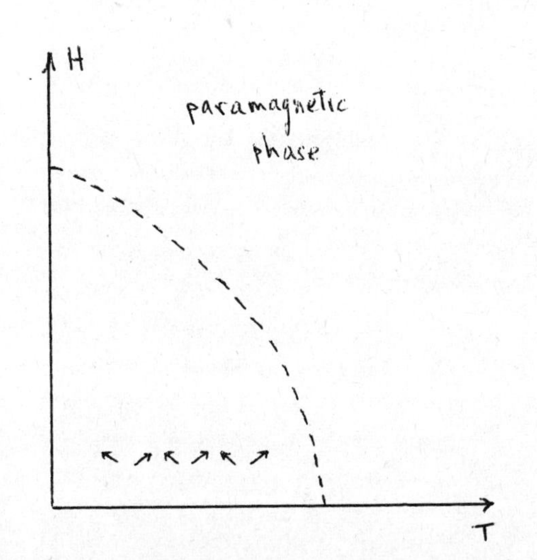

Fig.11. Phase diagram for isotropic antiferromagnet.

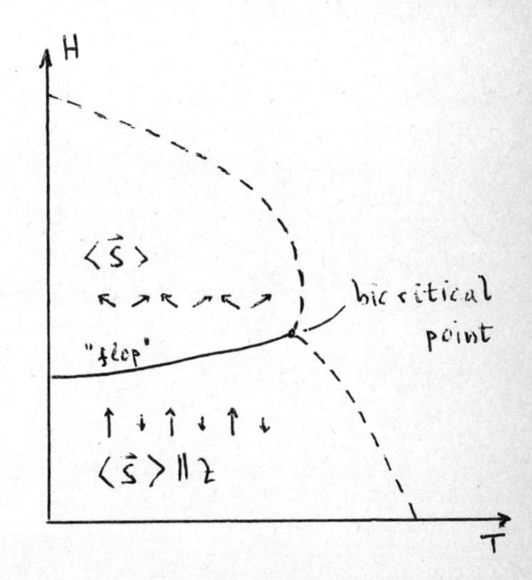

Fig.12. Phase diagram for anisotropic antiferromagnet.

to the uniform field, i.e. in the XY plane (plus a small paramagnetic component parallel to the field). The resulting phase diagram is then shown in Fig.11. As the field grows larger, the longitudinal paramagnetic component grows larger, and finally there appears a second order transition into the paramagnetic phase, at which the transverse antiferromagnetic order parameter approaches zero continuously (even at T = 0).

The situation for $g < 0$ is shown in Fig.12. For small values of H it still pays to have the staggered magnetization along the z-axis. As some finite value of H ($\propto\sqrt{g}$) the magnetic field term wins, and the spin flop transition occurs. As before, the spin flop line meets the two second order transitions at a bicritical point.

It is interesting to study the phase diagram for the case in which the magnetic field is not exactly aligned along the easy z-axis. Extending the phase diagram in the transverse field direction, one ends up with Fig.13.[16] The spin flop line now becomes a "shelf" within the ordered phase. Note that if the phase diagram is cut along the surface of this "shelf", the cut looks like Fig.10. The "bicritical" point thus becomes a "tetracritical" point.

It should thus be emphasized that there does not yet exist a systematic way of naming the various multicritical points. Some of the existing names indeed depend on the parameter subspace in which the particular points are being considered.

Finally, we comment that the phase diagrams of Figs.6 and 12 are to be expected for anisotropic antiferromagnets in the limits of large $|g|$ and small $|g|$, respectively. Intermediate values of $|g|$ may yield intermediate situations, like the one shown in Fig.14.[17] As $|g|$ decreases, the lines of tricritical points and of critical end points which occur in Fig.14 approach each other. They meet the line of bicritical points (from Fig.12, for varying g) at a new (higher order) multicritical point.[17] As $|g|$ increases, the critical end point moves to lower temperatures, and finally the flopped phase disappears, leaving the phase diagram of Fig.6.

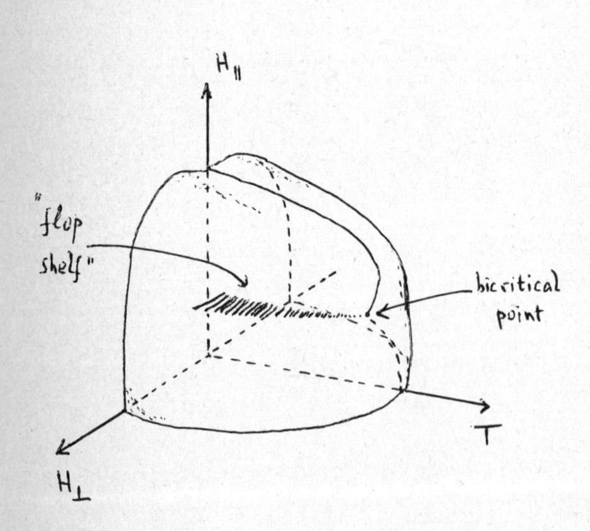

Fig.13. Bicritical point for skew magnetic field.

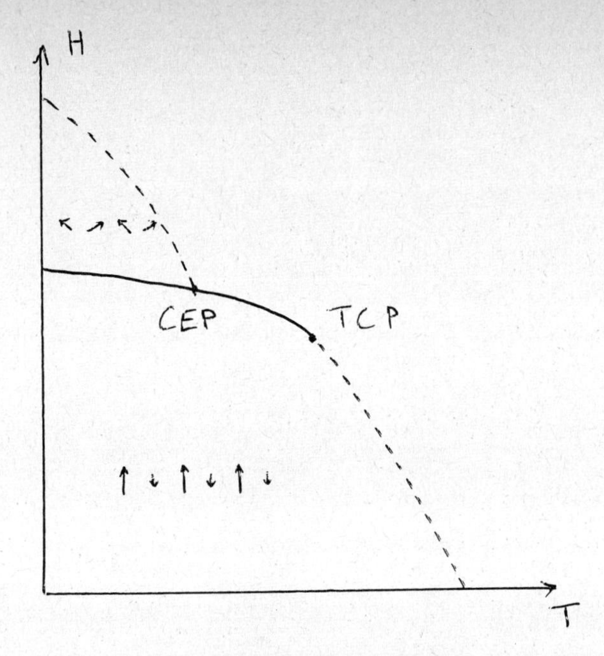

Fig.14. Phase diagram for anisotropic antiferromagnet with intermediate anisotropy.

III. LANDAU THEORY: TRICRITICAL SCALING

The basic assumption of the Landau (Mean Field) theory is that the Helmholtz free energy density A is an analytic function of the order parameter M, the temperature T and any other variable. If we are not too far away from a second order transition into the disordered phase, so that M is not too large, we may expand A in powers of M,

$$A = A_o + \frac{1}{2} A_2 M^2 + \frac{1}{4} A_4 M^4 + \frac{1}{6} A_6 M^6 + \quad . \; . \; . \; . \quad . \tag{21}$$

We included only even powers of M, assuming the symmetry $M \longrightarrow - M$ (dictated by time reversal symmetry for magnets). The actual value of M is determined by minimizing A with respect to M, i.e. by the equation

$$\frac{\partial A}{\partial M} = A_2 M + A_4 M^3 + A_6 M^5 + \; . \; . \; . \; . \; = 0 \; . \tag{22}$$

Clearly, if all the coefficients are positive then the only acceptable minimum is the paramagnetic one, $M = 0$. If $A_2 < 0$ and all other coefficients remain positive, then we have two minima at $M \simeq \pm (- A_2/A_4)^{\frac{1}{2}}$. Thus, a second order transition is identified at the point $A_2 = 0$, $(A_4 > 0,\ A_6 > 0,\ . \; . \; .)$. Assuming that this transition occurs as a result of variation in temperature, we may thus write

$$A_2 = a_2 (T - T_c) \quad , \qquad a_2 > 0. \tag{23}$$

If the free energy is also a function of other (non-ordering) fields, e.g. the pressure or the chemical potential, then A_2, and therefore T_c will also depend on them. Representing such a variable by g, we thus find a critical line in the T-g plane, at $T = T_c(g)$. At this transition we have

$$|M| \propto |T - T_c(g)|^{\beta} \quad , \quad \beta = 1/2 \; . \tag{24}$$

The shape of the function $A(M)$, for $A_4 > 0$ and for various values of T, is illustrated in Fig.15. The single minimum which exists at $M = 0$ for $T > T_c$ is seen to split into two minima, which continuously move away from the origin as T decreases below T_c.

Note that in the expression $|M| \simeq |A_2/A_4|^{1/2}$ we have ignored the term with A_6, in Eq.(22). However, a simple perturbation expansion shows that to leading order in A_6 one has

$$|M| = \left|\frac{A_2}{A_4}\right|^{\frac{1}{2}} \left(1 + \frac{1}{2} A_6\, A_2 \,/\, A_4{}^2 + \; . \; . \; . \; . \; . \right) \; . \tag{25}$$

In this expansion, A_6 always appears in the combination A_6A_2/A_4^2. Thus, higher powers of A_6 also imply higher powers of A_2. These do not affect the leading power, $|A_2|^{\frac{1}{2}}$, which will be observed asymptotically close to T_c. Terms involving A_6 are thus corrections to the asymptotic form. They are important in determining how close to T_c does one have to go in order to observe this asymptotic behavior. It is useful to note that in fact the solution of Eq.(22) (truncated after A_6) can always be written in the scaling form

$$|M| = \left|\frac{A_2}{A_4}\right|^{\frac{1}{2}} \; m \; (A_6A_2/A_4^2). \quad (26)$$

As long as $A_4 > 0$, the function $m(x)$ is analytic in x, and may be expanded in powers of it.

The situation becomes quite different when $A_4 < 0$. The qualitative shapes of the function A(M), assuming $A_6 > 0$, are now shown in Fig.16. One now encounters two additional minima (in addition to the one at M = 0), which move downwards in energy as the temperature is lowered. These minima reach the line A = 0 at a positive value of A_2, i.e. at a temperature $T_o > T_c$ (g). For $T < T_o$, these minima have $A < 0$, and thus one expects a discontinuous (first order) transition from M = 0 into one of these other minima when T crosses T_o. To find these new minima, we now solve Eq.(22) without neglecting A_6. The solutions are given by

$$M^2 = [-A_4 \pm (A_4^2 - 4A_2A_6)^{\frac{1}{2}}]/2A_6. \quad (27)$$

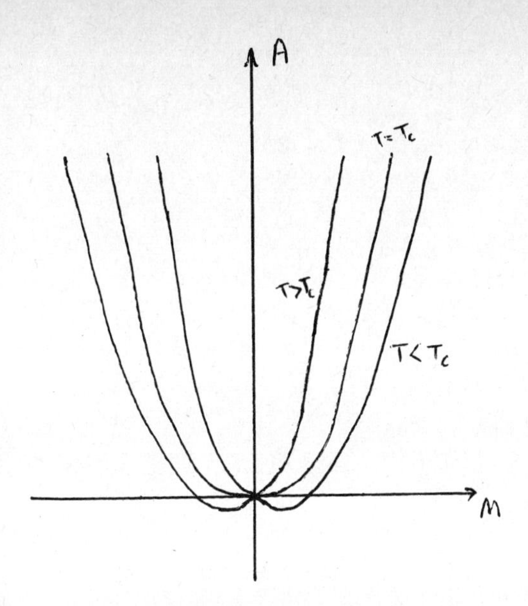

Fig.15. A(M) for $A_4 > 0$.

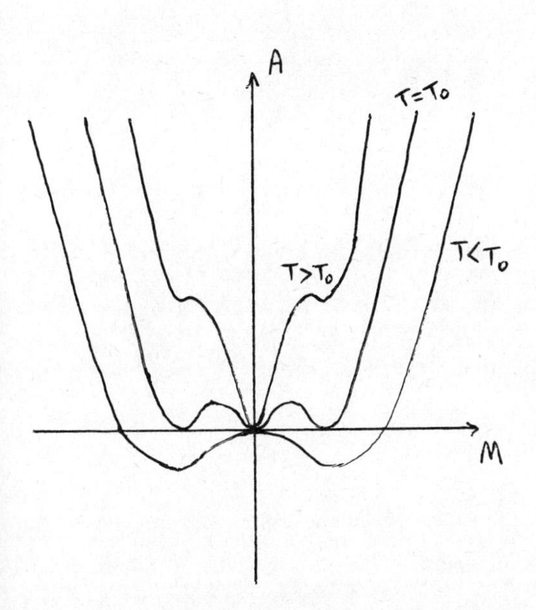

Fig.16. A(M) for $A_4 < 0$.

The first order transition will actually take place when $A(M) = 0$. Solving this equation together with Eq.(22) we find that T_o is given by

$$A_2 = \frac{3}{16} \frac{A_4^2}{A_6} \quad . \tag{28}$$

At the first order transition, the discontinuity (from M to zero) is

$$(\Delta M)^2 = - \frac{3}{4} \frac{A_4}{A_6} \quad . \tag{29}$$

Clearly, Eqs.(28) and (29) apply only for $A_4 < 0$. The point $T = T_c$, $A_4 = 0$ thus separates the critical line from a first order line, and may thus be identified as the tricritical point. The two equations $A_2 = 0$ and $A_4 = 0$ clearly determine the values of T and g at which this tricritical point will occur in the T-g plane.

Fig.17 shows the resulting phase diagram in the A_2 - A_4 plane: the critical line is given by $A_2 = 0$, $A_4 > 0$, and the first order line is the parabola given by Eq.(28) for $A_4 < 0$.

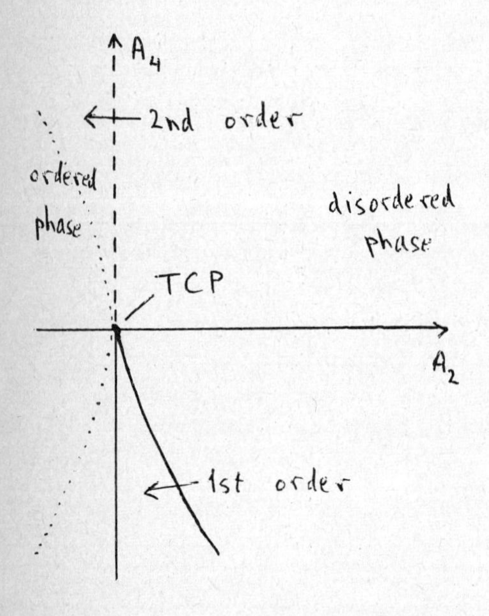

Fig.17. Tricritical phase diagram, A_2 - A_4 plane.

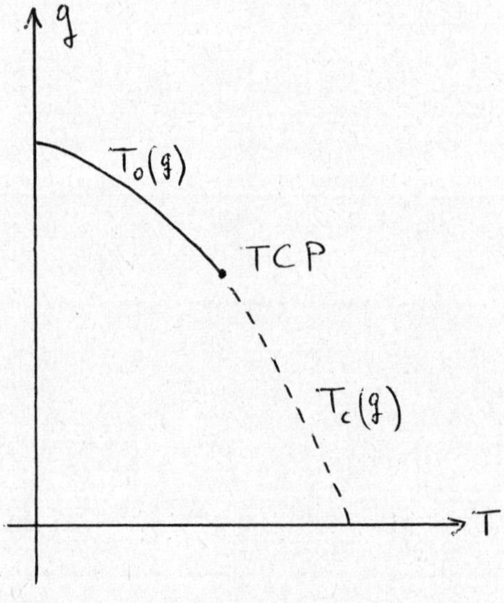

Fig.18. Tricritical phase diagram, T - g plane.

At the tricritical point, $A_4 = 0$, Eq.(27) yields

$$|M| = |A_2/A_6|^{\beta_t} \quad , \quad \beta_t = 1/4 \quad . \tag{30}$$

One thus expects new critical exponents when one performs measurements near the tricritical point! Note that Eqs.(28) and (29) also define new critical exponents, which may be generally written as '

$$(T_o - T_t) \propto A_4^{1/\psi} \quad , \quad \psi = \frac{1}{2} \quad . \tag{31}$$

$$\Delta M \propto |A_4|^{\beta_u} \quad , \quad \beta_u = \frac{1}{2} \quad . \tag{32}$$

The exponent ψ is called "the shift exponent".

We now make a few comments about scaling. Eq.(27) may be rewritten in the form

$$|M| = \left|\frac{A_2}{A_6}\right|^{\frac{1}{4}} \tilde{m} \left(\frac{A_4}{2|A_2|^{\frac{1}{2}} A_6^{\frac{1}{2}}}\right) \quad . \tag{33}$$

with

$$\tilde{m}(x) = [- x + (x^2 - \text{sign } A_2)^{\frac{1}{2}}]^{\frac{1}{2}} \quad . \tag{34}$$

Clearly, $\tilde{m}(0) = 1$ for $A_2 < 0$, and the result reduces to Eq.(30). Moreover, for finite values of A_4, the argument of $\tilde{m}$ in Eq.(33) becomes very large as $|A_2| \to 0$. For large x we have

$$\tilde{m}(x) = \left(- \frac{\text{sign } A_2}{2x} + 0\left(\frac{1}{x^2}\right)\right)^{\frac{1}{2}} \quad . \tag{35}$$

Substitution in Eq.(33) now recovers Eq.(25). Note also that Eqs.(33) and (26) are related via

$$m(x) = \tilde{m}(1/2|x|^{\frac{1}{2}})\,(2|x|^{\frac{1}{2}})^{\frac{1}{2}} \quad . \tag{36}$$

The function $\tilde{m}(x)$ [or $m(x)$] thus serves to describe the <u>crossover</u> from Eq.(24) to Eq.(30). The asymptotic tricritical behavior (30) will be observed only when the variable in $\tilde{m}$ in Eq.(33) is not too large, so that $\tilde{m}(x)$ may be treated as a constant. The condition is thus

$$\frac{A_4}{2|A_2|^{\frac{1}{2}}A_6^{\frac{1}{2}}} << 1 \quad . \tag{37}$$

There exists a parabola in the A_2 - A_4 plane, $A_2 \propto A_4^2/A_6$, which separates between the "tricritical" and the "critical" regimes. This parabola is also shown in Fig.17. It is therefore clear that Eq.(30) will be observed along any experimental trajectory which does not approach the tricritical point tangentially to the lines $T = T_o(g)$ or $T = T_c(g)$!

The fact that M depends on A_4 only through the combination $A_4/(2|A_2|^{\frac{1}{2}}A_6^{\frac{1}{2}})$ has further consequences. Since the first order transition at $T = T_o(g)$ signals a singularity in M, it must result from a singularity in the dependence of $\tilde{m}$ on its variable. Indeed, Eq.(28) tells us that $\tilde{m}(x)$ has a singularity at $x = (4/3)^{\frac{1}{2}}$.

Substituting Eq.(33) into Eq.(21), adding a magnetic field term $-\mu_o HM$, one easily checks that A may be written in the scaling form

$$A = \frac{|A_2|^{3/2}}{A_6^{\frac{1}{2}}} \quad \mathcal{A}\left(\frac{A_4}{2|A_2|^{\frac{1}{2}} A_6^{\frac{1}{2}}} \; , \; \frac{H\, A_6^{1/4}}{|A_2|^{5/4}}\right) \quad . \tag{38}$$

Written with general exponents, this becomes

$$A = \left|T - T_t\right|^{2-\alpha_t} \quad \mathcal{A}\left(\frac{\tilde{g}}{|T-T_t|^{\phi_t}} \; , \; \frac{H}{|T-T_t|^{\Delta_t}}\right) \quad , \tag{39}$$

with $\alpha_t = 1/2$, $\phi_t = 1/2$, $\Delta_t = 5/4$. The other tricritical exponents now follow by scaling, $\delta_t = 5$, $\gamma_t = 1$, etc. The "scaling field" $\tilde{g}$ in Eq.(39) has been defined to be proportional to A_4. It is thus measured along the tangent to the critical line at the tricritical point.

The exponent ϕ is called the "crossover exponent". As was explained before, it determines the crossover between the tricritical and the critical regimes. In most cases one has $\psi = \phi$.

As we did following Eq.(25), it is now easy to include A_8, A_{10}, etc. in our discussion. For example, A_8 will always appear in the combination $A_8\, A_2^{1/2}/\, A_6^{3/2}$. Its contributions to thermodynamic singular functions will thus become negligible when $|A_2| \longrightarrow 0$, and will only represent corrections to the leading singularities. In order to reach a point on the critical line one had to set $H = 0$ and $T = T_c(g)$, i.e. two variables. The tricritical point is reached only after a third variable, i.e. A_4, is appropriately set equal to zero. One can now generalize the discussion, and consider the situation when A_6 changes sign. Straightforward algebra shows that when $A_6 < 0$ one arrives at the phase diagram shown in Fig.3. The point $H=A_2=A_4=A_6=0$ (see Fig.4) is now a critical point of fourth order, because four variables have to be set in order to reach it. One can now easily generalize our discussion, identify the exponent $\beta_4 = 1/6$, write scaling functions, etc. . The generalization to higher order critical points is also obvious.

Finally, we consider the relevance of the above tricritical calculations to realistic experimental systems. From the Ginzburg criterion,[19] we should compare the fluctuation in the magnetization of a volume ξ^d (ξ is the correlation length, given in the Landau theory by $\xi \propto |T - T_c|^{-\nu}$, $\nu = 1/2$), $<\Delta M^2> = \xi^d \, k \, T \, \chi \propto |T - T_c|^{-d\nu-\gamma}$ to the square of the magnetization itself, $M^2 \propto \xi^{2d} \, |T - T_c|^{2\beta} \propto |T - T_c|^{2\beta-2d\nu}$. The fluctuations become important when $<\Delta M^2> \gg M^2$, i.e. when

$$d\nu < \gamma + 2\beta \quad , \tag{40}$$

with the Landau values of ν, γ and β. For the critical point, $2\nu = \gamma = 2\beta = 1$ and we have $d < 4$. For the tricritical point $2\beta = 1/2$, and thus <u>the upper critical dimensionality is equal to three</u>. The corrections to the theory described above, due to fluctuations, at three dimensions, are expected to be very weak! As we shall see below, these corrections involve only powers of $\log|T - T_t|$. Similarly, the upper critical dimensionality for the critical point of fourth order is 8/3.

IV. LANDAU THEORY: OTHER CASES

The general rule in the Landau theory is to expand the free energy density in powers of the relevant order parameters. We now outline the derivation, within this theory, of the bicritical and the Lifshitz point phase diagrams.

IV.1. Bicritical Point

Following the Hamiltonians of Eqs.(17), (18) and (19), we may write (for an n-component classical spin problem), to quartic order,[12]

$$A = \frac{1}{2} r_o |\vec{M}|^2 + u |\vec{M}|^4 + \frac{1}{2} g [M_1^2 - \frac{1}{n-1} (M_2^2 + \ldots + M_n^2)] +$$

$$+ v \sum_{\alpha=1}^{n} M_\alpha^4 \quad . \tag{41}$$

In order to conform with the notation in the literature, we have now replaced A_2 by r_o and $\frac{1}{4}A_4$ by u.

We now proceed to minimize A with respect to $\vec{M}$. The direction of M enters only into the last two terms in Eq.(41). As noted before, both of these terms prefer ordering of $\vec{M}$ along a cubic axis when $v < 0$. For $g < 0$ we thus have

$$A = \frac{1}{2} (r_o + g) M_1^2 + (u + v) M_1^4 \ , \tag{42}$$

and a second order transition is predicted to occur at $r_o = - g$, provided that $u + v > 0$. Similarly, for $g > 0$ we have e.g.

$$A = \frac{1}{2} (r_o - \frac{g}{n-1}) M_2^2 + (u + v) M_2^4 \ , \tag{43}$$

and the continuous transition will occur at $r_o = g/(n-1)$ (with $u + v > 0$). This yields the phase diagram of Fig.9. Generally, we may describe the two critical lines by

$$T_c(g) - T_c(0) = A_{\pm}|g|^{1/\psi} \quad . \tag{44}$$

Here we found $\psi = 1$ and $A_+/A_- = - 1/(n-1)$ (the $+$ and $-$ signs stand for $g > 0$ and $g < 0$). Note that the line $u = -v$ is <u>tricritical</u>. One must add sixth order terms when $u < -v$.

In the case $v > 0$, the g and the v terms compete. In this case, we write

$$M_1 = M \cos\theta, \ M_2 = M \sin\theta \ m_2, \ \ldots, \ M_n = M \sin\theta \ m_n \ , \tag{45}$$

with

$$\sum_{\alpha=2}^{n} m_\alpha^2 = 1 . \tag{46}$$

The minimum with respect to m_α yields $m_\alpha^2 = 1/(n-1)$. Eq.(41) now becomes

$$A = \frac{1}{2}\left(r_o - \frac{g}{n-1}\right) M^2 + u\, M^4 + \frac{ng}{2(n-1)} M^2 \cos^2\theta +$$

$$+ v\, M^4 \left\{\cos^4\theta + \frac{1}{n-1}\sin^4\theta\right\} . \tag{47}$$

Differentiating with respect to θ and demanding that $\partial A/\partial\theta = 0$, we find three solutions,

$$\begin{aligned} &(I) \quad \cos\theta = 0, \\ &(II) \quad \cos^2\theta = (1 - ng/4vM^2)/n, \\ &(III) \quad \sin\theta = 0. \end{aligned} \tag{48}$$

Clearly, solution II is possible only if

$$- 4(n-1)\, vM^2/n \leq g \leq 4vM^2/n . \tag{49}$$

We now differentiate A with respect to M, and find that $(\partial A/\partial M) = 0$ if $M = 0$ (disordered phase) or if M has the following expressions corresponding to the above three solutions:

$$\begin{aligned} &(I) \quad M^2 = -\frac{1}{4}\left(r_o - \frac{g}{n-1}\right) / [u + v/(n-1)] , \\ &(II) \quad M^2 = -\frac{1}{4}\, r_o / (u + v/n) , \\ &(III) \quad M^2 = -\frac{1}{4}(r_o + g) / (u + v) . \end{aligned} \tag{50}$$

A study of the second derivatives of A now identifies the regions in the $g - r_o$ (or $g - T$) plane in which each solution represents the minimum. This finally yields the phase diagram shown in Fig.10, with the two internal lines given by $r_o = -(nu + v)g/v$ for $g > 0$ and $r_o = (nu + v)g/(n-1)v$ for $g < 0$. Note that these results apply only for $(nu + v) > 0$. Some of the transitions become first order for $nu + v < 0$, and the line $nu + v = 0$ is thus also tricritical.

Within the context of the Landau theory, the exponents describing the order parameters at all the transitions involved in the bicritical phase diagram are all equal to $\beta = 1/2$. The crossover phenomena, which we saw in the tricritical situation, are therefore less striking. However, it is still instructive to convince oneself that the free energy may be written in the general scaling form

$$A(T, g, v) = (T - T_b)^{2-\alpha_b} A\left(\frac{g}{(T-T_b)^{\phi_g}}, \frac{v}{(T-T_b)^{\phi_v}}\right), \quad (51)$$

with (in the Landau theory) $\alpha_b = 0$, $\phi_g = 1$, $\phi_v = 0$.

The situation for the anisotropic antiferromagnet is more complicated.[20] Since the exchange term in Eq.(20) favors an antiferromagnetic ordering, we must use a staggered magnetization, $\vec{M}^\dagger$, as our order parameter. On the other hand, the magnetic field H couples to the uniform magnetization, M. The appropriate Landau expansion thus contains powers of $M^\dagger_\parallel$, $M^\dagger_\perp$, $M_\parallel$ and $M_\perp$ ($\parallel$ is along the magnetic field, and $\perp$ is perpendicular to it). The coupling between $\vec{M}^\dagger$ and $\vec{M}$ arises through biquadratic terms, e.g. $|\vec{M}|^2 \, |\vec{M}^\dagger|^2$ and $(\vec{M}\cdot M^\dagger)^2$. Non-zero values of H generate non-zero values of M ($\propto H$). When these are substituted back into the free energy, we end up with an expansion in powers of $M^\dagger_\parallel$ and $M^\dagger_\perp$. The coefficients of $(M^\dagger_\parallel)^2$ and of $(M^\dagger_\perp)^2$ are now shifted by amounts of order H^2. Finally, this resulting expansion can be brought exactly into the form (41) discussed above. The only effect of the field is thus to move the axes, and to turn Fig.9 into Fig.12. Similar mappings are possible for all the bicritical situations.

IV.2. Lifshitz Point

The order parameter for an incommensurate phase is related to the Fourier component of the magnetization at wave vector $\vec{q}_0$, e.g. Eq.(16). Denoting this component by

$$M(\vec{q}_0) = \frac{1}{\sqrt{N}} \sum_i e^{-i\,\vec{q}_0\cdot\vec{r}_i} \langle S_i \rangle , \quad (52)$$

the free energy may be written as

$$A = \frac{1}{2} A_2(T, \vec{q}_0) \, |M(\vec{q}_0)|^2 + \frac{1}{4} A_4 \, |M(\vec{q}_0)|^4 + \dots \, . \quad (53)$$

In principle, we should consider a functional of all the $M(\vec{q}_0)$'s, and minimize with respect to all of them. In writing Eq.(53) we assumed that only one mode is important.

The crucial point concerning the Lifshitz point is now the dependence of A_2 on

$\vec{q}_o$. Expanding A_2 in powers of the components of $\vec{q}_o$, for a uniaxially anisotropic situation, we have

$$A_2 = a_2(T - T_c) + e_{\parallel}\, q^2_{o,\parallel} + e_{\perp}\, |\vec{q}_{o,\perp}|^2 + f\, q^4_{o,\parallel} + \dots \quad . \tag{54}$$

If $e_{\parallel}$ and $e_{\perp}$ are positive, then it is clear that $\vec{q}_o = 0$ is a minimum of A_2 and therefore of A. For $e_{\parallel}$, $e_{\perp} > 0$ we thus return to the usual ferromagnetic ordering, with $M(\vec{q}_o = 0)$ as the order parameter. If $e_{\parallel} < 0$, then $q_o = 0$ is no longer a minimum. If $|e_{\parallel}|$ is not too large, an alternative minimum is now found at

$$q^2_{o,\parallel} = - e_{\parallel}/2f \quad , \qquad \vec{q}_{o,\perp} = 0. \tag{55}$$

Fixing $\vec{q}_o$ at the value given in Eq.(55), we now find a second order transition in $M(\vec{q}_o)$, which occurs at

$$(T - T_c) \simeq \frac{e_{\parallel}^{\,2}}{4fa_2} \quad . \tag{56}$$

The free energy must thus be considered as a function of the two variables T and $e_{\parallel}$. The borderline between the ferromagnetic phase and the incommensurate phase, occurring at $T = T_c$, $e_{\parallel} = 0$, may now be identified as the Lifshitz point (Fig.8).

Again, Eqs. (55) and (56) may be generalized as

$$|q_{o,\parallel}| \propto |\kappa - \kappa_L|^{\beta_k} \quad , \qquad \beta_k = \frac{1}{2} \; , \tag{57}$$

$$|T - T_L| \propto |K - K_L|^{1/\psi} \quad , \qquad \psi = 2 \; , \tag{58}$$

and the free energy can be shown to depend on $e_{\parallel}$ only through the scaled variable $e_{\parallel}/|T - T_L|^{\phi}$, with $\phi = \psi = \frac{1}{2}$.

V. RENORMALIZATION GROUP AND SCALING

The basic idea of the renormalization group is that short range effects are irrelevant in the vicinity of a second order transition, where the correlation length diverges. Given a Hamiltonian H, one thus divides the N degrees of freedom $\{S_i\}$ of the problem into two groups: N/b^d $(b > 1)$ degrees of freedom, $\{S_i'\}$, which represent the (relatively) longer range features of the problem are kept, and the partition function trace is performed over the remaining ("short range") $N(1-b^{-d})$ degrees of freedom, $\{S_i''\}$. The former N/b^d degrees of freedom may for example represent the new "cell" spins, when the lattice is subdivided into "cells" of b^d spins. This is used for real space renormalization group transformations. Alternatively, they may represent the long-wavelength Fourier components of the spin variables, as used in the momentum space renormalization group.

Irrespective of the details, the result of the above trace is that the partition function is now represented as a trace only over the remaining N/b^d degrees of freedom,

$$Z = \operatorname*{Tr}_{\{S_i\}} e^{-\overline{H}\{S_i\}} = \operatorname*{Tr}_{\{S_i'\}} \left[\operatorname*{Tr}_{\{S_i''\}} e^{-\overline{H}} \right] . \tag{59}$$

(We denote $\overline{H} = \beta H$). Defining a new "Hamiltonian" via

$$e^{-\overline{H}_1\{S_i'\}} \equiv \operatorname*{Tr}_{\{S_i''\}} e^{-\overline{H}\{S_i', S_i''\}} , \tag{60}$$

we have

$$Z = \operatorname*{Tr}_{\{S_i'\}} e^{-\overline{H}_1} . \tag{61}$$

The nearest neighbor distance among the "new" spins $\{S_i'\}$ is larger, by a factor b, from the original nearest neighbor distance. It is thus convenient to rescale all distances in $\overline{H}_1$ by a factor b. As we shall see below, one sometimes also rescales the spin variables. The resulting Hamiltonian, $\overline{H}'$, defines the renormalization group (RG) transformation,[21-24]

$$\overline{H}' = R\,\overline{H} . \tag{62}$$

If one calculates the correlation length ξ', using $\overline{H}'$, then it is related to ξ via

$$\xi' = \xi/b . \tag{63}$$

Similarly, since the partition function and therefore the total free energy is unchanged, we have for the free energy density

$$F = b^{-d} F' \quad . \tag{64}$$

A <u>fixed point</u> of (62) is defined as a Hamiltonian $\overline{\mathcal{H}}^*$ which satisfies

$$\overline{\mathcal{H}}^* = R\,\overline{\mathcal{H}}^* \quad . \tag{65}$$

Combining (63) with (65) it is clear that $\overline{\mathcal{H}}^*$ must have $\xi = 0$ or $\xi = \infty$. The latter cases represent a <u>critical point</u>. We thus concentrate on the vicinity of such fixed points. Setting the magnetic field (for a magnetic system) equal to zero, and assuming that both $\overline{\mathcal{H}}$ and $\overline{\mathcal{H}}'$ depend on a single variable, T and T', resepctively, Eq.(62) has the form

$$T' = f(T) \quad . \tag{66}$$

The fixed point, T*, is now given by the solution of $T^* = f(T^*)$. Ignoring the trivial case ($\xi = 0$), T* thus corresponds to $\xi = \infty$, i.e. T* represents the critical point. Linearization of (66) about T* now yields

$$T' - T^* = f'(T^*)\,(T - T^*) \quad . \tag{67}$$

Combining this with Eq.(63), and with

$$\xi = A(T - T^*)^{-\nu} \quad , \quad \xi' = A(T' - T^*)^{-\nu} \quad , \tag{68}$$

now yields

$$f'(T^*) = b^{\lambda_t} \quad , \tag{69}$$

with $\lambda_t = 1/\nu$. The renormalization group recursion relation (66) thus determines the exponent ν!

In general, the Hamiltonian must be characterized by many parameters, g_1, g_2, ... After finding a (non-trivial) fixed point in this large parameter space, g_1^*, g_2^*, ... , and linearizing about it, we have

$$g_i' - g_i^* = \sum_j L_{ij}\,(g_j - g_j^*) \quad . \tag{70}$$

Diagonalization of the matrix L will now yield new variables, $\{\tilde{g}_i\}$, which are linear combinations of the $(g_i - g_i^*)$'s, such that

$$\tilde{g}_i' = \Lambda_i\,\tilde{g}_i = b^{\lambda_i}\,\tilde{g}_i \quad . \tag{71}$$

The "eigenvectors" $\tilde{g}_i$ are now called "scaling fields".[24] Note that the exponential form $\Lambda_i = b^{\lambda}$ is dictated by the semi group nature of the transformation.

There exist three types of eigenvalues: If $\Lambda_i > 1$ $(\lambda_i > 0)$, $\tilde{g}_i$ is called "relevant". Clearly, $\tilde{g}_i$ becomes larger under the iteration of the RG transformation. If $\Lambda_i < 1$ $(\lambda_i < 0)$, $\tilde{g}_i$ is called "irrelevant". Such variables decay to zero under the RG transformation. Finally, if $\Lambda_i = 1$ $(\lambda_i = 0)$ then $\tilde{g}_i$ is "marginal". One must go to higher order (non linear) terms in Eq.(71) in order to see how $\tilde{g}_i$ varies under the transformation.

Setting all the relevant variables equal to zero (and assuming there are no marginal ones), the Hamiltonian will "flow" under the RG transformation towards the fixed point. Since at the fixed point $\xi = \infty$, we must have $\xi = \infty$ for all the points from which this point is reached. Thus, the surface on which the relevant variables vanish is identified as the critical surface of the problem. Since all the Hamiltonians on this surface flow to the same fixed point, all of them will have the same critical exponents.

At a fixed point which describes an ordinary critical point we expect only two relevant variables, i.e. the temperature $t = (T - T_c)/T_c$, with eigenvalue $b^{\lambda t}$, and the ordering field h, with eigenvalue $b^{\lambda h}$. Ignoring all other variables, Eq.(64) assumes the form

$$F(t, h) = b^{-d} F(b^{\lambda_t} t, b^{\lambda_h} h) \quad . \tag{72}$$

Repeating the transformation ℓ^* times, until $b^{\lambda_t \ell^*} t = t_o = 0(1)$, Eq.(72) becomes

$$F(t, h) = (t/t_o)^{d/\lambda_t} F(t_o, h(t/t_o)^{-\lambda_h/\lambda_t}) \quad , \tag{73}$$

which is identical with the usual scaling form

$$F(t, h) = t^{2-\alpha} f(h/t^{\Delta}), \tag{74}$$

with $2-\alpha = d/\lambda_t = d\nu$, $\Delta = \lambda_h/\lambda_t$.

The general form of Eq.(72) is

$$F(t, h, \{\tilde{g}_i\}) = b^{-d} F(b^{\lambda_t} t, b^{\lambda_h} h, \{b^{\lambda_i} \tilde{g}_i\}) \quad , \tag{75}$$

and thus (after ℓ^* iterations)

$$F(t, h, \{\tilde{g}_i\}) = t^{2-\alpha} f(\frac{h}{t^{\Delta}} \; , \; \{\frac{\tilde{g}_i}{t^{\phi_i}}\}) \quad , \tag{76}$$

with

$$\phi_i = \lambda_i/\lambda_t = \nu \lambda_i \quad . \tag{77}$$

If $\tilde{g}_i$ is irrelevant then $\phi_i < 0$, and $\tilde{g}_i t^{-\phi_i} = \tilde{g}_i t^{|\phi_i|}$ becomes very small as $t \to 0$. In most cases, f is an analytic function of $\tilde{g}_i$ near $\tilde{g}_i \to 0$, and the main effect of $\tilde{g}_i$ is to yield corrections to scaling, e.g.

$$F(t, 0, \tilde{g}_i) = t^{2-\alpha} [f(0, 0) + f_i'(0, 0) \tilde{g}_i t^{\omega_i}] \quad , \tag{78}$$

with $\omega_i = |\phi_i|$. This was the effect of the variable A_6 near the Landau critical behavior, see e.g. Eq.(25). In some special cases, f is singular in $\tilde{g}_i$. In these cases, $\tilde{g}_i$ will be called a "dangerous irrelevant variable".[25] An example was A_4 in Eq.(26). In both cases, it suffices to know the behavior of f for small $\tilde{g}_i$.

If $\tilde{g}_i$ is relevant, and is not equal to zero, then the Hamiltonian will "flow" away from the fixed point. The fixed point is thus reached only if more than two variables (t and h) are set at special values. Whenever there exist more than the two basic relevant variables, the fixed point represents a multicritical point. In Eq.(76), the combination $\tilde{g}_i/t^{\phi_i}$ (with $\phi_i > 0$) now grows larger as $t \to 0$. The critical properties of the multicritical point will be expected to be observed only for

$$\tilde{g}_i/t^{\phi_i} \ll 1 \quad . \tag{79}$$

The crossover exponent ϕ_i thus determines the shape of the crossover region [see also the discussion following Eq.(37)].

As $\tilde{g}_i$ grows larger, the Hamiltonian flow sometimes goes to an alternative (more "stable") fixed point, with different critical properties. The new critical behavior, which implies a singularity of F at a finite value of t (t measures the distance from the fixed point of the multicritical point), must now result from a singularity of f as function of $\tilde{g}_i/t^{\phi_i}$. If $f(0, y)$ is singular at $y = y_o$, this identifies the location of the new critical line at $\tilde{g}_i/t_c^{\phi_i} = y_o$, i.e.

$$t_c \propto \tilde{g}_i^{\,1/\psi_i} \quad , \quad \psi_i = \phi_i \quad . \tag{80}$$

In some other cases, the flow never reaches a stable fixed point with an infinite correlation length. In many cases this implies a first order transition, which occurs on a line whose $\tilde{g}_i$-dependence is also given by Eq.(80). We shall return to examples of this situation below.

VI. CONTINUOUS SPINS, WILSON'S RENORMALIZATION GROUP AND THE GAUSSIAN MODEL

In the remaining parts of these lectures we are going to limit ourselves to a specific method of performing the renormalization group, i.e. the one performed in momentum space.

VI.1. Continuous Spin Model

The method is based on the continuous spin model, which may be written (in the Ising case) as

$$Z = \prod_i \left(\int_{-\infty}^{\infty} d\, S_i \right) e^{-\overline{H}\{S_i\}} \quad , \tag{81}$$

with the spins S_i varying continuously in the range $-\infty < S_i < \infty$. One usually also performs a coarse graining of the space coordinates, so that S_i is replaced by a spin density $S(\vec{x})$. For a simple Ising model, at zero magnetic field, $\overline{H}\ \{S(\vec{x})\}$ is usually written as[21]

$$\overline{H}\ \{S(\vec{x})\} = \int d^d x\ \{\frac{1}{2} r\, S(\vec{x})^2 + \frac{1}{2} e\ (\vec{\nabla} S)^2 + u\, S^4 + ...\} \quad , \tag{82}$$

where the dots indicate higher powers of S or of $\vec{\nabla}S$, and where r is assumed to be linear in the temperature.

The simplest justification of (82) is that it is a power series in S and in $\vec{\nabla}S$. Such an expansion lies in the basis of all the Ginzburg-Landau theories.[19] A more direct derivation of (82) starts with a discrete Ising Hamiltonian,

$$\overline{H} = -\frac{1}{2} \sum_{i,j} K_{ij}\, S_i\, S_j \quad , \tag{83}$$

and then writes the partition function in the form

$$Z = \underset{\{S_i=\pm 1\}}{\mathrm{Tr}}\ e^{-\overline{H}} = \prod_i \left(\int_{\infty}^{\infty} d\, S_i\ e^{-w(S_i)} \right) e^{-\overline{H}} \quad . \tag{84}$$

For the discrete Spin case, the weight function $w(S_i)$ is given by

$$e^{-w(S_i)} = \delta\ (S_i^2 - 1) = \lim_{u\to\infty} \sqrt{\frac{u}{\pi}}\ e^{-u(S_i^2-1)^2} \quad . \tag{85}$$

Keeping u finite now yields

$$w(S_i) = u\, S_i^4 - 2u\, S_i^2 + \text{const.} \quad . \tag{86}$$

Expanding S_j around S_i, $S_j \simeq S_i + \vec{r}_{ji} \cdot \vec{\nabla} S_i + \ldots$, finally yields Eq.(82).[21] A third derivation of Eq.(82) is based on the Hubbard-Stratonovitch transformation.[26]

Generalizing Eq.(86), we shall write

$$w(S_i) = \frac{1}{2} S_i^2 + u S_i^4 + O(S_i^6) \quad . \tag{87}$$

Fourier transforming the spins,

$$S(\vec{x}) = \frac{1}{N} \sum_{\vec{q}} e^{-i\vec{q}\cdot\vec{x}} \hat{S}_{\vec{q}} \quad , \quad \hat{S}_{\vec{q}} = \sum_{\vec{x}} e^{i\vec{q}\cdot\vec{x}} S(\vec{x}) \quad , \tag{88}$$

and defining

$$\hat{K}(\vec{q}) = \sum_{\vec{x}_{ij}} e^{i\vec{q}\cdot\vec{x}_{ij}} K_{ij} = \hat{K}(0) - e\,\vec{q}^{\,2} + O(\vec{q}^{\,4}) \quad , \tag{89}$$

we end up with Eq.(82), where

$$r = 1 - \hat{K}(0) = 1 - \hat{J}(0)/kT \quad . \tag{90}$$

In the Fourier transformed variables, changing sums into integrals, Eq.(82) reads

$$\bar{H} = \frac{1}{2} \int_{\vec{q}} (r + e\vec{q}^{\,2}) \, |\hat{S}_{\vec{q}}|^2 + u \int_{\vec{q}_1} \int_{\vec{q}_2} \int_{\vec{q}_3} \hat{S}_{\vec{q}_1} \hat{S}_{\vec{q}_2} \hat{S}_{\vec{q}_3} \hat{S}_{-\vec{q}_1-\vec{q}_2-\vec{q}_3} + \ldots \, , \tag{91}$$

where $\int_{\vec{q}}$ denotes $(2\pi)^{-d} \int d^d q$ over the Brillouin zone of the system. For simple cubic systems, this implies $0 < |q_i| < \Lambda$, $i = 1, 2, \ldots, d$, with $\Lambda = \pi/a$ (a is the lattice constant). The trace in the partition function is now done over the new variables $\hat{S}_{\vec{q}}$.

VI.2. Wilson's Renormalization Group

The Fourier transform $\hat{S}_{\vec{q}}$ represents fluctuations with wave vector $\vec{q}$, i.e. wavelength $\lambda = 2\Pi/|\vec{q}|$. The large values of $\vec{q}$ thus represent short wavelengths. In the vicinity of T_c, where the correlation length ξ is very large, such fluctuations are expected to be unimportant. We now follow Wilson[21] in dividing the Brillouin zone into two regimes, an "internal" one with $|q_i| < \Lambda/b$ and an "external" one with $\Lambda/b < |q_i| < \Lambda$. The variables $\hat{S}^{>}_{\vec{q}}$, in the "external" regime, are now integrated out of the partition function, leaving Z as an integral over $S^{<}_{\vec{q}}$ (in the "internal" regime):

$$Z = \int \prod_{\vec{q}}^{<} d\,\hat{S}_{\vec{q}}^{<} \; e^{-\bar{H}_1\{\hat{S}_{\vec{q}}^{<}\}} \quad . \tag{92}$$

We next use the fact that both $\vec{q}$ and $\hat{S}_{\vec{q}}$ are "dummy" variables in (92). We thus rescale both of them, defining $\sigma_{\vec{q}}$ via

$$\hat{S}_{\vec{q}/b} = \zeta\,\sigma_{\vec{q}} \quad . \tag{93}$$

The rescaling of $\vec{q}$ by a factor b brings the reduced Brillouin zone back into its original size. The spin rescale factor ζ will be chosen below. Finally, we expand the resulting Hamiltonian in powers of $\sigma_{\vec{q}}$ and of $\vec{q}$, and identify the new coefficients r', e', u', etc. This defines our recursion relations.

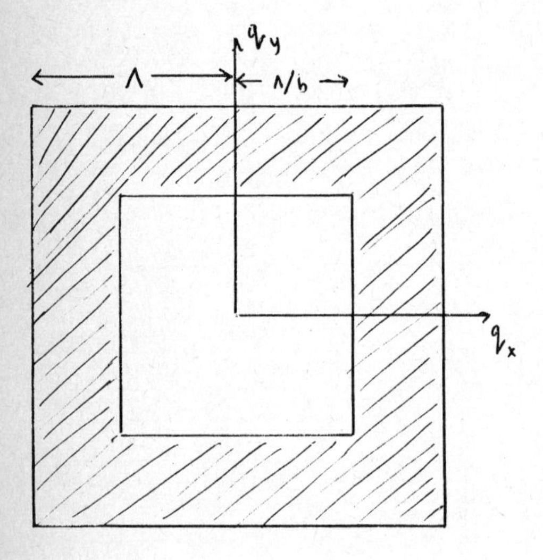

Fig.19. Reduced Brillouin Zone.

VI.3. Gaussian Model

It is very instructive to study the simple "Gaussian" model, in which one maintains only the quadratic terms in Eq.(91). This model is exactly soluble, and one can later check the relevance of the higher order terms. In terms of the weight function $w(S_i)$, this is eqivalent to replacing the spin distribution function by the Gaussian $\exp(-\frac{1}{2}S_i^2)$ [instead of the double peaked Eq.(85)].

The Gaussian model partition function is thus

$$Z = \int \prod_{\vec{q}}^{<} d\,\hat{S}_{\vec{q}} \; e^{-\frac{1}{2}\int_{\vec{q}}(r+e\vec{q}^{\,2})|\hat{S}_{\vec{q}}|^2} \quad . \tag{94}$$

Integration over the external regime now simply introduces a multiplicative constant, and we have

$$\bar{H}_1\{\hat{S}_{\vec{q}}^{<}\} = \frac{1}{2}\int_{\vec{q}}^{<} (r + e\vec{q}^{\,2})\,|S_{\vec{q}}^{<}|^2 \quad . \tag{95}$$

The substitution (93) now turns this into

$$\overline{H}'\{\sigma_{\vec{q}}\} = \frac{1}{2}\zeta^2 b^{-d} \int_{\vec{q}} (r + e\vec{q}^2/b^2)\,|\sigma_{\vec{q}}|^2 \quad , \tag{96}$$

and we easily identify

$$r' = \zeta^2 b^{-d} r \quad , \tag{97}$$

$$e' = \zeta^2 b^{-d-2} e \quad . \tag{98}$$

We now turn to the choice of ζ. It should be emphasized right away that _any_ choice of ζ will yield correct results. However, some choices are more convenient than others. For example, the choice $\zeta^2 < b^d$ implies that both r and e are _irrelevant_. However, the resulting fixed point, at $r = e = 0$, is not very useful, and the physics must be extracted from the way the irrelevant variables decay to zero. Similarly, the choice $\zeta^2 > b^{d+2}$ implies that both r and e are relevant near the fixed point $r = e = 0$. By our definitions, this fixed point represents a _multi_critical point rather than a simple critical point. Indeed, we shall see below that such a choice is appropriate for a _Lifshitz point_. The choice $b^d < \zeta^2 < b^{d+2}$ is also not very convenient: Although the fixed point $r = e = 0$ has only one relevant variable, r, the irrelevant variable e cannot be ignored since it contains the information on correlations. Finally, we are led to the choice

$$\zeta^2 = b^{d+2} \quad , \tag{99}$$

for which

$$r' = b^2 r \quad , \tag{100}$$

$$e' = e \quad . \tag{101}$$

These recursion relations have the _Gaussian fixed point_,

$$r = 0 \;, \qquad e = \text{const.} \tag{102}$$

Since $r \propto (T - T_c)$, by Eq.(90), we now identify $\lambda_t = 2$ [Eq.(72)], i.e.

$$\nu = 1/\lambda_t = 1/2 \quad . \tag{103}$$

The scaling relation $d\nu = 2 - \alpha$ [Eqs.(75), (76)] now yields

$$\alpha = (4 - d)/2 \quad . \tag{104}$$

In order to find the other exponents we must introduce a magnetic field term,

$$\overline{H}_h = h \hat{S}_{\vec{q}=0} \quad . \tag{105}$$

Under our transformation, this becomes

$$\overline{H}_h' = h \zeta \sigma_{\vec{q}} \quad , \tag{106}$$

hence

$$b^{\lambda_h} = \zeta = b^{(d+2)/2} \quad . \tag{107}$$

It is easy to derive from Eq.(73) the scaling relations

$$\beta = (d - \lambda_h)/\lambda_t \quad , \quad \gamma = (2\lambda_h - d)/\lambda_t \quad ,$$

yielding for the Gaussian model

$$\beta = (d - 2)/4 \quad , \quad \gamma = 1 \quad . \tag{108}$$

It should be emphasized, however, that the Gaussian model has no "ordered" phase, since the integral in Eq.(94) diverges for $r < 0$. One must include higher order terms (e.g. $S(\vec{x})^4$) to stabilize this integral.

The choice (99) for ζ^2 may be "justified" by a comment about correlation functions. The Fourier transform of the spin-spin correlation, $\langle \hat{S}_{\vec{q}} \hat{S}_{\vec{p}} \rangle$, is expected by translational invariance to have the form

$$\langle \hat{S}_{\vec{q}} \hat{S}_{\vec{p}} \rangle = G(\vec{q})\, \delta\, (\vec{q} + \vec{p}) \quad . \tag{109}$$

At T_c, the exponent η is defined via $G(\vec{q}) \propto |\vec{q}|^{-2+\eta}$. Performing the renormalization group transformation, Eq.(93), we now find that

$$\zeta^2 \langle \sigma_{\vec{q}} \sigma_{\vec{p}} \rangle = b^{d+2-\eta} G(\vec{q})\, \delta\, (\vec{q} + \vec{p}). \tag{110}$$

Since at T_c the correlations should be invariant of the length scale, we conclude that

$$\zeta^2 = b^{d+2-\eta} \quad . \tag{111}$$

Together with Eq.(99), this implies that in the Gaussian model one has $\eta = 0$. Indeed, this can be directly checked from the exact result

$$\langle \hat{S}_{\vec{q}} \hat{S}_{\vec{p}} \rangle = \frac{\delta(\vec{q} + \vec{p})}{r + e|\vec{q}|^2} , \qquad (112)$$

as well as from the exponents in Eqs.(103) and (108) and scaling relations.

Different choices of ζ^2 would imply that e is not invariant under the renormalization group transformation. This would result in an additional rescaling of the function $G(\vec{q})$:

$$G'(\vec{q}) = b^{d+2-\eta} \zeta^{-2} G(\vec{q}) , \qquad (113)$$

implying Eq.(98). For example, the choice $b^d < \zeta^2 < b^{d+2}$ implied that e is an irrelevant variable. However, e cannot be ignored, since at T_c it appears in the denominator of $G(\vec{q}) = 1/e|\vec{q}|^2$. A non-trivial rescaling of e thus generates a rescaling of the whole function G. Such variables are called "dangerous irrelevant variables".[25]

VII. LANDAU AND LIFSHITZ POINT THEORIES

We now turn to the higher order terms in Eq.(91), and consider their relevance near the Gaussian fixed point (102).

VII.1. Landau Theory for d > 4

We start with the quartic spin term,

$$\overline{H}_u = u \int d^d x \, S(\vec{x})^4 \quad . \tag{114}$$

In the Fourier transformed variables (88), this becomes

$$\overline{H}_u = u \int_{\vec{q}_1} \int_{\vec{q}_2} \int_{\vec{q}_3} \hat{S}_{\vec{q}_1} \hat{S}_{\vec{q}_2} \hat{S}_{\vec{q}_3} \hat{S}_{-\vec{q}_1-\vec{q}_2-\vec{q}_3} \quad . \tag{115}$$

Replacing each $\hat{S}$ by $\hat{S}^< + \hat{S}^>$ ($\hat{S}_{\vec{q}}^<$ is non-zero only for $|q_i| < \Lambda/b$, and $\hat{S}_{\vec{q}}^>$ is non-zero only for $\Lambda/b < |q_i| < \Lambda$), the integrand becomes

$$\hat{S}^< \hat{S}^< \hat{S}^< \hat{S}^< + 4\, \hat{S}^< \hat{S}^< \hat{S}^< \hat{S}^> + 6\, \hat{S}^< \hat{S}^< \hat{S}^> \hat{S}^> +$$

$$+ 4\, \hat{S}^< \hat{S}^> \hat{S}^> \hat{S}^> + \hat{S}^> \hat{S}^> \hat{S}^> \hat{S}^> \quad . \tag{116}$$

We now expand $\exp(\overline{H}_u)$ in powers of $\overline{H}_u$, and integrate over the $\hat{S}^>$'s. Up to first order in u, this yields

$$\overline{H}_{1,u} = u \int_{\vec{q}_1}^< \int_{\vec{q}_2}^< \int_{\vec{q}_3}^< \hat{S}^<_{\vec{q}_1} \hat{S}^<_{\vec{q}_2} \hat{S}^<_{\vec{q}_3} \hat{S}^<_{-\vec{q}_1-\vec{q}_2-\vec{q}_3}$$

$$+ 6\, u \int_{\vec{q}}^< \hat{S}_{\vec{q}} \hat{S}_{-\vec{q}} \int_{\vec{p}}^> \frac{1}{r + e\vec{p}^2} + \text{const.} \tag{117}$$

We now use Eq.(93), and end up with

$$u' = \zeta^4 b^{-3d} u + O(u^2) , \tag{118}$$

$$r' = \zeta^2 b^{-d} [r + 12\, u \int_{\vec{p}}^> \frac{1}{r + e\vec{p}^2} + O(u^2)] , \tag{119}$$

$$e' = \zeta^2 b^{-d-2} e + O(u^2) \quad . \tag{120}$$

By the rules explained above, we choose ζ^2 via Eq.(99), ending up with Eq.(101).

With this choice, Eq.(118) becomes

$$u' = b^{4-d} u + 0 (u^2) . \tag{121}$$

Thus, u is irrelevant for $d > 4$. The Gaussian fixed point (102) is therefore expected to describe the correct critical behavior for dimensionalities $d > 4$.

However, it turns out that the exponents α and β, as given by Eqs.(104) and (108), do not describe the correct physics once $u > 0$. In order to see this in a simplified way, we iterate the recursion relations (118) and (119) ℓ times, until $|r(\ell)| >> 1$, so that the fluctuations are negligible. At that point, we use the Landau theory for the free energy

$$F(\ell) = \frac{1}{2} r(\ell) M^2(\ell) + u(\ell) M^4(\ell) . \tag{122}$$

Since $u(\ell) > 0$, we can follow the discussion of Sec.III and find that $M \propto [|(r(\ell)|/u(\ell)]^{\frac{1}{2}}$, and $F(\ell) \propto |r(\ell)|^2/u(\ell)$. Using Eqs.(118) and (119) for $r(\ell)$ and $u(\ell)$, and Eq.(64) for $F(\ell)$, the original free energy density becomes

$$F \propto b^{-d\ell} |b^{2\ell}r|^2/[u\ b^{(4-d)\ell}] = \frac{r^2}{u} . \tag{123}$$

Thus, $F \propto |T - T_c|^2$ and $\alpha = 2$, rather than Eq.(104). A similar argument yields $\beta = \frac{1}{2}$ instead of Eq.(108). All the critical exponents thus have their <u>Landau theory</u> values, rather than those of the Gaussian model. The mechanism which led to this result is the fact that $u(\ell)$, which appeared in the <u>denominator</u> of $F(\ell)$, had a non-trivial rescaling, which affected that of $F(\ell)$. The variable $u(\ell)$ is thus a <u>dangerous irrelevant variable</u>.[25]

We could have avoided the need to worry about dangerous irrelevant variables, by choosing ζ differently. Since we realized that u is going to be important, we could have chosen ζ so that

$$u' \equiv u . \tag{124}$$

By Eq.(118), this implies

$$\zeta = b^{3d/4} , \tag{125}$$

and hence

$$e' = b^{(d-4)/2} e , \tag{126}$$

$$r' = b^{d/2} [r + 0 (u)] . \tag{127}$$

For $d > 4$, this implies that $e(\ell) \to \infty$, so that the spatial variation of $S(\vec{x})$ tends to disappear, and we may use a uniform value $S(\vec{x}) \equiv M$. Using the same arguments as before,

$$F \propto b^{-d\ell} \, |b^{d\ell/2} \, r \, |^2 \, / \, u \quad = \quad r^2/u \quad . \tag{128}$$

Similar arguments may apply to the tricritical point for $d > 3$ (see below).

VII.2. Lifshitz Points

We now return to Eq.(94), and add higher order terms in the gradients, e.g.

$$H_4 \;=\; -\frac{1}{2}\, e_4 \int_{\vec{q}} |\vec{q}|^4 \, |\hat{S}_{\vec{q}}|^2 \quad . \tag{129}$$

By the same rules as before, we find that

$$e'_4 \;=\; \zeta^2 \, b^{-d-4} \; e_4 \quad . \tag{130}$$

Thus, if we use the choice (99) then

$$e'_4 \;=\; b^{-2} \, e_4 \tag{131}$$

and e_4 is irrelevant. It is for this reason that all the details concerning the lattice structure are usually irrelevant.

However, the fixed point (102) is useful only if $e > 0$. If $e = 0$ then the leading behavior of the correlation function is $G(\vec{q}) = 1/e_4|\vec{q}|^4$, and e_4 becomes a dangerous irrelevant variable!

It is therefore convenient to choose

$$\zeta^2 \;=\; b^{d+4} \quad , \tag{132}$$

so that

$$e'_4 \;\equiv\; e_4 \quad . \tag{133}$$

The recursion relations for r and e now become

$$r' \;=\; b^4 \, r \quad , \tag{134}$$

$$e' \;=\; b^2 \, e \quad , \tag{135}$$

so that both are relevant. The fixed point

$$r = e = 0, \quad e_4 = \text{const.} \tag{136}$$

thus represents a multicritical point. In the vicinity of this multicritical point the free energy density has the form

$$F(r,e) = b^{-d} \; F(b^4 r, \, b^2 e) \quad . \tag{137}$$

Setting $b^4 r = 1$, this yields

$$F(r,e) = r^{d/4} \, F(1, \, e/r^{\frac{1}{2}}) \; , \tag{138}$$

and we identify $\alpha = (8-d)/4$ and $\phi = \frac{1}{2}$. Similarly, $\nu = 1/\lambda_r = 1/4$. These exponents will be observed as long as $e/r^{\frac{1}{2}} \ll 1$. One can now add H_u, Eq.(114), and see that the fixed point (136) describes the multicritical point for $d > 8$.

The problem discussed here is very similar to that of the Lifshitz point, discussed in Sec.IV.2. The only difference is that here we allowed the coefficients of all the quadratic terms in the momentum to go to zero [rather than only one of them, as in Eq.(54)]. In that case, the correlations at T_c are described by

$$G(\vec{q}) \;=\; 1/(e_\perp q_\perp^2 \;+\; f\, q_\parallel^4 \,) \quad . \tag{139}$$

In order to preserve both terms under the renormalization group, one chooses to rescale the momenta via [9]

$$q_\parallel \longrightarrow \sqrt{b}\; q_\parallel \quad , \quad q_\perp \longrightarrow b\; q_\perp \quad . \tag{140}$$

This implies that $d^d q \longrightarrow d^d q / b^{d-\frac{1}{2}}$, and thus

$$r' \;=\; \zeta^2 \, b^{-d+1/2} r \quad , \tag{141}$$

$$f' \;=\; \zeta^2 \, b^{-d-3/2} \, f \quad , \tag{142}$$

$$e'_\perp \;=\; \zeta^2 \, b^{-d-3/2} \, e_\perp \quad . \tag{143}$$

The choice

$$\zeta^2 \;=\; b^{d+3/2} \tag{144}$$

now leaves both f and $e_\perp$ invariant, while

$$r' \;=\; b^2 \, r \quad . \tag{145}$$

Adding the quartic term (115) now yields

$$u' = \zeta^4 b^{-3d+3/2} u = b^{9/2-d} u , \tag{146}$$

from which one deduces that the Gaussian Lifshitz fixed point

$$r = e_{\parallel} = 0, \quad e_{\perp} = \text{const.}, \quad f = \text{const.} \tag{147}$$

describes the critical properties of the Lifshitz point for $d > 9/2$.

VIII. THE ε-EXPANSION

VIII.1. Non-trivial Fixed Point

For $d < 4$, Eq.(121) shows that u is a relevant variable. One thus must go to the next order in the perturbation expansion in u. The term of order u^2 involves a product of eight spins. Simple observation now shows that only the product $(\hat{S}^< \hat{S}^< \hat{S}^> \hat{S}^>)(\hat{S}^< \hat{S}^< \hat{S}^> \hat{S}^>)$ will contribute to u'. The resulting recursion relation turns out to be

$$u' = b^{4-d} [u - 36 u^2 \int_{\vec{p}}^{>} \frac{1}{(r+e\vec{p}^2)^2} + O(u^3)] . \tag{148}$$

Clearly, the second term in Eq.(148) tends to slow down the increase in u, which results from the prefactor b^{4-d}. However, this slowing down is meaningful only if the two effects are of comparable magnitude, which happens only if u is of order $\varepsilon = (4-d)$. This is the origin of the studies at $d = 4-\varepsilon$, with ε treated as a small parameter. It then turns out that u "flows" to a fixed point, of order ε. A study of the vicinity of that fixed point yields the ε-expansions for the critical exponents.

For convenience, one now renormalizes the spins and the momenta so that both e and Λ become equal to unity, and one performs a first iteration which brings the cubic Brillouin zone into a spherical one.[21,23] One next replaces b by $e^{\delta\ell}$, and rewrites Eq.(148) in the differential form

$$\frac{du}{d\ell} = \varepsilon u - B u^2 + \ldots. , \tag{149}$$

with $B = 36K_4 = 9/2\pi^2$, where K_d is the area of a d-dimensional unit sphere divided by $(2\pi)^d$, $K_d^{-1} = 2^{d-1} \pi^{d/2} \Gamma(d/2)$. Similarly, Eq.(119) now becomes

$$\frac{dr}{d\ell} = 2r + \frac{A u}{1+r} + \ldots , \tag{150}$$

with $A = 12K_4$.

Eq.(149) is now easily solved, to yield

$$u(\ell) = u e^{\varepsilon\ell} / Q(\ell) , \tag{151}$$

where

$$Q(\ell) = 1 + (e^{\varepsilon\ell} - 1) u/u^* \tag{152}$$

and

$$u^* = \frac{\varepsilon}{B} = \frac{\varepsilon}{36K_4} = \frac{2\pi^2\varepsilon}{9} \quad . \tag{153}$$

For large ℓ, $u(\ell)$ thus approaches the fixed point u^*. For small ℓ, $u(\ell)$ grows as $e^{\varepsilon\ell}$, as implied by the scaling of the relevant variable u near the Gaussian fixed point $u^* = 0$. Eq.(151) thus describes the crossover from the Gaussian to the non-trivial Ising fixed point.[27]

To order ε, Eq.(150) has the fixed point $r^* = -\frac{1}{2} A\, u^*$. In fact, it is convenient to define a new temperature scaling field (to leading order)

$$t(\ell) = r(\ell) + \frac{1}{2} A\, u(\ell) \quad . \tag{154}$$

It can then be shown that

$$t(\ell) = t(0)\, e^{2\ell} \,/\, Q(\ell)^{A/B} \quad . \tag{155}$$

When $t(0) = 0$, t remains equal to zero and the Hamiltonian flows to the fixed point (153). When $t(0) \neq 0$ then $t(\ell)$ grows larger. The Hamiltonian flows in the r-u plane are shown in Fig.20. The line $t(0) = 0$ is now identified as the critical line. For u=0, the flow of $t(\ell)$ is characterized by the Gaussian thermal

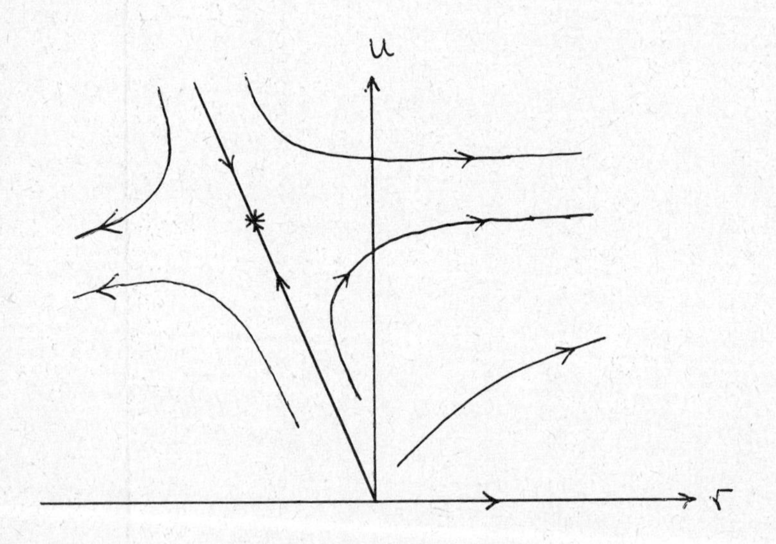

Fig.20. Hamiltonian RG flows.

exponent $\lambda_t = 2$. For $u \neq 0$, $Q(\ell)$ approaches $e^{\varepsilon\ell}\, u/u^*$, so that

$$t(\ell) \longrightarrow t(0)\, e^{(2-\varepsilon A/B)\ell} \tag{156}$$

and we identify

$$\lambda_t = 2 - \frac{A}{B}\varepsilon = 2 - \frac{1}{3}\varepsilon , \tag{157}$$

or

$$\nu = 1/\lambda_t = \frac{1}{2}[1 + \frac{1}{6}\varepsilon + O(\varepsilon^2)] . \tag{158}$$

VIII.2. n-Vector Case

Thus far, we have limited the discussion to the Ising case, in which the spin variable has a single component $S(\vec{x})$. In the more general Heisenberg-like case, the spins are n-component vectors $\vec{S} \equiv (S^1, S^2, \ldots, S^n)$, and the exchange interaction involves the scalar product $(\vec{S}_i \cdot \vec{S}_j)$. In this case, Eq.(82) is replaced by

$$\overline{H} = \int d^d x \{\frac{1}{2} r |\vec{S}|^2 + \frac{1}{2} e |\vec{\nabla}\vec{S}|^2 + u |\vec{S}|^4 + \ldots\} . \tag{159}$$

In Fourier language, Eq.(91) becomes

$$\overline{H} = \frac{1}{2} \int_{\vec{q}} (r + e|q|^2) |\hat{\vec{S}}_{\vec{q}}|^2 + u \int_{\vec{q}_1} \int_{\vec{q}_2} \int_{\vec{q}_3} (\hat{\vec{S}}_{\vec{q}_1} \cdot \hat{\vec{S}}_{\vec{q}_2})(\hat{\vec{S}}_{\vec{q}_3} \cdot \hat{\vec{S}}_{-\vec{q}_1-\vec{q}_2-\vec{q}_3}) . \tag{160}$$

The only effect of these changes is to alter the various combinatoric coefficients in the recursion relations. In particular, A and B in Eqs. (149) and (150) now become

$$A = (n + 2)\, K_4 , \qquad B = (n + 8)\, K_4 . \tag{161}$$

The critical exponent ν therefore becomes

$$\nu = \frac{1}{2}[1 + \frac{(n+2)}{2(n+8)}\varepsilon + O(\varepsilon^2)] . \tag{162}$$

Since the recursion relation of e remains unchanged to order - u, Eq.(111) implies that $\eta = O(\varepsilon^2)$. Using these two exponents and scaling relations one can derive all the other exponents. The ε-expansions for the various exponents have been carried out to order - ε^3. Some additional information, on the behavior of the coefficients of the high order terms in u, made it possible to obtain very accurate values at d = 3.[28]

VIII.3. Logarithmic Corrections

The recursion relations which we discussed above were based on a perturbation expansion in u and in ε. Therefore, they become more and more accurate for smaller and smaller ε. At d = 4, or $\varepsilon = 0$, these recursion relations become exact. At d = 4, Eq.(121) shows that u is a marginal variable. One therefore must go to the non-linear higher order terms in order to determine how u "flows" under the RG transformation. Eq.(149) can then be exactly solved, to yield the leading behavior

$$u(\ell) = \frac{1}{B(\ell_o + \ell)} , \tag{163}$$

where ℓ_o is determined by the initial value u(0). The Gaussian fixed point $u^* = 0$ is thus stable at d = 4, but the "flow" into it is slower than exponential.

Eq.(163) also results from Eq.(151), if we simply let $\varepsilon \to 0$. In the same limit, Eq.(155) becomes

$$t(\ell) \propto t(0)\ e^{2\ell} / (\ell + \ell_o)^{A/B} . \tag{164}$$

By Eq.(63), $\xi(\ell) = \xi e^{-\ell}$. Iterating until ℓ^*, so that $t(\ell^*)$ and $\xi(\ell^*)$ become of order unity, we can then eliminate ℓ^* and find that

$$\xi \propto t^{-\frac{1}{2}}\ |\ell n\ t|^{A/2B} . \tag{165}$$

For the n-vector model, this implies

$$\xi \propto t^{-\frac{1}{2}}\ |\ell n\ t|^{(n+2)/2(n+8)} . \tag{166}$$

Up to higher order corrections, this result is exact. Similar results are always expected when the linear terms in the recursion relations vanish, i.e. at "upper critical dimensionalities" (d = 4 is the highest dimensionality at which fluctuations affect the critical exponents for usual critical points).

For the Ising model with dipole-dipole interactions, the quadratic term in Eq.(91) is replaced by [29,30]

$$\int_{\vec{q}} [r + e\vec{q}^2 + g\ (\frac{q_z}{q})^2]\ |\hat{S}_{\vec{q}}|^2 . \tag{167}$$

Very roughly, it turns out that one can treat the angular variable $\cos\theta = (q_z/q)$ as an additional coordinate axis in momentum integrals, so that the behavior of this model at $d = 3$ is the same (to leading order) as that of the non-dipolar case at $d = 4$. This model was indeed predicted to have logarithmic corrections. [29,30] The experimental verification of these predictions[31] was one of the triumphs of the RG theory.

IX. RESULTS FOR MULTICRITICAL POINTS

We now review the RG results for the multicritical points discussed in earlier chapters.

IX.1. Tricritical Point

As mentioned above, Eq.(151) describes the crossover from the Gaussian fixed point to the non-trivial Ising (or Heisenberg) fixed point. For $d < 4$, one must set both t and u to be equal to zero in order to flow to the Gaussian fixed point. Having two (or three, including the ordering field) relevant variables, the Gaussian fixed point is thus identified as a multicritical point. Remembering the similarity between u and A_4 of Eq.(21), it is natural to expect that the Gaussian fixed point represents the tricritical point. Indeed, if we start with $u < 0$ then the recursion relation (149) yields a flow to more and more negative values of u, without ever reaching a fixed point. We thus iterate until ℓ^*, with $r(\ell^*)=O(1)$, and consider the resulting Landau-like free energy, Eq.(122). For $u(\ell) < 0$, this free energy is meaningless unless we include the sixth order term

$$\bar{H}_6 = u_6 \int d^d x \, S(\vec{x})^6 \quad . \tag{168}$$

Following the same routine as in Sec.VII.1, we now see that

$$u'_6 = \zeta^6 b^{-5d} u_6 + \dots \quad . \tag{169}$$

With Eq.(99), this yields

$$u'_6 = b^{6-2d} u_6 + \dots \quad . \tag{170}$$

Thus, u_6 is irrelevant near the Gaussian fixed point for $d > 3$, and

$$u_6(\ell) \simeq u_6 \, e^{(6-2d)\ell} \quad . \tag{171}$$

Adding $u_6(\ell) M^6(\ell)$ to Eq.(122), we can now use the results of Sec.III. In particular, transitions for $u < 0$ can be shown to be first order, and the point $u = 0$ can be identified as the tricritical point [more accurately, u_6 also contributes to the recursion relation for u, and therefore the tricritical point occurs at $u = -c(n)\, u_6$.[32]]. Substituting Eq.(171) and its analogs for $r(\ell)$ and $u(\ell)$ into Eq.(38) we then see that at the tricritical point $(u = 0)$, u_6 is a "dangerous irrelevant variable" which turns the Gaussian exponents (104) and (108) into their Landau theory counterparts $\alpha_t = 1/2$ and $\beta_t = 1/4$.

From Eq.(171) we now conclude that the upper critical dimensionality for the tricritical point is $d = 3$. At $d = 3$, the addition of the terms of order u_6^2 and $u_4 u_6$ to the recursion relations yields logarithmic corrections. For $d=3-\varepsilon$ one can find a fixed point, with $u_6^* = O(\varepsilon)$, and derive ε-expansions for the tricritical exponents.

IX.2. Lifshitz Point

Sec.VII.2 contained a discussion of two types of Lifshitz points. For the isotropic case, in which $e = 0$ [in Eq.(91)], ζ was chosen by Eq.(132). With this choice, the recursion relation for u, Eq.(118), becomes

$$u' = b^{8-d} \left[u - 36u^2 \int_{\vec{p}}^{>} \frac{1}{(r+e_4\vec{p}^4)^2}\right] . \tag{172}$$

All the arguments of Ch.VIII can now be repeated, ending up with an expansion of the critical exponents in powers of $\varepsilon = 8 - d$.

Similarly, the choice (144) for the anisotropic Lifshitz point yields expansions in $\varepsilon = 9/2 - d$.[9]

IX.3. Bicritical Point

As explained in Sec.II.2, the bicritical point results from a breaking of the rotational invariance via quadratic spin terms, as in Eq.(18). In the continuous spin model, this amounts to replacing Eq.(160) by the more general[22,23]

$$\overline{H} = \frac{1}{2} \int_{\vec{q}} \sum_{\alpha} (r_\alpha + e_\alpha|\vec{q}|^2) |\hat{S}^\alpha_{\vec{q}}|^2 + \int_{\vec{q}_1}\int_{\vec{q}_2}\int_{\vec{q}_3} \sum_{\alpha\,\beta} u_{\alpha\beta} \hat{S}^\alpha_{\vec{q}} \hat{S}^\alpha_{\vec{q}} \hat{S}^\beta_{\vec{q}} \hat{S}^\beta_{-\vec{q}_1-\vec{q}_2-\vec{q}_3} . \tag{173}$$

If we choose our recursion relations so that

$$e_\alpha' \equiv e_\alpha = 1 \tag{174}$$

for all α, i.e. we rescale all spin components by the same factor

$$\zeta_\alpha^2 = \zeta^2 = b^{d+2} , \tag{175}$$

then Eq.(119) is replaced by

$$r'_\alpha = b^2 [r_\alpha + 4 \sum_\gamma u_{\alpha\gamma} \int_{\vec{q}}^{>} \frac{1}{r_\gamma + q^2} + 2 u_{\alpha\alpha} \int_{\vec{q}}^{>} \frac{1}{r_\alpha + q^2}] . \quad (176)$$

Similarly, Eq.(148) is replaced by n^2 equations for the $u_{\alpha\beta}$'s. We next concentrate on the vicinity of the isotropic fixed point, at which

$$u_{\alpha\beta} \equiv u^* = \frac{\varepsilon}{4K_4(n+8)} . \quad (177)$$

Substituting this value in Eq.(176), and linearizing about the isotropic fixed point $r^* = -\frac{1}{2} K_4(n+2) u^*$, we find the n linearized recursion relations

$$\Delta r'_\alpha = b^2 [\Delta r_\alpha - 4u^* (\int_{\vec{q}}^{>} q^{-4}) (\sum_\gamma \Delta r_\gamma + 2\Delta r_\alpha)] . \quad (178)$$

These recursion relations have two eigenvalues. One, corresponding to the isotropic eigenvector $\Delta r_\alpha \equiv \Delta r$, is non-degenerate and is equal to $b^{1/\nu}$, with ν given by Eq.(162). The second eigenvalue, b^{λ_g}, with

$$\lambda_g = 2 - 8K_4 u^* = 2 - \frac{2\,\varepsilon}{n+8} , \quad (179)$$

has the (n-1) anisotropic eigenvectors with $\sum_\gamma \Delta r_\gamma = 0$ [compare with Eq. (18)]. All of these anisotropies are thus <u>relevant</u>, with the <u>crossover exponent</u>

$$\phi = \frac{\lambda_g}{\lambda_t} = 1 + \frac{n}{2(n+8)} \varepsilon + 0\,(\varepsilon^2) . \quad (180)$$

There are several ways to study where the Hamiltonian flows to, when the anisotropy g is not equal to zero. A simple way is based on the fact that for the anisotropy (18), if $g > 0$, the variable r_1 will grow faster than the variable $r_2 = r_3$. After ℓ^* iterations we may have $r_1(\ell^*) = 1$ and $r_2(\ell^*) \ll 1$. At this point one may integrate the spin variable S^1 out of the partition function, and remain with an effective Hamiltonian which contains only the two spin components S^2 and S^3. The coefficients in this effective Hamiltonian, e.g. u_{eff}, will be renormalized via the integration over S^1. Apart from this, the effective Hamiltonian now represents a two component spin Hamiltonian. The crossover is thus from the Heisenberg (n=3) fixed point to that of the XY (n=2) model.

It is important to comment here that in some cases, the renormalized effective coefficients change sign or move out of the region of attraction of a stable fixed point. In these cases, the order of the transition may change via tricritical points.[32]

X. THE CUBIC PROBLEM

We devote this last chapter to a discussion of the effects of the cubic symmetry breaking Hamiltonian, Eq.(19).[23] A simple power counting shows that the leading term in the recursion relation for v is similar to that for u, i.e.

$$v' = \zeta^4 b^{-3d} v + \ldots\ldots \quad . \tag{181}$$

When $u = 0$, the Hamiltonian decouples into n independent Ising-like Hamiltonians. Therefore, we expect the recursion relation for v to reduce to that given in Eq.(148), with v replacing u. By symmetry, we do not expect a term of order u^2 in the recursion relation for v (cubic symmetry should not be generated by purely isotropic terms). It is also not expected to find terms of order v^2 in the recursion relation for u. The only quadratic terms we need to worry about are thus of order uv. Finally, the recursion relations for u and v become

$$\frac{du}{d\ell} = \varepsilon u - 4K_4 [(n+8) u^2 + 6 u v] + \ldots. \quad , \tag{182}$$

$$\frac{dv}{d\ell} = \varepsilon v - 4K_4 [12 uv + 9 v^2] + \ldots. \quad . \tag{183}$$

These recursion relations are easily found to have four fixed points, i.e. the Gaussian fixed point,

$$u^*_G = v^*_G = 0 , \tag{184}$$

the isotropic Heisenberg fixed point,

$$u^*_H = \frac{\varepsilon}{4K_4(n+8)} , \quad v^*_H = 0 , \tag{185}$$

the decoupled Ising fixed point,

$$u^*_I = 0 , \quad v^*_I = \frac{\varepsilon}{36K_4} , \tag{186}$$

and the new cubic fixed point,

$$u^*_C = \frac{\varepsilon}{12K_4 n} , \quad v^*_C = \frac{\varepsilon(n-4)}{36K_4 n} . \tag{187}$$

For $n < 4$, the flows in the u-v plane are shown in Fig.21. The isotropic fixed point is found to be the only stable one, with exponents

$$\lambda_u = -\varepsilon + O(\varepsilon^2) \quad , \qquad \lambda_v = \frac{n-4}{n+8}\,\varepsilon + O(\varepsilon^2) \quad . \tag{188}$$

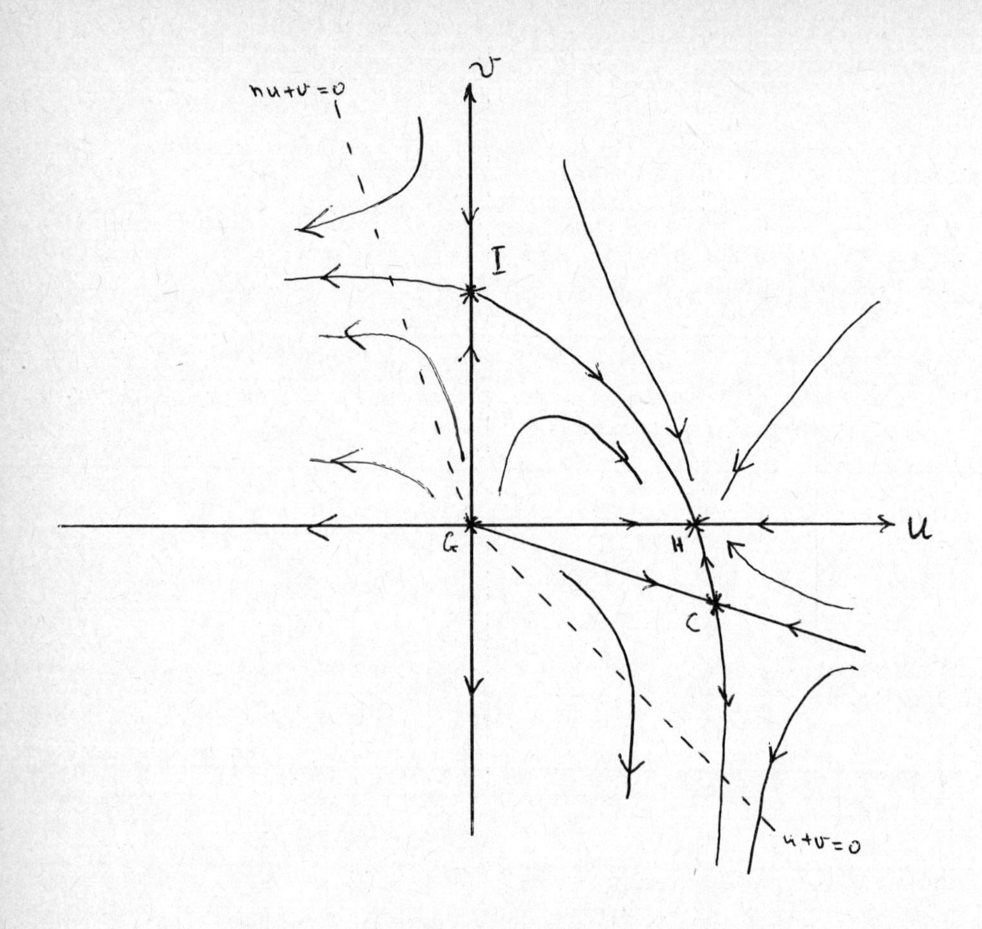

Fig.21. Hamiltonian flows in u-v plane.

All the Hamiltonians which flow to the isotropic Heisenberg fixed point will thus have second order transitions, with the asymptotic exponents having the same values as if the cubic symmetry was absent. However, we should note that the cubic interaction is also a dangerous irrelevant variable! At any finite distance below T_c, the cubic term will remain finite, although it will be rescaled by $t^{|\phi_v|}$, with $\phi_v = \lambda_v/\lambda_t$. Thus, the system will have well defined easy axes (along diagonals for $v > 0$ and along axes for $v < 0$) and the transverse susceptibility will remain finite. It is only asymptotically close to T_c that these cubic effects become negligible. Since $|\phi_v|$ is very small, one never reaches this asymptotic limit in practice.

Another effect of the cubic term is to yield a tetracritical point, as shown in Fig.10. The only effect of the fluctuations is that the two critical lines which bound the intermediate ordered phase approach each other tangentially as $T \to T_c$, somewhat faster than the other two lines.[12]

It remains to discuss the flows which do not approach the isotropic fixed point. Except for the special flows which go to I or to C, these flows go away, to larger negative values of u or v. Eventually, they cross the Landau stability lines $u+v = 0$ or $nu+v = 0$ [see discussion following Eq.(50)]. Matching to Landau theory after ℓ^* iterations then shows that the transitions become first order. Thus, there exist transitions which are driven by the fluctuations to become first order! [32,33]

The Hamiltonians which flow to I or to C are now identified as tricritical, since they separate between second order and first order transitions.

Similar structures are expected for many systems in which the symmetry is broken via quartic spin terms.[8]

CONCLUSION

The aim of this series of lectures has been to review several examples of multicritical points, using a phenomenological description, a Landau theory analysis, a scaling theory and a renormalization group analysis. Although we were limited to a small number of examples, it is hoped that the reader is now able to follow the current literature on the subject and to appreciate the large variety of possible multicritical points.

ACKNOWLEDGEMENTS

The author is grateful to Professors F. J. W. Hahne and C. A. Englebrecht for their warm hospitality in Stellenbosch, and to the CSIR, who organized the summer-school. Comments from the other lecturers at the school are also gratefully acknowledged. The work described here was supported in part by the U.S.-Israel Binational Science Foundation.

REFERENCES

1. M. Wortis, Phys. Lett. 47A, 445 (1974).
2. M. Blume, Phys. Rev. 141, 517 (1966); H. W. Capel, Physica 32, 966 (1966).
3. For a generalized version, see M. Blume, V. J. Emery and R. B. Griffiths, Phys. Rev. A4, 1071 (1971).
4. R. J. Griffiths, Phys. Rev. B7, 545 (1973).
5. e.g. R. J. Birgeneau, G. Shirane, M. Blume and W. C. Koehler, Phys. Rev. Lett. 33, 1098 (1974).
6. M. Blume, L. M. Corliss, J. M. Hastings and E. Schiller, Phys. Rev. Lett. 32, 544 (1974).
7. M. E. Fisher and W. Selke, Phys. Rev. Lett. 44, 1502 (1980).
8. D. Mukamel and S. Krinsky, Phys. Rev. B13, 5065 (1976).
9. R. M. Hornreich, M. Luban and S. Shtrikman, Phys. Rev. Lett. 35, 1678 (1975).
10. A. Aharony and A. D. Bruce, Phys. Rev. Lett. 33, 427 (1974).
11. M. E. Fisher and D. R. Nelson, Phys. Rev. Lett. 32, 1350 (1974).
12. A. D. Bruce and A. Aharony, Phys. Rev. B11, 478 (1975).
13. D. Mukamel, Phys. Rev. B14, 1303 (1976).
14. A. Aharony, Ann. Israel Phys. Soc. 2, 13 (1978).
15. A. Aharony, Ferroelectrics 24, 313 (1980).
16. M. E. Fisher, AIP Conf. Proc. 24, 273 (1975).
17. S. Galam and A. Aharony, J. Phys. C13, 1065 (1980); 14, 3603 (1981).
18. E. K. Riedel and F. J. Wegner, Phys. Rev. Lett. 29, 349 (1972).
19. e.g. L. P. Kadanoff et al., Rev. Mod. Phys. 39, 395 (1967).
20. J. M. Kosterlitz, D. R. Nelson and M. E. Fisher, Phys. Rev. B13, 412 (1976).
21. K. G. Wilson and J. Kogut, Phys. Reports 12C, 74 (1974).
22. M. E. Fisher, Rev. Mod. Phys. 46, 597 (1974).
23. A. Aharony, in Phase Transitions and Critical Phenomena, edited by C. Domb and M. S. Green (Academic, N.Y., 1976), Vol. 6, p. 357.
24. F. J. Wegner, ibid, p. 8.
25. M. E. Fisher, in Renormalization Group in Critical Phenomena and Quantum Field Theory: Proceeding of a Conference, edited by J. D. Gunton and M. S. Green (Temple University, 1973), p. 65.
26. J. Hubbard, Phys. Lett. A39, 365 (1972).
27. J. Rudnick and D. R. Nelson, Phys. Rev. B13, 2208 (1976).
28. J. C. Le Guillou and J. Zinn-Justin, Phys. Rev. Lett. 39, 95 (1977) and references.
29. A. I. Larkin and D. E. Khmel'nitzkii, Zh. Eksp. Teor. Fiz. 56, 2087 (1969) [Sov. Phys. -JETP 29, 1123 (1969)].
30. A. Aharony, Phys. Rev. B8, 3363 (1973); 9, 3946 (E).
31. G. Ahlers, A. Kornblit and H. J. Guggenheim, Phys. Rev. Lett. 34, 1237 (1975).
32. D. Blankschtein and A. Aharony, Phys. Rev. Lett. 47, 439 (1981).
33. E. Domany, D. Mukamel and M. E. Fisher, Phys. Rev. B15, 5432 (1977).

Lectures on Disordered Systems*

Michael J. Stephen
Physics Department
Rutgers University
Piscataway, NJ 08854/USA

Contents

*Ten lectures presented at the South African school on Critical Phenomena held at Stellenbosch, January 1982.

I. PERCOLATION

1. Introduction

Percolation problems were introduced by Broadbent and Hammersley (1957). For reviews see Frish and Hammersley (1963), Shante and Kirkpatrick (1971), Essam (1972), Kirkpatrick (1973), Stauffer (1979) and Essam (1980).

Consider a crystal lattice in which some of the sites are occupied by particles, the remaining sites being vacant. Nearest neighbor particles are connected by bonds thus forming clusters. As the concentration of particles increases the average size of the clusters increases. Eventually we reach a concentration, the critical concentration, at which a cluster of particles extends from one side of the lattice to the other, i.e. there is an "infinite" cluster (for an infinite lattice). If the bonds are conductors of some type, the solid will be conducting or percolating in that current can now percolate from one side to the other side of the lattice.

It is clear that the dimension of our lattice is important. In 1-d an arbitrarily small concentration of vacancies causes the solid to be non-conducting so that the critical concentration is 1. In 2-d at low concentrations we have an infinite cluster of vacancies and finite clusters of sites. At the critical concentration an infinite cluster of occupied sites appears and this coincides with the disappearance of the infinite cluster of vacancies.

For theoretical purposes it is simpler to consider percolation processes on a regular lattice. We consider a regular array of sites, each site being connected to its neighbors (usually only nearest neighbors) by bonds of some kind. There are two classes of percolation problems known as site and bond percolation. In the site problem, the sites are occupied independently and at random with a probability p and empty with probability $q = 1-p$. Nearest neighbor occupied sites are connected by bonds. The percolation probability $P^{(s)}(p)$ is the probability that a site belongs to the infinite cluster of occupied sites. The critical concentration p_c is the value of p at which $P^{(s)}$ is first non-zero. For $p<p_c$ all clusters are finite. In the bond percolation problem the bonds are occupied independently and at random with probability p and empty with probability $q = 1-p$. The percolation probability $P^B(p)$ is again the probability that a site is connected to an infinite cluster of sites by occupied bonds and p_c^B is the critical concentration at which $P^B(p)$ is first non-zero. Some authors define p^B as the probability that a bond belongs to an infinite cluster of bonds.

One of the interesting problems in percolation theory is that of determining the distribution of cluster sizes in the lattice. Consider the bond percolation problem and let $Nn_s(p)$ be the average number of clusters of size s (i.e. containing s connected sites) at the concentration p on the lattice of N sites. Complete information on the cluster size distribution is contained in the generating function

$$F(p,h) = \sum_{s=1} n_s e^{-hs} \tag{1.1}$$

We will see that F is like a free energy and h (the variable in the generating function) like a magnetic field. The probability that a given site lies in a cluster of s sites per site is $P_s(p) = s\, n_s(p)$ and we define

$$A(p,h) = \sum_{s=1} P_s e^{-hs} \tag{1.2}$$

Clearly $A = -\frac{\partial F}{\partial h}$. Only finite clusters contribute to the sums in (1.1) and (1.2) and thus

$$\begin{aligned} A(p,o) &= 1 \qquad\qquad p<p_c \\ &= 1-P(p) \qquad p>p_c \end{aligned} \tag{1.3}$$

where P is the percolation probability. The moments of the cluster distribution are obtained by differentiating (1.2). Thus the average cluster size per site is

$$S(p) = -\left(\frac{\partial A}{\partial h}\right)_{h=o} \tag{1.4}$$

This diverges at the critical concentration and is analogous to the susceptibility in a magnetic problem. In the site problem Eq. (1.3) must be modified. For $p<p_c$, $A(p,o) = p$ and is the probability that a site is occupied while for $p>p_c$ $A(p,o) = p-P(p)$.

2. Magnetic models

Percolation theory has been used as a model for the behavior of alloys of magnetic and non-magnetic atoms. Antiferromagnetic alloys such as $KMn_pMg_{1-p}F_3$ show percolation behavior while $K_2Mn_pMg_{1-p}F_4$ and $Rb_2Mn_pMg_{1-p}F_4$ behave like two-dimensional percolating systems (Birgeneau et al. (1976). Clearly these alloys are examples of the site percolation problem, with magnetic atoms corresponding to occupied sites and non-magnetic atoms to empty sites. Assuming only nearest neighbor interactions, a magnetic phase transition can only occur if there is an infinite cluster of magnetic atoms. The finite clusters respond to a magnetic field and contribute a Curie-like term to the magnetic susceptibility. As $T\to 0°K$ the coupling between the atoms is infinitely strong and the magnetic properties of the alloy are determined by the cluster distribution. If the spins are Ising-like the spontaneous magnetization per site is P(p). A cluster of s sites behaves as a single spin and thus the magnetic moment per site is

$$M = P(p) + \sum_s P_s \tanh(Hs/kT) \tag{2.1}$$

The zero field susceptibility is given by

$$\chi = \frac{1}{kT}\sum_s sP_s = \frac{S(p)}{kT} \tag{2.2}$$

In a classical Heisenberg magnet tanh Hs/kT in (2.1) is replaced by (coth Hs/kT - kT/Hs). In addition in a Heisenberg magnet spin waves can exist in the infinite cluster and a further quantity of interest is the spin wave stiffness. The infinite cluster near p_c is rather tenuously connected and the energy of the spin deviations can be reduced by concentrating the changes in angle to narrow channels. The spin wave stiffness is then reduced in the percolating system and vanishes at p_c. The spin wave stiffness is directly related to the electrical conductivity of a random network and is discussed further below.

3. Inhomogeneous conductors

The problem of determining the conductivity or dielectric constant of a random mixture of insulating and metallic material, e.g. a cermet/ceramic metallic composite) or of a granular metal which consists of overlapping islands of metal mixed with insulator is an old one. We define a spatially varying conductivity $\sigma(\vec{r})$ which takes on values appropriate to the metal and insulator and solve for the potential distribution $V(\vec{r})$ which satisfies current conservation $\vec{\nabla}\cdot(\sigma(\vec{r})\vec{\nabla}V(\vec{r})) = 0$. It is more convenient to consider the problem on a lattice and look at this equation at the grid of lattice points. We then get Kirchoff's equation

$$\sum_j \sigma_{ij}(V_i - V_j) = 0 \tag{3.1}$$

where i, j are the lattice points and σ_{ij} is the conductivity between sites i and j. The grid of lattice points is chosen coarse enough so that the σ_{ij} are statistically independent. In this way we obtain a bond percolation problem in which the bonds are the conductors σ_{ij} and the sites are the nodes of the electrical network. If the conductors σ_{ij} have a probability p of having the value σ and 1-p of being zero then the network is conducting for $p>p_c$, the current being carried by the infinite cluster and is insulating for $p<p_c$. A related problem is that where the conducting elements are replaced by superconductors and the insulating regions by normal conductors. The conductivity is finite for $p<p_c$ when only isolated islands of superconductors occur and diverges at p_c where an infinite superconducting cluster forms.

4. Exact results

Comparison of different lattices allows certain inequalities to be established (Fisher and Essam (1961), Essam (1972))

(a) The percolation probability for the bond problem is greater than for the site problem

$$P^B(p) \geqslant P^S(p) \tag{4.1}$$

It can also be shown that P(p) is a non-decreasing function of p and thus on the same lattice

$$p_c^B \leqslant p_c^S \tag{4.2}$$

(b) The removal of certain edges or bonds from a lattice will not change p the fraction of occupied bonds or sites and will not lead to an increase in P(p). Thus if L is a lattice obtained from L^+ by removing certain bonds

$$P_{L^+}(p) \geqslant P_L(p) \quad , \quad p_{cL^+} < p_{cL} \tag{4.3}$$

Applying this result to the 2-d triangular, square and hexagonal lattices gives

$$p_{c\,\Delta} \leqslant p_{c\,\square} \leqslant p_{c\,\text{hexag}} \tag{4.4}$$

and in 3-d to the fcc, bcc and simpler cubic lattices gives

$$p_{c\ \text{s.c.}} \leqslant p_{c\,\text{bcc}} \leqslant p_{c\,\text{fcc}} \tag{4.5}$$

(c) The critical concentration in some 2-d lattices can be obtained using a transformation similar to the Kramers Wannier duality transformation for the Ising model (see Essam (1972)). Thus

$$p_c^S(\Delta) = \frac{1}{2}$$

$$p_c^B(\square) = \frac{1}{2}$$

$$p_c^B(\Delta) = 1-p_c^B(\text{hexag}) = 2\sin(\pi/18) \tag{4.6}$$

An approximate formula (Kirkpatrick 1973) for p_c^B in the bond problem that works well in any dimension d is

$$p_c^B = z^{-1}\left(\frac{d}{d-1}\right) \tag{4.7}$$

where z is the coordination number.

5. Bethe lattice

The bond and site percolation problems can be solved exactly on the Bethe lattice (Fisher and Essam 1961). We will not go into the complete solution but consider certain aspects of the bond percolation problem. With a little modification the results also apply to the site problem. Suppose R^z is the probability that beginning at a site (the origin) there is no path of connected bonds to the outer boundary i.e. $R^z(p)$ is the probability that the origin site is not part of the infinite cluster and the percolation probability is

$$P(p) = 1 - R^z(p) \tag{5.1}$$

For an infinite lattice R satisfies the relation

$$R = 1-p + pR^{z-1} \tag{5.2}$$

where the first term corresponds to the absence of a bond and the second term to the presence of a bond but with all the bonds attached to it not leading to infinity. One solution of (5.2) is $R = 1$ and this applies for $p<p_c$. For $p>p_c$ a second solution exists and it is easily obtained close to p_c by putting $R = (1-z^{-1}P)$ where P is small. Substituting in (5.2) and expanding in P gives

$$P[p(z-1)-1] = \frac{p}{2z}(z-1)(z-2)P^2 \tag{5.3}$$

Thus $p_c = \frac{1}{z-1}$ and close to p_c

$$P = \frac{2z(z-1)}{(z-2)}(p-p_c) \tag{5.4}$$

The one dimensional case is $z = 2$ and $p_c = 1$. Essam and Fisher have also determined the cluster distributions.

6. Critical exponents

In percolation problems we are interested in the cluster distribution functions $n_s(p)$. The behavior of systems close to a phase transition is usually described by critical exponents α, β, γ ... and close to the percolation point we define percolation exponents $\alpha_p\ \beta_p\ \gamma_p$ ...

$$\left(\sum_s n_s(p)\right)_{sing} \sim (p-p_c)^{2-\alpha_p} \qquad \text{(free energy)} \tag{6.1}$$

$$\left(\sum sn_s(p)\right)_{sing} \sim (p-p_c)^{\beta_p} \qquad \text{(order parameter)} \tag{6.2}$$

$$\left(\sum_s s^2 n_s\right)_{sing} \sim (p-p_c)^{-\gamma_p} \qquad \text{(suscept)} \tag{6.3}$$

$$\left(\sum_s s\, n_s(p)e^{-hs}\right)_{sing} \sim h^{1/\delta_p} \qquad p = p_c \qquad \text{(order parameter at } T_c) \tag{6.4}$$

$$\xi(p) \sim (p-p_c)^{-\nu_p} \qquad \text{correl. length} \tag{6.5}$$

The subscript sing. denotes the leading non-analytic part of a quantity. In order to justify these definitions and analogies we turn to a formulation of the percolation problem in terms of a spin problem.

7. Potts model

A great deal is known about the critical behavior of spin systems and it is very useful to relate the percolation problem to a spin problem. The spin problem is the Potts model and its relation to bond percolation was first pointed out by Kasteleyn and Fortuin (1969). It is also possible to relate the site percolation problem to a different form of the Potts model.

The Potts model is a generalization of the Ising model. In the Ising model a spin is placed on each site of the lattice and each spin can take on two values generally taken to be ± 1. In the Potts q state model each spin can take on q possible values. Nearest neighbor spins have any energy $-\varepsilon$ if they are in the same state and 0 if they are in different states. If $\varepsilon > 0$ this leads to an ordered state at low temperatures. If λ_i is the Potts spin on site i the partition function is

$$Z = \mathrm{Tr}\, e^{K \sum_{(ij)} \delta_{\lambda_i,\lambda_j} + h \sum_i \delta_{\lambda_i,1}} \tag{7.1}$$

where $K = \varepsilon/kT$, the first sum is over all nearest neighbor pairs of spins and we have included a magnetic field h which acts on spins in the state $\lambda = 1$. The trace indicates a sum on the q values of each spin.

We will show that the Potts model in the limit $q \to 1$ is related to the bond percolation problem and to obtain this relation we write Z as

$$Z = \mathrm{Tr} \prod_{(ij)} [1 + v\delta_{\lambda_i,\lambda_j}] e^{h \sum_i \delta_{\lambda_i,1}} \tag{7.2}$$

where $v = e^K - 1$. The product is over all nearest neighbor pairs and we can regard the factors 1 and $v\delta_{\lambda_i \lambda_j}$ in the square bracket as representing the absence or presence of a bond connecting sites i and j. In order to weight the configurations correctly we choose

$$v = \frac{p}{1-p} \tag{7.3}$$

which relates the concentration p in the percolation problem to the temperature in the Potts model with low concentrations corresponding to high temperatures.

When the product in (7.2) is multiplied out we generate all possible arrangements of bonds on the lattice. Owing to the factors $\delta_{\lambda_i,\lambda_j}$ the spins on the sites of a connected cluster must all be the same and on taking the trace on λ a cluster of s sites and b bonds is weighted by a factor

$$v^b [e^{hs} + q - 1] \tag{7.4}$$

In a given configuration G of B(G) bonds let $N n_s(G)$ be the number of clusters of s sites. Then

$$Z = \sum_G v^{B(G)} \prod_s [e^{hs} + q - 1]^{N n_s(G)} \tag{7.5}$$

where N is the number of lattice sites. For q=1

$$Z = (1-p)^{-Nz/2}\, e^{hN} \tag{7.6}$$

where z is the coordination number. Differentiating (7.5) with respect to q and setting q=1 we find the "free energy" per site

$$\begin{aligned} F(p,h) &= \frac{1}{N}\left(\frac{\partial}{\partial q} \ln Z\right)_{q=1} \\ &= \sum_s n_s\, e^{-hs} \end{aligned} \tag{7.7}$$

where n_s is the average number of clusters of size s per site. This is the required generating function. To summarize: the bond percolation problem corresponds to the unphysical q=1 limit of the Potts model. This is a form of the well known replica trick.

8. Mean field theory

The relation between bond percolation and the Potts model enables us to obtain a mean field theory quite easily (Stephen (1977)). We begin with the partition function of the Potts model and introduce a representation for the spins in which $\lambda=1, \omega, \ldots \omega^{q-1}$ where $\omega = e^{2\pi i/q}$ is a qth root of unity. We can then write

$$q\delta_{\lambda_i,\lambda_j} = 1+\sum_{r=1}^{q-1} \lambda_i^r \lambda_j^{q-r} \tag{8.1}$$

$$q\delta_{\lambda,1} = 1+\sum_{r=1}^{q-1} \lambda^r \tag{8.2}$$

The Hamiltonian for the Potts model is

$$H/kT = -\frac{K}{q} \sum_{(ij)} \sum_{r} \lambda_i^r \lambda_j^{q-r} - \frac{h}{q} \sum_{i} \sum_{r} \lambda_i^r - \frac{KNz}{2q} - \frac{hN}{q} \ . \tag{8.3}$$

By differentiation of the partition function with respect to h the magnetization per site is $\frac{1}{q}$ + M where

$$M = \frac{1}{q} < \sum_{r=1}^{q-1} \lambda_i^r > \tag{8.4}$$

The cluster generating function is

$$A(p,h) = -\frac{1}{N}\left[\frac{\partial^2}{\partial q\, \partial h} \ln Z\right]_{q=1} = -\left[\frac{\partial}{\partial q}\left(\frac{1}{q} + M\right)\right]_{q=1} = 1-\left(\frac{\partial M}{\partial q}\right)_{q=1} \tag{8.5}$$

and is thus determined by the magnetization of the Potts model.

In the mean field theory we introduce an order parameter $R = <\lambda^r>$. We choose the symmetrical solution so that R is independent of the index r (r=1 ... q-1). Omitting constants the mean field Hamiltonian is

$$H_{MF}/kT = -\frac{1}{q}(xR + h) \sum_{r=1}^{q-1} \lambda^r \tag{8.6}$$

where x = zK = -z ln (1-p). Using this Hamiltonian the self consistency condition $R = <\lambda^r>$ becomes

$$R = \frac{e^{xR + h} - 1}{e^{xR + h} + q - 1} \tag{8.7}$$

We are interested in these formulae when q = 1 and thus setting q=1 we find

$$A(p,h) = 1-R \tag{8.8}$$

$$R = 1- e^{-xR-h} \tag{8.9}$$

Eq. (8.9) is more conveniently written in terms of A

$$A = e^{-x-h} e^{xA} \tag{8.10}$$

which is easily solved by iteration

$$A(p,h) = \sum_{s=1}^{\infty} \frac{(sx)^{s-1}}{s!} e^{-s(h+x)} \tag{8.11}$$

and the cluster distribution

$$P_s = \frac{(sx)^{s-1}}{s!} e^{-sx} \tag{8.12}$$

This solution applies for $p<p_c$.

The average size of clusters diverges at the percolation point and this serves to locate the transition point as $x_c = -z\ln(1-p_c) = 1$ or $p_c = 1-e^{-1/z}$. Close to critical point we approximate P_s for large s and small $1-x$ by

$$P_s \simeq \frac{1}{(2\pi s^3)^{1/2}} e^{-\frac{s}{2}(1-x)^2} \tag{8.13}$$

Close to p_c Eq. (8.9) can be expanded in powers of R:

$$R(1-x) + \frac{x^2}{2} R^2 = h \quad . \tag{8.14}$$

The following results are then easily derived

(i) For $p>p_c$ and $h=0$

$$R = P(p) \text{ and } P(p) = 2(1-x)^{\beta_p} \quad , \quad \beta_p = 1 \tag{8.15}$$

(ii) For $p<p_c$ $R = hS(p)$ where the average cluster size

$$S(p) = (1-x)^{-\gamma_p} \quad , \quad \gamma_p = 1 \tag{8.16}$$

(iii) For $p = p_c$

$$R \sim h^{1/\delta_p} \qquad \delta_p = 2 \tag{8.17}$$

Note that the form of the mean field equation (8.14) is different from that for a magnet because the non-linear term is R^2 and not M^3 as is a magnet. This also leads to different mean field exponents for percolation.

The full equation (8.9) can also be solved to find the distribution of finite clusters above p_c. Firstly the solution of (8.9) for h=0 determines the percolation probability P

$$P = 1- e^{-xP} \tag{8.18}$$

For $x<1$ (i.e. $p<p_c$), $P=0$ (we reject the solution of (8.18) with $P<0$) and for $x>1$ (i.e. $p>p_c$) a positive solution $P>0$ exists (see (8.15). The distribution of finite clusters is determined by A (Eq. (8.8)) and it is convenient to put

$$A(p,h) = A_F(1-P) \quad , \quad p>p_c \tag{8.19}$$

Then using (8.18) Eq. (8.10) can be written

$$A_F = e^{-x(1-P)-h} e^{x(1-P)A_F} \tag{8.20}$$

This is of exactly the same form as (8.10) with x replaced by $x(1-P) < 1$. The mean field theory thus gives the same form for the distribution of finite clusters, Eq. (8.11), for p both above and below p_c.

9. Scaling theories

One of the most important recent developments in theory of critical phenomena has been that of scaling theories and that close to the critical temperature there is a characteristic length scale for correlations. As the critical temperature is approached all quantities scale with this length. These same ideas can be applied to percolation problems and using the relation between the bond percolation problem and the Potts model the identification of various quantities is straightforward. The "free energy" F(p,h) has a singular part near p_c.

$$F_{sing}(p,h) \sim |p-p_c|^{2-\alpha_p} f_\pm(h/|p-p_c|^{\Delta_p}) \tag{9.1}$$

where we use a subscript p on the exponents to indicate that they are percolation exponents and the subscripts ± apply for $p>p_c$ and $p<p_c$ respectively. The function f is not universal but in accordance with the universality hypothesis it can be made so by introducing two further parameters q_o and q_1. The form

$$F_{sing}(p,h) = q_o|\varepsilon|^{2-\alpha_p} f_\pm(h|\varepsilon|^{\Delta_p}) \tag{9.2}$$

where $\varepsilon = q_1(\frac{p_c-p}{p_c})$ is universal except for the parameters q_o and q_1 which must be determined for each model.

The cluster generating function is then given by

$$A_{sing}(p,h) = q_o|\varepsilon|^{2-\alpha_p-\Delta_p} f_\pm'(h/|\varepsilon|^{\Delta_p}) \tag{9.3}$$

This form implies that all the critical exponents can be expressed in terms of the two exponents α_p and Δ_p. Thus

$$S(p) = C_\pm(1-\frac{p}{p_c})^{-\gamma_p} \tag{9.4}$$

where

$$\gamma_p = 2\Delta_p - 2 + \alpha_p \quad ; \quad C^\pm = q_o q_1^{-\gamma_p} f_\pm''(o) \tag{9.5}$$

Similarly

$$P(p) = B(\frac{p}{p_c} - 1)^{\beta_p} \tag{9.6}$$

where

$$\beta_p = 2-\alpha_p-\Delta_p \quad , \quad B = q_o q_1^{\beta_p} f_+'(o) \tag{9.7}$$

Exactly at p_c

$$P(p_c,h) = Eh^{1/\delta_p} \tag{9.8}$$

where

$$\frac{1}{\delta_p} = \frac{2-\alpha_p}{\Delta_p} - 1 \quad , \quad E = q_o \lim_{z\to\infty} z^{-1/\delta_p} f_\pm'(z) \quad . \tag{9.9}$$

A further test of universality is provided by the amplitude ratio

$$R = C_+^{1/\delta_p} E^{-1} B^{1-1/\delta_p} \tag{9.10}$$

which should be universal due to the cancellation of the non-universal parameters q_o and q_1 (Stauffer (1976)). The results of series expansions (mainly in 2-d) are consistent with this result.

Finally we discuss the correlation length exponent defined by $\xi \sim |p-p_c|^{-\nu_p}$ and hyperscaling. We suppose that the concentration of large clusters n_s is inversely proportional to the volume that they occupy i.e. ξ^d. Thus Eq. (9.1) for F should be proportional to ξ^d and this leads to the hyperscaling relation

$$2-\alpha_p = d\nu_p \tag{9.11}$$

We can find the form of P_s by taking the Laplace transform with respect to h of (9.3) which gives

$$P_s \sim s^{-\frac{\beta_p}{\Delta_p}} g(s|p-p_c|^{\Delta_p}) \tag{9.12}$$

This form for the distribution of cluster sizes has been well established by computer simulations in 2-d (Leath and Reich (1978)).

The mean field values of the exponents follow by comparison of these results with those of the previous section and are $\alpha_p = -1$ $\beta_p = 1$, $\gamma_p = 1$ $\Delta_p = 2$ and it can further be shown that $\nu_p = \frac{1}{2}$. This latter result together with the hyperscaling relation enable us to determine the upper critical dimension for percolation as $d^* = 6$.

It is also possible to construct a field theory for the Potts model and hence for the bond percolation problem. The field theory has a Hamiltonian of the form

$$H/kT = \int d^d x [\frac{1}{2} r_o \sum_{r=1}^{q-1} z_r^*(x) z_r(x) - \frac{u}{3!} \sum_{r_1 r_2 r_3} z_{r_1} z_{r_2} z^*_{r_1+r_2} + O(z^4)]$$

where z_r is a complex field with q-1 components. An important difference with the field theory of the Ising or Heisenberg models is the appearance here of terms cubic in the fields and this implies that classical mean field behavior is only obtained in 6 or more dimensions. This has led to expansions in $\varepsilon = 6-d$ for the critical exponents (Harris et al. 1975) with results such as

$$\gamma_p = 1+ \frac{\varepsilon}{7} \quad , \quad \beta_p = 1- \frac{\varepsilon}{7} \qquad \gamma_p = \frac{1}{2} + \frac{5\varepsilon}{84} \quad . \tag{9.14}$$

10. Lattice animals

We have discussed the cluster distribution $n_s(p)$, the average number of clusters of s sites per site. The probability that a given site lies in a cluster of s sites is then $P_s = sn_s$ per site. We can look at this in a different way by focussing our attention on the given site and asking how many ways we can form a cluster of s sites with perimeter t including the given site. We denote this number by $s\,g_{st}$. The factor s arises because the given site could be any one of the s sites in the cluster. The perimeter t is the number of empty sites adjacent to the occupied sites in the site problem or unoccupied bonds adjacent to the cluster in the bond problem. Translations of a cluster are not counted as distinct clusters but rotations do give rise to distinct clusters. Thus

$$n_s = \sum_t g_{st} p^s (1-p)^t \tag{10.1}$$

The g_{st} have been called lattice animals by Domb and are more basic than the cluster distribution n_s. This form has been used by Kunz and Souillard (1978) to obtain some rigorous results for n_s for large s. We present here a simplified and non-rigorous discussion of their results. Assume that for sufficiently large s, at fixed p, the cluster numbers n_s decay like $\ln n_s \sim -s^{\zeta}$. The exponent $\zeta = \zeta(p)$ is not a critical exponent in the usual sense since it is defined for all p, not only close to p_c. For p near 1 only the most compact configurations with the smallest perimeter occur in (10.1). The minimum perimeter varies as $s^{1-1/d}$ for large s. Thus

$$\zeta(p\to 1) = 1-\frac{1}{d} \tag{10.2}$$

It is plausible that this result holds for all $p>p_c$. For $p\to 0$ all animals in (10.1) get an equal weight because $(1-p)^t$ can be approximated by 1 and n_s is proportional to the total number $\sum_t g_{st}$ animals. Numerical results from series suggest that this number increases as $s^{-\theta}\lambda^s$. Thus for $p\to 0$ $\ln n_s \sim s \ln p\lambda$ and

$$\zeta(p\to 0) = 1 \tag{10.3}$$

Using these results we can say something about the average number of perimeter sites t_s for a cluster of s sites. Thus differentiating (10.1) with respect to p we find

$$t_s = \frac{1-p}{p}\, s - \frac{1-p}{n_s}\left(\frac{\partial n_s}{\partial p}\right) \tag{10.4}$$

For $p>p_c$ from (10.2) the second term on the right is s^{ζ} as $s\to\infty$ and thus

$$\lim_{s\to\infty} \frac{t_s}{s} = \frac{1-p}{p} \qquad p>p_c \tag{10.5}$$

For $p<p_c$ t_s is again proportional to s for large s but the constant of proportionality is different because $\zeta=1$. This proportionality of the perimeter to the system size has led Domb (1976) to describe the clusters as ramified. Holes or missing bonds in the interior of the material and dangling ends are responsible for this proportionality to the system size.

Not much progress has been made in determining the form of g_{st}. It is simpler to follow the reverse path and infer the scaling form of g_{st} from that of n_s (see Essam 1980).

11. Fractals

The word fractal was introduced by Mandelbrot (1977) to describe objects with fractal dimension d' smaller than the Euclidean dimensionality d of the underlying lattice or space. Roughly speaking if the mass or size s of a system varies as $(\text{length})^{d'}$ then d' is called the fractal dimensionality and can be different from d.

Studies of computer generalized clusters show two important features which suggest that ideas of fractals may be appropriate to describe percolation clusters. These features are

(1) The texture appears to be self similar: the coarse features are roughly similar to the features of small portions of the same figure which have been enlarged. This self similarity suggest that a single parameter may be used to characterize the texture.

(2) Studies of clusters in which L/ξ is kept constant by increasing sample size L as $p\to p_c$ suggests that the self similarity is independent of $p-p_c$. The coarse features (scale ξ) remain similar while increasing L and ξ simply adds finer details to the texture.

A number of different definitions have been proposed for the fractal dimensionality of a percolating cluster. For example (Reatto and Rastelli (1972)) the average number of sites in a cluster $\bar{s}$ varies as $(p-p_c)^{-\beta\delta}$. Since we may regard $\bar{s}$ as proportional to the mass of a cluster we equate $\bar{s} \sim \xi^{d'}$. This gives a fractal dimensionality

$$d' = \frac{\beta\delta}{\nu} = \frac{d}{1+\delta^{-1}} = d - \frac{\beta}{\nu} \qquad (11.1)$$

In conclusion the fractal dimension of droplet-ramified clusters are useful catch words (Stauffer 1979).

II. RANDOM MAGNETS

12. Harris criterion and nodes and links picture

I now want to discuss the behavior of random magnetic alloys, i.e. alloys of a mixture of magnetic and non-magnetic atoms, e.g. $Rb_2Mn_pMg_{1-p}F_4$. In dilute magnets we have the problem of treating simultaneously the effects of thermal fluctuations and quenched in disorder. We first consider the question of whether a random magnetic alloy will show a single sharp T_c or some sort of distribution of local critical temperatures and hence a rounded transition. A second question is if it shows a sharp T_c is the critical behavior the same as that for the pure system.

Harris (1974) introduced an argument to show that if the specific heat exponent $\alpha<0$ then a small amount of randomness will not affect the critical behavior while if $\alpha>0$ then some new type of behavior will occur.

The argument is as follows: divide the magnetic system into cells of dimension L. We choose $L\sim\xi$ so that the cells are sufficiently large to have a well defined average concentration of impurities but are weakly correlated. Let T_c be the average transition temperature and let us estimate the fluctuations in T_c from cell to cell. We expect from the central limit theorem that

$$\Delta T_c \sim (\text{no. of impurities in cell})^{-1/2} \sim L^{-d/2} \tag{12.1}$$

Thus

$$\Delta T_c \sim \xi^{-d/2} \sim |T_c - T|^{d\nu/2} \tag{12.2}$$

As long as $\Delta T_c < |T_c - T|$ the fluctuations in T_c will not round the transition and this requires $d\nu/2 > 1$ or using the scaling relation $d\nu = 2-\alpha$ we find

$$\alpha<0 \tag{12.3}$$

as the condition for a sharp T_c. Only the 3d Ising model is thought to have $\alpha>0$. Luther and Grinstein (1976) and Lubensky (1975) have studied magnetic systems near d=4 with small fluctuations in the exchange parameters using the renormalization group and concluded that when the Harris criterion is satisfied the fluctuations due to a small amount of randomness are irrelevant. For $\alpha>0$ they found a new fixed point with different critical exponents.

The magnetic alloy is like a site percolation problem and let p be the probability that a given site is occupied with a magnetic atom and 1-p that it is occupied with a non-magnetic atom. In a quenched system the transition temperature $T_c(p)$ decreases as p is decreased and eventually reaches zero at the critical value for percolation p_c. (See Fig. 12.1.) For $p<p_c$ there is no infinite cluster and no spontaneous magnetization is possible. The behavior of $T_c(p)$ at the two end points has been calculated for the Ising model by Domany (1975). Close to p=1

$$\frac{\partial T_c(p)}{\partial p} = 1.329\ T_c(1) \tag{12.4}$$

and close to p_c

$$e^{-2J/kT_c(p)} = (2\ln 2)(p-p_c) \tag{12.5}$$

The appropriate temperature variable in Ising systems near p_c and T=0°K is $e^{-2J/kT}$ where J is the exchange constant. This measures the probability of an excitation which costs an energy 2J from the ground state. For XY or Heisenberg systems with no gap in the excitation spectrum the appropriate temperature variable is T/J and we expect near the percolation point that $T_c(p) \sim J(p-p_c)^{\zeta}$, i.e. a power law behavior.

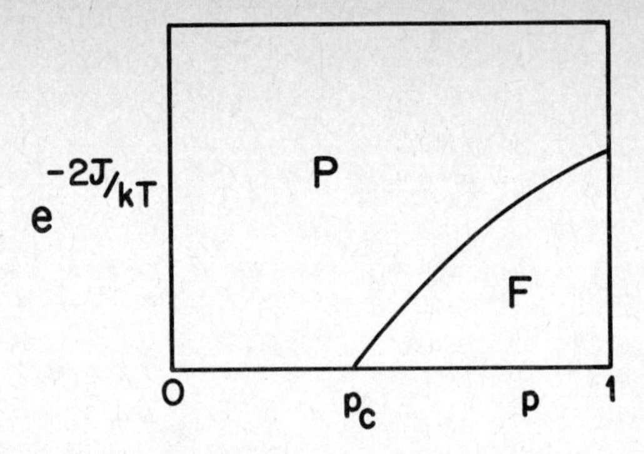

Fig. 12.1. Phase diagram for a random Ising magnet indicating the ferromagnetic (F) and paramagnetic (P) regions and the percolation point p_c.

Stauffer (1975) has argued that the point $p=p_c$, T=0 is a type of multicritical point and that the scaling fields for the Ising model are $\mu_2=p-p_c$ and $\mu_1=e^{-2J/kT}$. He proposed that the free energy should be a function of the scaled variable $\bar{\mu}_2^{\,\zeta}/\bar{\mu}_1$ where $\bar{\mu}_1$ and $\bar{\mu}_2$ are linear combinations of μ_1 and μ_2 and ζ is a crossover exponent.

A very useful picture of a random diluted lattice for discussing the magnetic phase transition has been proposed by de Gennes (1976) and by Skal and Shklovskii (1975). For p just above p_c the lattice can be viewed as a collection of nodes (compact clusters) which are connected by links which can be thought of as random paths (see Fig. 12.2). Two important lengths enter this picture. The distance

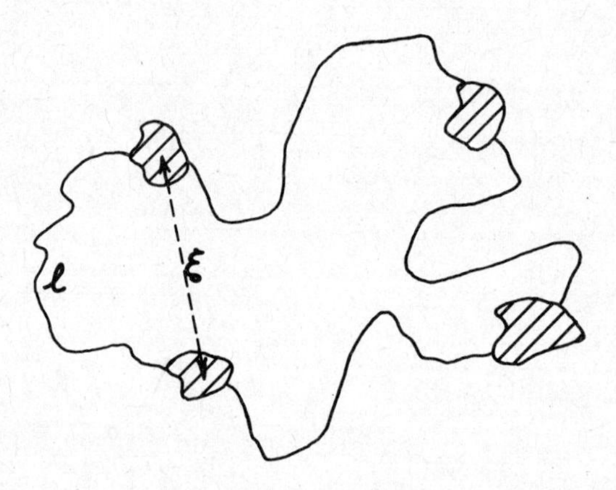

Fig. 12.2. Nodes and links picture of the infinite cluster.

between nodes is the percolation correlation length $\xi_p \sim (p-p_c)^{-\nu_p}$. The other length is the length ℓ of the random path between nodes. We introduce a new exponent ζ to describe the divergence of ℓ near p_c

$$\ell \sim |p-p_c|^{-\zeta} \tag{12.6}$$

We can place some reasonable bounds on ℓ. We would certainly expect ℓ to be greater than ξ_p. A reasonable guess for ℓ is that it is of order of the number of steps in a self avoiding walk between nodes (Lubensky (1977); Stanley et al (1976)). This gives

$$\xi_p \leqslant \ell \leqslant \xi_p^{1/\nu_s} \tag{12.7}$$

where ν_s is the correlation length exponent for the self avoiding walk. Eq. (12.7) yields the following relation for ζ

$$\nu_p \leqslant \zeta \leqslant \nu_p/\nu_s \tag{12.8}$$

Skal and Shklovskii (1975) have given a simple argument based on the nodes and links picture which gives $\zeta=1$. Suppose we remove a fraction $(1-\pi)$ of the bonds already present. The probability of breaking a link is $(1-\pi)\ell$ and at the critical concentration π_c $(1-\pi_c)\ell = 1$. At π_c we also have $p\pi_c = p_c$ and combining these results gives $\ell \sim (p-p_c)^{-1}$ and hence $\zeta=1$.

The argument that leads to scaling functions for the phase transition at $p=p_c$ $T=0$ is that on a length scale small compared to ℓ the lattice appears to be a collection of non-interacting contorted one-dimensional chains. On a scale large compared with ℓ the true d-dimensional nature of the lattice appears. We then compare the correlation length $\xi_1(T)$ of a one-dimensional system to ℓ. $\xi_1(T)$ can be calculated exactly and at low temperatures

$$\begin{aligned} \xi_1(T) &= e^{2J/T} && \text{Ising} \\ &= J/T && \text{XY, Heisenberg} \end{aligned} \tag{12.9}$$

If $\xi_1 \gg \ell$ the spins are ordered and we expect the magnetic properties to be determined by the percolation theory. The magnetization M is given by the percolation probability and the magnetic susceptibility is related to the mean cluster size:

$$M \sim P(p) \sim |p-p_c|^{\beta_p} \tag{12.10}$$

$$\chi T \sim S(p) \sim |p-p_c|^{-\gamma_p} \tag{12.11}$$

To obtain the behavior in other regimes we assume that the free energy and other thermodynamic functions can be expressed in scaling form

$$\frac{1}{T} G_{sing}(p-p_c,T,H) = |p-p_c|^{2-\alpha_p} F\left(\frac{H}{T|p-p_c|^{\Delta_p}}, \frac{\ell}{\xi_1(T)}\right) \tag{12.12}$$

where H is the external magnetic field (note that H/T is the appropriate variable near T=0) and α_p and Δ_p are percolation exponents. Other thermodynamic quantities scale in a similar manner, e.g. the correlation length

$$\xi(p-p_c,T) = |p-p_c|^{-\nu_p} g(\ell/\xi_1(T)) \tag{12.13}$$

Thus if $\ell >> \xi_1$ it follows from (12.12) and (12.13) that

$$\begin{aligned} M &\sim (\xi_1)^{-\beta_p/\zeta} \\ \chi T &\sim (\xi_1)^{\gamma_p/\zeta} \qquad\qquad \xi_1 << \ell \\ \xi &\sim (\xi_1)^{\nu_p/\zeta} \end{aligned} \tag{12.14}$$

and that the transition temperature satisfies

$$\xi_1(T_c(p)) = \ell \sim |p-p_c|^{-\zeta} \tag{12.15}$$

or

$$\begin{aligned} T_c(p) &\sim \frac{2J}{\zeta \ln(p-p_c)} \qquad \text{Ising} \\ &\sim J(p-p_c)^{\zeta} \qquad \text{Heisenberg} \end{aligned} \tag{12.16}$$

in agreement with (12.5).

The behavior of random magnets near the percolation threshold can be put in terms of the Potts model and the crossover exponent ζ has been calculated in powers of $\varepsilon = 6-d$ with the result $\zeta=1$ to all orders in ε for Ising systems (Stephen and Grest (1977), Wallace and Young (1978)). This is in good agreement with experiments on $Rb_2Mn_pMg_{1-p}F_4$ a random 2-d Ising system (Birgeneau et al (1976)).

The nodes and links picture works well in higher dimensions $d \geqslant 3$. It is obviously wrong in 2-d where $\nu_p = 1.365$ which exceeds ζ and is inconsistent with the model.

For Heisenberg magnets the situation is not so clear. The recent experimental data of Birgeneau et al (1976,1980) give $\zeta \sim 1.48$. Coniglio (1981) has argued that in the nodes and links picture it is the resistance between nodes which determines the propagation of correlations and that $\zeta = \zeta_R^{-1}$ where ζ_R is a resistivity critical exponent.

III. RANDOM CONDUCTORS

13. Random Resistor Networks

The problem of determining the conductivity of a random mixture of insulating and metallic material is an important one. As discussed above where we consider this problem on a lattice we get Kirchoff's equations

$$\sum_j \sigma_{ij}(V_i - V_j) = 0 \qquad (13.1)$$

The conductors σ_{ij} connecting nearest neighbor sites are statistically independent and in this way we obtain a bond percolation problem in which the conductors σ_{ij} are the bonds. If we have a mixture of two materials a and b where a is a good conductor and b is a poor conductor then we assume that σ_{ij} can take on two values: $\sigma_{ij} = \sigma_a$ with probability p and $\sigma_{ij} = \sigma_b$ with probability 1-p and $\sigma_a >> \sigma_b$. Thus p is the concentration of the good conductor. There are then two interesting limiting cases (Straley 1977)

(a) $\sigma_a > 0$, $\sigma_b = 0$ i.e. a metal-insulator mixture. For $p < p_c$ the macroscopic conductivity $\sum$ (response to a uniform applied electric field) is zero and for $p \gtrsim p_c$ we define a conductivity exponent t by

$$\sum \sim \sigma_a (p - p_c)^t \qquad p > p_c \qquad (13.2)$$

(b) $\sigma_a = \infty$, $\sigma_b > 0$ i.e. a supercondutor-normal metal mixture. For $p < p_c$ only finite islands of superconductors exist and $\sum$ is finite. At p_c an infinite superconducting cluster forms and $\sum$ becomes infinite. In this case we can define an exponent s describing the divergence of $\sum$ by

$$\sum \sim \frac{\sigma_b}{(p_c - p)^s} \qquad p < p_c \qquad (13.3)$$

The relations (13.2) and (13.3) are sketched in Fig. 13.1.

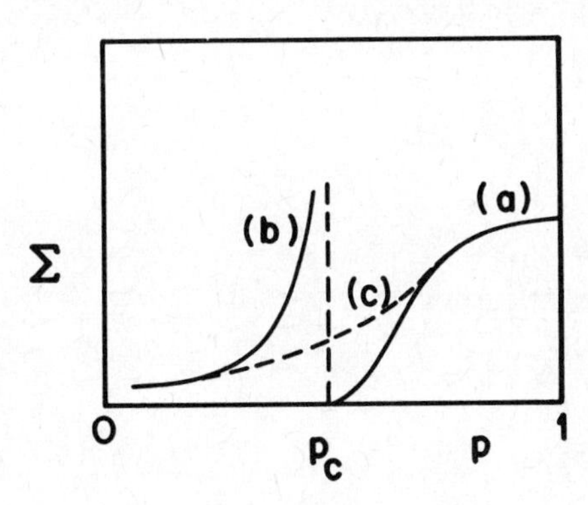

Fig. 13.1. Phase diagram of a random resistor network
(a) metal-insulator mixture $\sigma_a > 0$, $\sigma_b = 0$
(b) superconductor-normal metal mixture $\sigma_a = \infty$, $\sigma_b > 0$
(c) $\sigma_a \neq 0$, $\sigma_b \neq 0$ with $\sigma_a / \sigma_b >> 1$.

(c) In the more general case of a mixture of good and bad conductors the conductivity at p_c is finite and varies smoothly between the two limiting cases. In this case the conductivity depends on two variables $p-p_c$ and $\sigma_b/\sigma_a \ll 1$. We now make a scaling assumption (Straley 1977) about the dependence of the macroscopic conductivity on these two variables

$$\Sigma(\sigma_a,\sigma_b,p-p_c) = \sigma_a(p-p_c)^t f\left(\frac{\sigma_a(p-p_c)^{s+t}}{\sigma_b}\right) \tag{13.4}$$

This scaling form is assumed to hold for $(p-p_c) \ll 1$ and $\sigma_b/\sigma_a \ll 1$. The exponent of $p-p_c$ in the unknown scaling function f has been chosen so that the two limiting cases (a) and (b) above are correctly described. Thus if $\sigma_b=0$ and $f(\infty)>0$ we obtain case (a) and if $\sigma_a\to\infty$ $(p<p_c)$ we obtain case (b) provided $\lim_{x\to-\infty} f(x) \sim \frac{1}{x}$. In case (c), $p=p_c$, $p-p_c$ must cancel in (13.4) and this leads to

$$\Sigma(\sigma_a,\sigma_b,0) \sim \sigma_a^{1-u}\sigma_b^{u} \tag{13.5}$$

where $u = \frac{t}{s+t}$. In two dimensions duality arguments (Straley 1977) give s=t and $u = \frac{1}{2}$.

14. Effective Medium Theory

An effective medium theory for describing a random mixture of conductors was developed by Bruggeman (1935) and by Landauer (1978). The inhomogeneous medium is replaced by a uniform medium with conductors σ_m. The effective medium conductivity σ_m is chosen in such a way that the average effect of one of the real components embedded in the effective medium is zero. This theory is surprisingly accurate in 3-d except very close to p_c.

To see how it works consider the random resistor network with the good and bad conductors σ_a and σ_b with concentrations p and 1-p. In the medium suppose each bond of conductivity σ_m carries a current i_m. The voltage across it is given by

$$\sigma_m V_m = i_m \tag{14.1}$$

Now introduce a bond with conductivity σ_{AB} in the medium under conditions of constant current (see Fig. 14.1). The current through σ_{AB} is $i_m + i'$ say and the voltage across it is $V_m + V'$ so that

$$\sigma_{AB}(V_m + V') = i_m + i' \tag{14.2}$$

The excess current i' must flow from B to A through the effective medium by paths avoiding σ_{AB}. Thus (Fig. 14.1b) $G'_{AB}V' = -i'$ and from (14.2)

$$V' = \frac{\sigma_m-\sigma_{AB}}{\sigma_{AB}+G'_{AB}} \tag{14.3}$$

It is easy to show that $G'_{AB} = (\frac{z}{2}-1)\sigma_m$ where z is the lattice coordination number. We now require that $\langle V'\rangle = 0$ which gives the condition determining σ_m:

$$\left\langle \frac{\sigma_m-\sigma_{AB}}{\sigma_{AB}+(\frac{z}{2}-1)\sigma_m} \right\rangle = 0 \tag{14.4}$$

For the case of the mixture of good and bad conductors we get

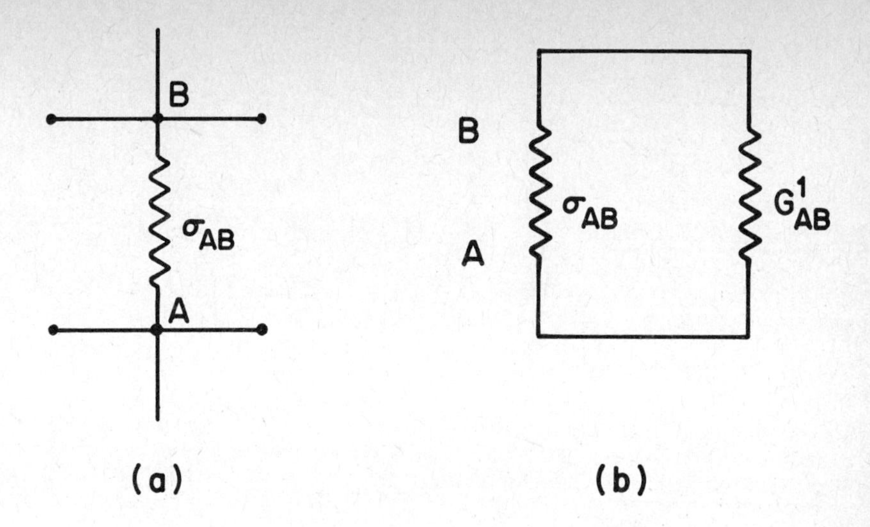

Fig. 14.1. (a) Impurity bond in the effective medium
(b) Equivalent circuit of (a)

$$p\,\frac{\sigma_m-\sigma_a}{\sigma_a+(\frac{z}{2}-1)\sigma_m} + (1-p)\,\frac{\sigma_m-\sigma_b}{\sigma_b+(\frac{z}{2}-1)\sigma_m} = 0 \tag{14.5}$$

In case (a) where $\sigma_b=0$ we find $\sigma_m = \sigma_a(pz-2)/(z-2)$. Thus $p_c = 2/z$ and the exponent $t=1$. In case (b) where $\sigma_a=\infty$ we find $\sigma_m = 2\sigma_b/(2-pz)$ and the exponent $s=1$. Exactly at p_c and if $\sigma_a \gg \sigma_b$ we get $\sigma_m \simeq (2\sigma_a\sigma_b/z-2)^{1/2}$ in agreement with the scaling prediction.

The effective medium theory can also be applied to the equivalent problem of determining the dielectric constant of an inhomogeneous medium (Landauer 1978). The result for the effective dielectric constant ε_m is

$$\frac{x_1(\varepsilon_1-\varepsilon_m)}{\varepsilon_1+2\varepsilon_m} + \frac{x_2(\varepsilon_2-\varepsilon_m)}{\varepsilon_2+2\varepsilon_m} = 0 \tag{14.6}$$

where x_1 and x_2 are the concentrations of the two components with dielectric constants ε_1 and ε_2. In deriving this result it is assumed that the average polarization of a spherical inclusion in the medium is zero.

The exponents t and s given by the effective medium theory are not correct. In d=2 t has been calculated using simulation methods (Kirkpatrick 1973), real space renormalization (Kirkpatrick 1977, Kogut and Straley 1978), and scaling arguments. These methods give values of t in the range 1 to 1.4. In d=3 t is about 1.7. It is expected that at d=6, the upper critical dimension for percolation, the conductivity exponents t and s should reach their classical values. These have been shown to be t=3, s=0 (log) from studies of the conductivity of a Bethe lattice (Stinchcombe 1974, Straley 1977b) and by a mean field theory type of calculation (Stephen 1978). The ε expansions for t and s in $\varepsilon = 6-d$ have also been determined with the results

$$t = 3-\frac{11}{42}\varepsilon \qquad s = \frac{11}{42}\varepsilon \quad . \tag{14.7}$$

The nodes and links picture of the infinite percolating cluster of de Gennes (1976) and Skal and Shklovskii (1975) is a useful one in which to discuss the conductivity of a random resistor network. We consider case (a) i.e. a mixture of conducting and insulating links. The current is carried by the infinite cluster. In this picture, after the dangling ends have been removed as they carry no current, the infinite cluster is composed of nodes connected by effectively one dimensional links (see Fig. 12.2). The distance between nodes is $\xi_p \sim (p-p_c)^{-\nu_p}$ the percolation correlation length. The resistance R between nodes is proportional to the length of the one-dimensional links ℓ and thus from (12.6) $R \sim \sigma_a^{-1}(p-p_c)^{-\zeta}$ where ζ is the same exponent as in the random Ising model. When an electric field E is applied to the network in Fig. 12.2 the voltage between links is $E\xi_p$ and the current in one link is $E\xi_p/R$. The number of links per unit area in d dimenstions is ξ_p^{1-d} and thus the current density J is

$$J \sim \frac{\xi_p^{2-d}}{R} E \tag{14.8}$$

and the macroscopic conductivity $\sum$ is

$$\sum \sim \frac{\xi_p^{2-d}}{R} \sim \sigma_a (p-p_c)^{(d-2)\nu_p + \zeta} \tag{14.9}$$

This model thus predicts

$$t = (d-2)\nu_p + \zeta \quad . \tag{14.10}$$

As discussed above ε expansion gives $\zeta=1$ and this is also obtained by real space renormalization group calculations in d=2. Thus in d=6 $\nu_p = 1/2$ and t=3 as expected. In d=2 $t=\zeta$ and this is in fair agreement with the values of t quoted above.

The nodes and links model can also be applied to discuss a mixture of superconducting and normal conductors. In this case the nodes are interpreted as superconducting islands which grow and eventually join up to form the infinite cluster as $p \to p_c$. The quantity R is interpreted as the resistance between nearest neighbor superconducting clusters and vanishes with an exponent $R \sim (p_c-p)^{\phi_1}/\sigma_b$. The conductivity exponent s is thus

$$s = \phi_1 - (d-2)\nu_p \tag{14.11}$$

In d=6 $\phi_1=2$ giving s=0 (log).

Scaling theories have also been applied to the frequency dependent conductivity and Hall effect (Skal and Shklovskii 1975).

The macroscopic conductivity is only defined for $p>p_c$. It is usually easier to calculate critical exponents in the disordered phase and it is therefore interesting to find some quantity defined for $p<p_c$ which can be related to the conductivity exponent t by scaling. One such quantity is the resistive susceptibility (Harris and Fisch 1977) defined by

$$\chi_R(\vec{r}-\vec{r}\,') = [R(r,r')C(r,r')]_{av} \tag{14.12}$$

where R(r,r') is the resistance between points r and r' in the same cluster and C(r,r') = 1 if r and r' are in the same cluster and is zero otherwise. In a nonrandom system χ_R is easily calculated

$$\chi_R(r-r') = \int d^d k \, \frac{1-e^{i\vec{k}\cdot(\vec{r}-\vec{r}\,')}}{\sum k^2} \sim (\sum a^{d-2})^{-1} \tag{14.13}$$

where the integral is over the first Brillouin zone and a is the lattice spacing. In a percolating system ($p>p_c$) the factor $C(r\,r')$ restricts r and r' to be in the infinite cluster and

$$\chi_R(r-r') \sim \frac{P^2(p)}{\Sigma\xi^{d-2}} \tag{14.14}$$

as $|r-r'|\to\infty$.

Below p_c we define the average resistance of a finite cluster

$$\chi_R = \sum_{r'} \chi_R(r-r') \sim (p_c-p)^{-\gamma_R} \tag{14.15}$$

which should diverge with an exponent (γ_R say). γ_R is related to the conductivity exponent t by the following argument: Assume $\chi_R(r-r')$ satisfies the usual homogeneity relation

$$\chi_R(r-r') \sim |r-r'|^{-d+\frac{\gamma_R}{\nu_p}} f\left(\frac{|r-r'|}{\xi_p}\right) \tag{14.16}$$

This form has been constructed to agree with (14.15). Comparing this scaling form with (14.14) gives

$$\gamma_R = \gamma_p + t-(d-2)\nu_p = \gamma_p + \zeta \tag{14.17}$$

χ_R can be calculated by series expansion or ε expansion (Harris and Fisch 1977).

We briefly discuss the relation between the conductivity Σ of a random resistor network and the spin wave stiffness in a random ferromagnet (Kirkpatrick (1973)). In a Heisenberg magnet with Hamiltonian $\mathcal{H} = -\sum_{(ij)} J_{ij}\vec{S}_i\cdot\vec{S}_j$ the linearized spin wave equations at $T = 0°K$ are

$$i\,\dot{S}_i^+ = 2S\sum_j J_{ij}(S_j^+-S_i^+) \tag{14.18}$$

where $S^+ = S_{xi} + iS_{yi}$ and S is the spin. The J_{ij} are random ferromagnetic coupling constants. It has been shown that for $p>p_c$ where an infinite cluster exists this equation has spin wave like excitations with a dispersion relation for small wave vectors k

$$\omega = 2JS\,D(p)k^2 \tag{14.19}$$

where D(p) is the spin wave stiffness. D(p) vanishes at p_c. Eqs. (14.18) are of exactly the same form as Kirchoff's equations for a random resistor network in the case where an a.c. source is applied at each site. Using a response function formation it is possible to express the conductivity in terms of the spectrum of low lying excitations which leads to the relation

$$\Sigma(p) = P(p)D(p) \tag{14.20}$$

where P is the percolation probability.

IV. SPIN GLASSES

15. Introduction

Spin glasses are random dilute magnetic alloys, e.g. CuMn or AuFe alloys are two well studied examples (Cannella and Mydosh (1972)) with magnetic ion concentrations of 10^{-3} to 10^{-1}. At higher concentrations conventional ferromagnetic or antiferromagnetic ordering is observed. These are metals and it is believed that the exchange coupling between the magnetic ions is mediated by the conduction electrons and is of the oscillating RKKY form (see Fig. 15.1). Insulating spin glasses are

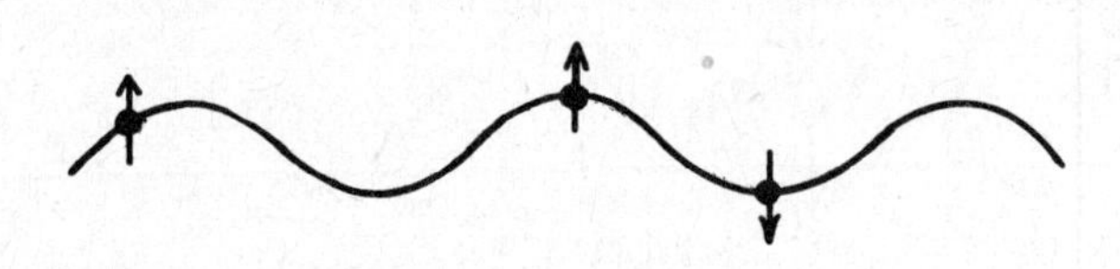

Fig. 15.1. Spins interacting via an RKKY interaction

also known, e.g. $Eu_xSr_{1-x}S$. In this case the n.n. exchange interaction $J_1>0$ is ferromagnetic while the second n.n. interaction J_2 is antiferromagnetic and $J_2 \simeq -\frac{1}{2}J_1$.

Experimentally spin glasses exhibit a cusp in the a.c. susceptibility at a temperature T_f called the freezing temperature. On the other hand the specific heat does not show any singularity at T_f but a rounded maximum. Below T_f the remanent magnetization decays slowly with time. The d.c. susceptibility below T_f depends on the conditions and is different if the sample is cooled in zero field and then a field is applied or if the sample is cooled in a small field ($\sim$5 gauss) and then the susceptibility is measured. A review of experimental results on spin glasses has been given by Ford (1982).

The questions that arise are (i) what is the nature of the low temperature state? (ii) is there a phase transition? (iii) is it a static phenomenon and what are the relevant time scales?

It is believed that spin glasses can be described by an exchange Hamiltonian

$$\mathcal{H} = \sum_{ij} J_{ij} \vec{S}_i \cdot \vec{S}_j \tag{15.1}$$

where the spins may have either Ising or Heisenberg symmetry. The exchange interactions J_{ij} are randomly positive and negative so that $[J_{ij}]_{av} = 0$ and $[J_{ij}^2]_{av} = J^2$ where the brackets $[\ldots]_{av}$ indicate an average over the impurity configurations. Such materials have been called spin glasses by Edwards and Anderson (EA) (1975). At temperatures $T<T_f$ they suggested that the spins become frozen so that for a given

impurity configuration $<S_i> \neq 0$ where the angular brackets indicate a thermal average. Because of the random signs of the exchange interaction the frozen ground states will not have all the spins parallel but the sign of $<S_i>$ will vary randomly from site to site with the result that in zero field there is no net magnetization

$$M = 1/N \sum_i <S_i> = [<S_i>]_{av} = 0 \qquad (15.2)$$

This led EA to introduce an order parameter for spin glasses

$$q = [<S_i>^2]_{av} \qquad (15.3)$$

which is the average of the square of the magnetization. The quantity q vanishes above T_f and is non-zero below. The meaning of the order parameter becomes clearer if we look at the spin-spin correlation function $<S_i S_j>$. These have the following properties:

(a) If we average over the random exchange interactions, spins on different sites will be uncorrelated, i.e. $[<S_i S_j>]_{av} = \delta_{ij}$ where we assume we are dealing with spins of unit length $S_i^2 = 1$. This result follows by making the transformation $S_i \to -S_i$ and $J_{ik} \to -J_{ik}$ for all sites k connected to i. This leaves the Hamiltonian (15.1) and the distribution of J_{ik} unchanged while the spin-spin correlation function changes sign showing that it must be zero.

(b) EA suggested that we should look at the spin-spin correlation function at a site as a function of time, i.e. $<S_i(o)S_i(t)>$. In the paramagnetic phase this will vanish for large t as the spin becomes uncorrelated with its initial orientation. Below the freezing temperature the spins become partly frozen and

$$\lim_{t\to\infty} <S_i(o)S_i(t)> = q \qquad (15.4)$$

is a measure of the order in the ground state. The ground state of the spins interacting via the random exchange interaction is not likely to be simple as in a ferromagnet or antiferromagnet. In fact we expect that there are many possible ground states of more or less the same energy i.e. there are many arrangements of the spins which lead to a low energy.

The order parameter q is related to the local susceptibility

$$\chi = \sum_j \chi_{ij} = \beta \sum_j [[<S_i S_j>]_{av} - [<S_i><S_j>]_{av}] \qquad (15.5)$$

The first term describing the long range spin correlations is important in understanding the susceptibility of ferromagnets or antiferromagnets. In the spin glass both terms vanish for $i \neq j$ and for $i=j$ the first term is 1 and the second term is the order parameter q (Fischer 1975). Thus

$$\chi = \beta(1-q) \qquad (15.6)$$

Other attempts to describe the hidden magnetic order in spin glasses include the concept of frustration (Toulouse 1977). Consider 4 spins at the corners of a square and suppose there is an odd number of antiferromagnet bonds. Then it is impossible to find a ground state in which all the exchange interactions are satisfied--there is always one spin which does not know which way to point i.e. it is frustrated. It is this frustration which gives rise to the expected large degeneracy of the ground state. A well known example of a fully frustrated spin system is the antiferromagnetic Ising model on a triangular lattice. This model has been solved exactly and does not exhibit a phase transition at finite temperature. The zero point entropy, which is a measure of the ground state degeneracy, is of order N.

16. The Replica Trick and Mean Field Theory

The partition function of the spin glass is

$$Z(\{J_{ij}\}) = \mathrm{Tr}\, e^{\beta \sum_{(ij)} J_{ij} S_i S_j} \tag{16.1}$$

for a given configuration of the exchange interactions. The average free energy is then

$$F = -kJ[\ln Z(\{J_{ij}\})]_{av} \tag{16.2}$$

This is called a quenched average and is the appropriate way to calculate the free energy of a disordered system with a fixed distribution of impurities. The other possibility is an annealed average in which Z is averaged. This would correspond to a physical situation in which the impurities are mobile and during a measurement take up all possible configurations.

The calculation outlined in (16.1) and (16.2) is difficult and for this reason the popular replica trick is often used. It is based on the formula

$$\lim_{n\to 0} \left[\frac{Z^n - 1}{n}\right]_{av} = [\ln Z]_{av} \tag{16.3}$$

Thus instead of considering Z we consider n replicas of the same system and calculate $[Z^n]_{av}$. In practice n is taken to be an integer and at the end of calculation n is set equal zero. There is a problem of analytic continuation from the integers n to zero which is discussed further below. The advantage of the method is that the impurity average is an annealed average and is easily accomplished. This is offset by having to deal with n replicas and the averaging introduces interactions between the replicas.

The n replicas are introduced by attaching an index α to each spin which runs from 1 to n. Thus

$$Z^n = \mathrm{Tr}\, e^{\beta \sum_{(ij)} J_{ij} \sum_{\alpha=1}^{n} S_i^\alpha S_j^\alpha + \beta H \sum_{i\alpha} S_i^\alpha} \tag{16.4}$$

where we have included a magnetic field H. The exchange interactions are assumed to have a Gaussian distribution

$$P(J_{ij}) = \left(\frac{1}{2\pi \bar{J}}\right)^{1/2} e^{-(J_{ij}-J_o)^2/2\bar{J}^2} \tag{16.5}$$

After averaging independently over all the J_{ij} we find (omitting constants)

$$\begin{aligned} Z(n) &= [Z^n]_{av} \\ &= \mathrm{Tr}\, e^{\beta J_o \sum_{(ij)} \sum_{\alpha} S_i^\alpha S_j^\alpha + \frac{1}{2}\beta^2 \bar{J}^2 \sum_{(ij)} \sum_{\alpha\neq\beta} S_i^\alpha S_i^\beta S_j^\alpha S_j^\beta} \end{aligned} \tag{16.6}$$

The averaging procedure thus mixes the replicas. In the mean field theory we focus attention on a single site and replace the effect of the neighboring spins by an average. This leads to introduce two order parameters

$$m_\alpha = \langle S^\alpha \rangle \quad , \quad q_{\alpha\beta} = \langle S^\alpha S^\beta \rangle \tag{16.7}$$

and a mean field Hamiltonian

$$-\beta \mathcal{H}_{mF} = K_1 \sum_{\alpha} m_\alpha S^\alpha + K_2 \sum_{\alpha\neq\beta} q_{\alpha\beta} S^\alpha S^\beta \tag{16.8}$$

where $K_1 = \beta J_o z$ and $K_2 = \frac{1}{2}\beta^2 \bar{J}^2 z$ and z is the coordination number. The self consistency conditions determining the order parameters (16.7) are

$$m_\alpha = \frac{\mathrm{Tr}\, S^\alpha e^{-\beta\mathcal{H}_{mF}}}{\mathrm{Tr}\, e^{-\beta\mathcal{H}_{mF}}} \tag{16.9}$$

$$q_{\alpha\beta} = \frac{\mathrm{Tr}\, S^\alpha S^\beta e^{-\beta\mathcal{H}_{mF}}}{\mathrm{Tr}\, e^{-\beta\mathcal{H}_{mF}}} \tag{16.10}$$

In the EA theory replica symmetry is assumed, i.e. $m_\alpha = m$, $q_{\alpha\beta} = q$. Close to the transition the right hand sides of (16.9) and (16.10) are expanded in m and q and after setting n=0 we obtain the mean field equations

$$\begin{aligned} m &= K_1 m - 2K_1K_2 qm \\ q &= K_2 q + K_1^2 - 8K_2^2 q^2 \end{aligned} \tag{16.11}$$

These equations have three types of solutions

(i) $m=q=0$ the paramagnetic phase

(ii) $m=0$, $q\neq 0$ the spin glass phase. From (16.11) $q = (K_2-1)/8K_2^2$ and thus the transition to the spin glass phase occurs at $kT_f = \sqrt{z/2}\, J$.

(iii) $m\neq 0$, $q\neq 0$ the ferromagnetic phase. From (16.11)

$$q = \frac{K_1-1}{2K_1K_2} \qquad m_o^2 = \frac{(1-K_2)(K_1-1)+4K_2(K_1-1)^2}{2K_1^4K_2} \tag{16.12}$$

and the transition occurs at $kT_c = zJ_o$. The transition between the ferromagnetic and spin glass phases occurs on the line where the spontaneous magnetization $m_o = 0$. All the transitions are second order. The phase diagram is sketched in Fig. 16.1.

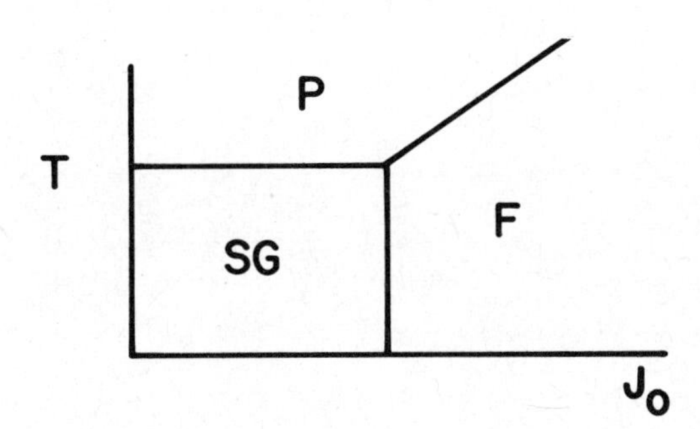

Fig. 16.1. Mean field phase diagram of a spin glass showing the spin glass (SG), ferromagnetic (F) and paramagnetic (P) phases.

17. Infinite Range Spin Glasses

In order to shed light on the EA mean field theory Sherrington and Kirpatrick (1975 SK) introduced the infinite range model. In this model the sum in (15.1) runs over all i and j so that each spin interacts with all the other N-1 spins. This procedure will make sense only if at the same time we weaken each individual interaction so that the sum of all exchange interactions with a given spin remains intensive on the average. Since this average might vanish fluctuations about the mean interaction strength must also be appropriately scaled. To do this we define

$$[J_{ij}]_{av} = J_o/N$$
$$[J_{ij}^2]_{av} = \tilde{J}^2/N \tag{17.1}$$

The infinite range pure ferromagnetic model ($\tilde{J}=0$) is soluble and its solution is identical with the Weiss molecular field theory. The infinite range spin glass has not been completely solved but is sufficiently interesting to discuss in some detail. We will consider the case of the Ising spin glass so that the spins in (15.1) have the values $S_i = \pm 1$. The distribution of J_{ij} is

$$P(J_{ij}) = \left(\frac{N}{2\pi\tilde{J}}\right)^{1/2} e^{-(J_{ij}-\frac{J_o}{N})^2 N/2\tilde{J}^2} \tag{17.2}$$

Proceeding exactly as in the previous section after averaging independently over all the J_{ij} we find (omitting constant terms of order N^o)

$$Z(n) = [Z^n]_{av}$$
$$= e^{\frac{\beta^2\tilde{J}^2 nN}{4}} \operatorname{Tr} e^{-\beta\mathcal{H}} \tag{17.3}$$

where

$$-\beta\mathcal{H} = \frac{\beta^2\tilde{J}^2}{4N}\sum_{\alpha\neq\beta}\left(\sum_i S_i^\alpha S_i^\beta\right)^2 + \frac{\beta J_o}{2N}\sum_\alpha\left(\sum_i S_i^\alpha\right)^2 + \beta H\sum_{i\alpha} S_i^\alpha \tag{17.4}$$

The averaging procedure mixes the replicas. The EA order parameter (15.3) is given by

$$q = \lim_{n\to 0} \langle S_i^\alpha S_i^\beta\rangle \tag{17.5}$$

where the angular brackets indicate a thermal average with respect to the replica Hamiltonian (17.4). We have also assumed a symmetry between the replicas.

For simplicity we will only consider the symmetrical case $J_o=0$, H=0. The general case is discussed in the papers of SK. The calculation of (17.3) is simplified by using the identity

$$e^{\lambda a^2} = \frac{1}{\sqrt{2\pi}}\int_{-\infty}^{\infty} dy\, e^{-\frac{1}{2}y^2 + (2\lambda)^{1/2}ay} \tag{17.6}$$

Then

$$\operatorname{Tr} e^{-\beta\mathcal{H}} = \left(\frac{N\beta\tilde{J}}{2\pi}\right)^{\frac{n(n-1)}{4}} \int_{-\infty}^{\infty}(dq)\, e^{-\frac{N}{2}\beta^2\tilde{J}^2\sum_{\alpha<\beta} q_{\alpha\beta}^2 + Nf(q_{\alpha\beta})} \tag{17.7}$$

where

$$e^{f(q_{\alpha\beta})} = \mathrm{Tr}\, e^{\beta^2\tilde{J}^2 \sum_{\alpha<\beta} q_{\alpha\beta} S^\alpha S^\beta} \tag{17.8}$$

The dependence on N has been explicitly displayed in (17.7) and suggests the use of the steepest descent method to evaluate the integrals. The integrand is stationary with respect to $q_{\alpha\beta}$ when

$$q_{\alpha\beta} = \frac{1}{\beta^2\tilde{J}^2}\frac{\partial f}{\partial q_{\alpha\beta}} = \frac{\mathrm{Tr}\, S^\alpha S^\beta e^{\beta^2\tilde{J}^2 \sum_{\alpha<\beta} q_{\alpha\beta} S^\alpha S^\beta}}{\mathrm{Tr}\, e^{\beta^2\tilde{J}^2 \sum_{\alpha<\beta} q_{\alpha\beta} S^\alpha S^\beta}} \tag{17.9}$$

which is the self consistency condition for $q_{\alpha\beta}$. The SK solution assumes that the saddle point in the $\frac{1}{2}n(n-1)$ dimensional space of the $q_{\alpha\beta}$ is the symmetrical one $q_{\alpha\beta} = q$. The following results are then easily obtained. Eq. (17.9) can be simplified by using (17.6) and the self consistent equation for q is (in the n=0 limit)

$$q = \frac{1}{\sqrt{2\pi}}\int_{-\infty}^{\infty} dz\, e^{-\frac{1}{2}z^2} \tanh^2\Omega \tag{17.10}$$

where $\Omega = \beta J_o m + \beta\tilde{J}q^{\frac{1}{2}}z + \beta H$ and we have included the magnetic field and asymmetry J_o. m is the magnetization and is given by

$$m = \frac{1}{\sqrt{2\pi}}\int dz\, e^{-\frac{1}{2}z^2} \tanh\Omega \tag{17.11}$$

The free energy per spin is given by (including the constant term in (17.3))

$$F = \frac{1}{2}J_o m^2 - \frac{1}{4}\beta^2\tilde{J}^2(1-q)^2 - \frac{1}{\sqrt{2\pi}\beta}\int dz\, e^{-\frac{1}{2}z^2} \ln(2\cosh\Omega) \tag{17.12}$$

From (17.10) it follows that the freezing temperature is determined by $\beta_f\tilde{J} = 1$ i.e. $kT_f = J$. Close to T_f $q \sim (T_f - T)$ and vanishes linearly at T_f. At low temperatures

$$q = 1 - \left(\frac{2}{\pi}\right)^{\frac{1}{2}} \frac{kT}{\tilde{J}} - \frac{1}{\pi}\left(\frac{kT}{\tilde{J}}\right)^2 \ldots \tag{17.13}$$

and it can be shown that the zero point entropy per spin is $S_o = -k/2\pi$, an unacceptable result.

It was pointed out by Almeida and Thouless (1978 AT) that the assumed saddle point $q_{\alpha\beta} = q$ in the SK solution is actually unstable. This was shown by explicitly considering fluctuations around the point $q_{\alpha\beta} = q$.

The SK solution is stable in a sufficiently large magnetic field. The AT results shows that the assumed symmetry amongst the replicas in the spin glass is incorrect below T_f. A wide variety of replica symmetry breaking schemes have been proposed and some will be discussed below. It should be noted that the Fischer relation (15.6) is satisfied by the SK solution but if the replica symmetry is broken this relation will no longer be valid.

18. TAP Mean Field Equations

Thouless, Anderson and Palmer (1977) (TAP) have attempted to improve on the mean field equations by using the Bethe-Peierls-Weiss (BPW) approximation which we briefly describe. The BPW approximation consider a spin and its nearest neighbors and replaces the effects of the rest of the spins by a field h_i acting on the neighbors. This approximation takes into account some of the effects of short range correlations. The Hamiltonian for the central spin S_o and its neighbors S_i is thus

$$\mathcal{H}_{BPW} = -\sum_i J_{oi} S_o S_i - \sum_i h_i S_i \tag{18.1}$$

Using the notation $g_{oi} = \tanh \beta J_{oi}$ and $t_i = \tanh\beta h_i$ it is not difficult to show that

$$\langle S_o \rangle = \tanh \left[\sum_i \tanh^{-1}(g_{oi} t_i)\right] \tag{18.2}$$

$$\langle S_i \rangle = \frac{1}{1-g_{oi}^2 t_i^2} \left[t_i(1-g_{oi}^2) + g_{oi}(1-t_i^2)\langle S_o \rangle\right] \tag{18.3}$$

where the angular brackets indicate a thermal average with respect to the BPW Hamiltonian (18.1). We define an effective field h_o acting on the central spin by

$$h_o = \beta^{-1} \sum_i \tanh^{-1}(g_{oi} t_i) \tag{18.4}$$

In the long range case where J_{oi} is small this reduces to

$$h_o = \sum_i J_{oi} \tanh\beta h_i \tag{18.5}$$

which is the equation of the Mean Random Field model.

A better approximation leads to the TAP equations. For convenience let $m_o = \langle S_o \rangle$ and $m_i = \langle S_i \rangle$. Again for small J_{oi} (the long range case) (18.2) is replaced by

$$m_o = \tanh\left(\sum_i \beta J_{oi} t_i\right) \tag{18.6}$$

and (18.3) is expanded in powers of J_{oi} giving $m_i = t_i + \beta J_{oi}(1-t_i^2)m_o + O(J^2)$. This can be solved for t_i giving

$$t_i = m_i - \beta J_{oi}(1-m_i^2)m_o + O(J^2) \tag{18.7}$$

Substituting in Eq. (18.6) gives the TAP equations for the random magnetizations

$$m_o = \tanh\left[\beta \sum_i J_{oi} m_i - \beta^2 \sum_i J_{oi}^2 (1-m_i^2) m_o\right] \tag{18.8}$$

The first term on the right is the usual Weiss molecular field term. The second is cavity field or Onsager term and represents the effect of the central spin on the neighbors. TAP have given analytic solutions of (18.8) near T_c and, with some further assumptions, numerical results near T=0 in good agreement with Monte Carlo results. These equations have been rederived by Sommers (1978) who showed that the TAP equations admit a solution involving the EA order parameter and an anomaly of the linear response (a violation of the Fischer relation). The Sommers solution is also not satisfactory. It is unstable like the SK solution but gives a good value for the ground state energy $E_o = -\frac{\sqrt{2}}{\pi}\tilde{J}$ (compared with the Monte Carlo result $E_o = -.765\tilde{J}$). The Sommers solution has a positive entropy but decreasing like $e^{-C/T}$ instead of the expected T^2 dependence.

As an example of replica symmetry breaking we consider a procedure introduced by Bray and Moore (1978) and de Dominicis and Garel (1979). It is equivalent to the Sommers solution and gives a positive entropy but is still unstable in the sense of Almeida and Thouless. The procedure is as follows:

(i) Introduce 2 replicas, $S_i^{(1)}$ and $S_i^{(2)}$, on each site with the following Hamiltonian

$$\mathcal{H} = - \sum_{(ij)} J_{ij}(S_i^{(1)}S_j^{(1)} + S_i^{(2)}S_j^{(2)}) \tag{18.9}$$

The EA order parameter is $q = [<S_i^{(1)}S_i^{(2)}>]_{av}$. The calculation of the free energy is performed using the replica trick. We replicate each of the initial 2 replicas m times giving $n = 2m$ replicas $S_i^{(1)\alpha}$, $S_i^{(2)\alpha}$ $\alpha = 1 \ldots m$. The EA order parameter is

$$q = \lim_{m\to 0} [<S_i^{(1)\alpha}S_i^{(2)\alpha}>]_{av} \tag{18.10}$$

We can also consider 2 other parameters

$$r_1 = [<S_i^{(1)\alpha}S_i^{(2)\beta}>]_{av} \qquad \alpha\neq\beta$$

$$r_2 = [<S_i^{(1)\alpha}S_i^{(1)\beta}>]_{av} = [<S_i^{(2)\alpha}S_i^{(2)\beta}>]_{av} \qquad \alpha\neq\beta \tag{18.11}$$

It can be shown that a stationary value of the free energy is consistent with $r_1 = r_2$.

(ii) We now extend this model by taking not 2 replicas but p replicas on each site $S_i^{(a)}$ $a=1 \ldots p$. We then replicate this system m times to get pm replicas $S_i^{(a)\alpha}$ $\alpha=1 \ldots m$. The EA order parameter is

$$q = [<S_i^{(a)\alpha}S_i^{(b)\alpha}>]_{av} \quad , \qquad a\neq b \tag{18.12}$$

The other parameter is defined by

$$r = <S_i^{(a)\alpha}S_i^{(b)\beta}> \qquad \alpha\neq\beta \tag{18.13}$$

This is a minimal procedure for breaking the symmetry between replicas.

The free energy can be calculated straightforwardly as in the SK solution in the limit $m\to 0$ for arbitrary p. It turns out that the zero point entropy and energy are given by

$$S_o(p) = S_{SK}/p \quad , \quad E_o(p) = E_{SK} \tag{18.14}$$

where SK stands for Sherrington-Kirkpatrick. Thus to get zero entropy requires $p=\infty$. It can be shown that the entropy decreases exponentially at low temperatures. The stability of this solution has been investigated by de Dominicis and Garel (1979) and it has been found to be unstable.

More elaborate replica symmetry breaking schemes have been introduced by Parisi (1979). It is difficult to motivate these schemes from a physical viewpoint although they appear to give a reasonable account of the thermodynamics of the infinite range spin glass model. For this reason we now turn to the dynamical theory.

19. Dynamical Theory of Spin Glasses

A dynamical theory of the infinite range spin glass has been introduced by Sompolinsky and Zippelius (1981) and by Sompolinsky (1981). This theory is interesting from a physical point of view. The static thermodynamic properties predicted by the theory are similar to those of Parisi (1979).

As discussed above the spin glass at low temperatures is characterized by the large number of states of low energy. Because of the frustration effects it is possible to construct many states of almost the same energy. These states can differ by reversing the sign of large blocks of spins. In a dynamical theory the existence of these states will give rise to very slow relaxation processes. Sompolinsky assumes that the time dependence of the order parameter is characterized not by a single (macroscopic) relaxation time but rather by a distribution of many large relaxation times, t_x, all of which become infinite in the thermodynamic limit. These relaxation times are parametrized in decreasing order by a parameter $0<x<1$. Thus t_o is longest time scale (the purely static limit) and t_1 is the shortest one. In general if $x>x^1$ $t_{x^1}/t_x \to \infty$ as $N\to\infty$. We then envisage a phase space with many ground states separated by macroscopic energy barriers b_i. These barriers give rise to relaxation times $t_i \sim e^{b_i/T}$ such that for $b_i<b_j$, $t_i/t_j<<1$. If the height of the barriers $b_i \sim A_i N^{\alpha}$ where the A_i vary continuously we will have $\lim_{N\to\infty} t_i = \infty$ and $\lim_{N\to\infty} t_i/t_j \to 0$ if $A_i<A_j$. With these properties the thermodynamic properties of the spin glass phase will be dependent on the distribution of relaxation times.

The EA order parameter was defined in (15.4) as the time persistent part of the average spin-spin correlations. We now define a generalized order parameter

$$q(x) = [<S_i(o)S_i(t_x)>]_{av} \tag{19.1}$$

which measures the amount of correlations which have not decayed at the time scale t_x. Thus $q(x)$ is expected to be a monotonic increasing function of x with a maximum value $q(1) = q_{EA}$ which is the frozen correlation measured in a finite time.

Another order parameter is the susceptibility measured at a frequency $\omega_x \sim t_x^{-1}$.

$$\chi(x) = \mathrm{Re}\chi(\omega_x) = \int_o^{t_x} \chi(t)dt \tag{19.2}$$

It is convenient to write this in the form

$$T\chi(x) = (1-q_{EA}) + \Delta(x) \tag{19.3}$$

The order parameter $\Delta(x)$ is the slow response due to overturning of large clusters and is a decreasing function of x with its maximum value $\Delta(o)$ corresponding to the purely static susceptibility $\chi(o)$.

The order parameter $q(x)$ and $\Delta(x)$ are assumed to be continuous as they are sums of a large number of contributions from a broad range of time scales ranging from the static to finite time limits. This implies that

$$q(x) \to 0 \quad \text{as} \quad x\to 0 \tag{19.4}$$

in zero field, showing the comple decay of spin-spin correlations at the longest time scale and

$$\Delta(x) \to 0 \quad \text{as} \quad x\to 1 \tag{19.5}$$

reflecting the validity of the Fischer relation and linear response theory at the short time scales. The Sompolinsky theory thus predicts that the time dependent susceptibility on a time scale large compared to microscopic processes but small compared to macroscopic ones, approaches a quasi-equilibrium value $T\chi(1) = 1-q_{EA}$. The susceptibility then very slowly relaxes to its true equilibrium value $T\chi(o)=1-q_{EA}+\Delta(o)$. The order parameter Δ provides a clearer definition of the spin glass transition than q as it measures the response of the system to a field. No estimates have yet been made of the time scales or the rate at which slow relaxation occurs. In principle it is possible to measure q_{EA} and $\Delta(o)$ as functions of T from measurements of the susceptibility. Low field measurements in spin glasses (Guy (1975) and Nagata, Keesom and Harrison (1979)) do exhibit a slow relaxation of the susceptibility from a non-equilibrium value χ_1 towards an equilibrium value χ_o. The above discussion has been for long ranged systems. In 3d short ranged systems relaxation times may be finite even in the thermodynamic limit.

V. LOCALIZATION

20. Electrons in a random potential

In this section we are going to discuss the properties of electrons in a random potential. In solid state physics we generally learn that a conduction electron in a metal sees a more or less periodic potential. The resulting electron states in the metal are modulated plane waves or Bloch states. We will refer to such states which extend throughout the metal as extended states. In order to understand the electrical conductivity of metals it is necessary to modify this picture to include the scattering of the electrons by impurities (elastic scattering) or by phonons (inelastic scattering). This scattering is usually treated as a perturbation on the motion of the electron by writing a Boltzmann equation with a collision term or if you are more sophisticated by the Kubo formula. The effects of scattering are often lumped into a collision time and this leads to the well known Drude formula for the conductivity $\sigma = ne^2\tau/m$. It is essential in order for a metal to have a non-zero conductivity as $T\to 0°K$, that the eigenstates of the electrons in the presence of scattering extend throughout the system. If they do not an electron initially in a particular region of space cannot diffuse out and there will be no conductivity.

In contrast to the above situation Anderson (1958) argued in his classic paper "Absence of Diffusion in Certain Random Lattices" that it was natural to expect that for sufficiently strong disorder the electron eigenstates would be localized in regions where the potential is favorable. Away from this region the wavefunction would drop off (perhaps exponentially).

It is difficult to vary the amount of disorder and it is easier to think of varying the Fermi energy. At low energies the states are localized while at high energies the states are extended. Separating these two regions Anderson suggested is an energy, E_c, the mobility edge where the transition from localized to extended electronic states occurs. If the Fermi energy $E_F<E_c$ at $T=0°K$ only localized states are occupied and the material is an insulator at $T=0°K$. If $E_F>E_c$ the extended states are occupied and the material is a metal. The nature of this transition is of considerable current interest.

This problem has some features not present in percolation models of conductivity. In percolation the motion is classical and a particle can only enter those regions which are classically accessible. In the quantum mechanical case this is no longer so because a particle may tunnel from one region to another. An interesting consequence is that it is not possible to have localized and extended states coexisting at the same energy. Suppose we had a localized state in some region of the material and in some other region there was a band of extended states, the energy of the localized state falling within the band. Some small matrix element would always exist between the localized state and the extended states and owing to the precise degeneracy would mix the localized states with the extended states.

Some experimental systems that are being investigated are
(i) Granular metals - A metal is evaporated on a substrate in an atmosphere of oxygen so that we end up with a mixture of metal and oxide. The concentration of metal can be controlled.
(ii) Mosfets - These are made by coating the surface of a semiconductor with oxide. A metallic electrode is then formed on the oxide and a voltage applied. This causes the conduction band to bend and electrons populate the surface states. The electrons can move parallel to the surface so the metal is effectively 2 dimensional. The disorder is provided by the rough surface and the concentration of electrons can be varied by varying the applied voltage.
(iii) Other systems studied are cermets (ceramic metal mixtures) and thin wires and thin films which are discussed below.

21. Minimum Metallic Conductivity

An interesting approach to this problem mainly developed by Mott and coworkers (1974) is the concept of minimum metallic conductivity or maximum metallic resistance. We have the picture of electrons propagating in a periodic lattice occasionally making collisions with impurities. Two important lengths enter this picture

(i) The de Broglie wavelength of the electrons

$$\lambda = \hbar/p_F = k_F^{-1} \tag{21.1}$$

where p_F and k_F are the Fermi momentum and wave vector respectively. The phase of the electron wavefunction changes by 2π in a distance $2\pi\lambda$.

(ii) The mean free path $\ell = v_F\tau$. The electron wavefunction loses phase coherence in the distance ℓ. It is argued that the above picture loses its meaning when the mean free path becomes comparable with the wavelength λ. This leads to the Yoffe-Regel criterion that when $\ell \sim \lambda$ i.e. $k_F\ell \sim 1$ a transition from metallic to insulating behavior will occur.

When this condition is combined with the relation between the conductivity and the mean free path $\sigma = ne^2\tau/m$ some very interesting results are obtained. It is interesting to consider separately the case of 2d and 3d metals.

(i) 2d - The electron density $n = k_F^2/2\pi$ and thus

$$\sigma = \frac{e^2}{2\pi\hbar}(k_F\ell) \tag{21.2}$$

According to the Yoffe-Regel criterion the minimum metallic conductivity is $\sigma_{min} = e^2/2\pi h$ independent of the material and depending only on fundamental constants. $e^2/2\pi\hbar \sim 4\ 10^{-5}$ ohm^{-1} which corresponds to a maximum metallic resistance of 25,000 ohms. Thin metallic films have been studied by Bishop et al (1981) who found that high temperature extrapolations for the high resistance curves all tended to converge to about 30,000 ohms.

(ii) 3d - The electron density $n = k_F^3/3\pi^2$ and

$$\sigma = \frac{e^2}{3\pi^2\hbar} k_F(k_F\ell) \tag{21.3}$$

which leads to a minimum metallic conductivity $\sigma_{min} \sim \frac{e^2}{3\pi^2\hbar} k_F$ which is inversely proportional to the de Broglie wavelength. This now depends on the material. If we take $k_F^{-1} \sim 10^{-8}$ cm, the interparticle spacing, the maximum metallic resistivity $\rho_{max} = \frac{3\pi^2\hbar}{e^2} a \sim 10^{-3}$ ohm cm. Mott suggested that when this value of the resistivity was reached the material would make a first order transition to the insulating state. It is now generally believed that the transition is continuous.

22. Models

It is useful to introduce a few models which we can use in the further discussion of localization.

(i) Anderson tight binding model. In this model we have regular lattice. The amplitude of the wavefunction on site i is a_i and satisfies the Schrödinger equation

$$Ea_i = \varepsilon_i a_i + \sum_j V_{ij} a_j \tag{22.1}$$

The ε_i are the energies of the state on sites and V_{ij} is the hopping matrix element usually taken to be non-zero only between nearest neighbor sites. The ε_i may be random (diagonal disorder) or the V_{ij} may be random (off diagonal disorder) or both may be random.

As an example let us assume that the ε_i are independently and uniformly distributed over the range -W/2 to W/2 and $V_{ij} = V$ for nearest neighbors only. The ratio W/V measures the degree of disorder. A rough estimate can be made of the rate at which the amplitude of a localized eigenstate falls off from its maximum value. Consider an electron which in the limit $V \to 0$ would be localized at a site whose energy ε_i is near the center of the band. Then the amplitude on a neighboring site j is, to lowest order of perturbation theory, $a_j \sim V/(\varepsilon_j - \varepsilon_i)$. Since the denominators can range between ±W/2 a typical value is W/4. Since there are z neighbors, where z is the coordination number, the rate of fall off of the amplitude is determined by the smallest denominator which is of order W/4z. The condition for the convergence of the perturbation theory is then

$$\frac{4zV}{W} < 1$$

The amplitude would be expected to fall off exponentially.

More generally we would expect a phase diagram of the form in Fig. 22.1.

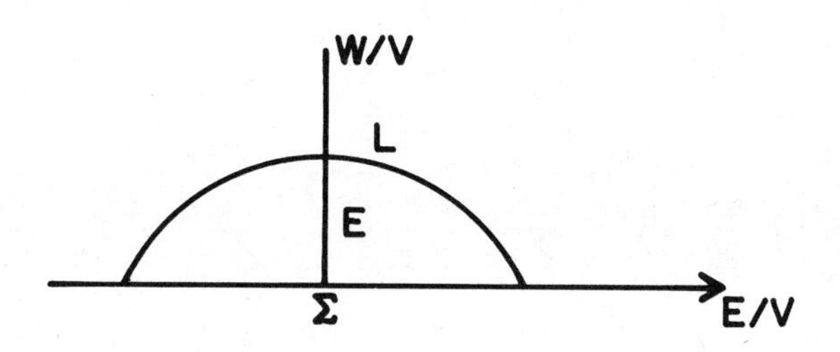

Fig. 22.1. Phase diagram for the Anderson model showing extended states E and localized states L

(ii) Schrödinger equation with random potential. The one electron Schrödinger equation is

$$[-\frac{\hbar^2}{2m}\nabla^2 + V(r)]\psi(r) = E\psi(r) \tag{22.2}$$

The potential is usually taken to be a sum of random placed potentials with possibly also random strengths

$$V(\vec{r}) = \sum_i V_i(\vec{r}-\vec{R}_i) \tag{22.3}$$

We hope that the exact nature of the model and the form of the disorder are unimportant, i.e. we hope that there is universality near the mobility edge in the same way that the critical behavior near 2nd order phase transitions is universal. For this reason we will generally use the Anderson model in our discussion.

What quantities should we look at to decide whether we have localized or extended states? This information should be contained in the Green's function

$$G_{ii}(E_\pm) = \sum_\alpha \frac{|a_i^{(\alpha)}|^2}{E_\pm - E_\alpha} \tag{22.4}$$

where $a_i^{(\alpha)}$ is the amplitude of the eigenstate with energy E_α at site i. $E_\pm = E \pm i\eta$ where η is a small real positive quantity and ± refers to the retarded or advanced

Green's function. G is the Green's function for a given configuration of the impurities and its analytic properties in E should be different for localized or extended states. For localized states G will have poles on the real axis (perhaps densely distributed) and the residue at a pole gives the squared amplitude of the state at site i. For extended states G will have a cut on the real axis. (An exception is the case of the Anderson model with infinite range interactions where in the absence of randomness G has poles at E=0, NV but the states are extended over all N sites).

If G is averaged over the distribution of random elements it is believed that the distinction between localized states and extended states is lost. Thus $\overline{G}_{ii}$ does not have any singularity at the mobility edge and analytically has a cut along the real E axis. The discontinuity in $\overline{G}$ across the real axis is the average density of states which is believed to be smooth at the mobility edge. This has not been proved rigorously. There is one model in which it can be worked out in detail (the Lloyd model) in which the distribution of site energies in the Anderson model is a Lorentzian

$$P(\varepsilon_i) = \frac{\gamma/\pi}{\varepsilon_i^2+\gamma^2} \tag{22.5}$$

The Green's function has its poles either in the upper half or lower half plane depending on whether we take $E_\pm$ in (22.4). The Green's function can then be averaged over ε_i by closing the ε_i contour either in the lower or upper half planes. This corresponds to replacing $\varepsilon_i \to \mp i\gamma$ wherever it occurs. From (22.1) the average Green's function satisfies

$$(E\pm i\gamma)\overline{G}_{ij} + \sum_k V_{ik}\overline{G}_{kj} = \delta_{ij} \tag{22.6}$$

which is easily solved by taking Fourier transforms. For a d dimensional hypercubic lattice

$$\overline{G}_{rs} = \frac{1}{N}\sum_{\vec{k}} \frac{e^{i\vec{k}\cdot(\vec{r}-\vec{s})}}{E\pm i\gamma + 2V(\cos k_x + \cos k_y \ldots)} \tag{22.7}$$

It is not difficult to show that $\overline{G}$ and the density of states have no singularities. The Lorentz distribution is of course very special but the same result is believed to be true for other distributions.

Recently Wegner (1981) has proved rigorously that the average density of states is positive under conditions that the distribution of site energies is finite. This is important because it rules out theories of the mobility edge in which there is a singularity in the density of states at the mobility edge, e.g. it was thought that the average density of states may vanish at the mobility edge.

On the other hand the d.c. conductivity is given by

$$\sigma = \frac{e^2}{6\pi m^2}\frac{1}{V}\sum_{kk'}(k\cdot k')\,\overline{G(kk'E+i\eta)G(k'kE-i\eta)} \tag{22.8}$$

The average now involves retarded and advanced Green's functions and the above method doesn't work.

23. One-dimensional Systems

The effects of disorder on the eigenstates depends in an important way on dimensionality. It is generally accepted that in 1d systems all states are localized if there is an arbitrarily small amount of disorder. A review of 1d systems has been given by Ishii (1973).

A rough argument for localization in 1d due to Borland (1963) goes as follows. We take the Anderson tight binding model and ask the following question: We begin at one end with amplitudes a_o and a_1 and ask how the mean square amplitudes behave a distance L along the chain. This is easy to work out in detail. We rewrite the Schrodinger equation in transfer matrix form (assuming V=1)

$$\begin{pmatrix} a_{i+1} \\ a_i \end{pmatrix} = \begin{pmatrix} E-\varepsilon_i & -1 \\ 1 & 0 \end{pmatrix} \begin{pmatrix} a_i \\ a_{i-1} \end{pmatrix} = T^{(i)} \begin{pmatrix} a_i \\ a_{i-1} \end{pmatrix} \tag{23.1}$$

We can iterate this relation beginning at one end:

$$\begin{pmatrix} a_{L+1} \\ a_L \end{pmatrix} = T^{(L)} \ldots T^{(1)} \begin{pmatrix} a_1 \\ a_o \end{pmatrix} = M_L \begin{pmatrix} a_1 \\ a_o \end{pmatrix} \tag{23.2}$$

On the right we have a product of random matrices and a theorem due to Furstenburg tells us that under almost all conditions the amplitudes increase exponentially, i.e. $|a_L|^2 \sim e^{L/L_o}$ where L_o is related to the localization length. The states we have constructed in this way are not eigenstates but Borland argued that an eigenstate could be constructed by matching the magnitudes and slopes of exponentially growing states from each end. This matching would only be possible for certain values of the energy. These arguments have been made more rigorous (see Ishii 1973). The initial value problem can also be solved in higher dimensions (Stephen 1981) but the matching problem is more complicated as the eigenstates are not ordered in increasing energy.

The calculation of the resistance of a 1d system is closely related to the above initial value problem. The resistance ρ_L of a segment of length L can be expressed in terms of the transmission T_L and reflection $R_L = 1-T_L$ coefficients (Landauer 1970)

$$\rho_L = \frac{2\pi\hbar}{e^2}\left(\frac{R_L}{T_L}\right) = \frac{2\pi\hbar}{e^2}\left(\frac{1}{T_L} - 1\right) \quad . \tag{23.3}$$

The inverse transmission coefficient can in turn be expressed in terms of the product of transfer matrices introduced in Eq. (23.2). At the band center (E=0)

$$\rho_L = \frac{\pi\hbar}{2e^2}\left(\mathrm{Tr}\, M_L^+ M_L - 2\right) \tag{23.4}$$

A slightly more complicated formula holds away from the band center. Calculating the average of (23.4) is simple because each scattering center may be averaged independently. The result is that the average resistance increases exponentially

$$\bar{\rho}_L \sim e^{L/L_o} \quad , \quad (L>L_o) \tag{23.5}$$

Thus "Ohms law" for thin wires should read

$$\rho_{L_1+L_2} = \rho_{L_1}\rho_{L_2} \tag{23.6}$$

We all know that Ohms law applies to wires. The reason that we do not generally see these effects is because measurements are made at finite temperatures. The above calculation only took elastic scattering into account and only applies at T=0°K. At finite temperatures inelastic scattering processes can occur and an electron in a localized state can be inelastically scattered into a nearby localized state and eventually make its way to the other end of the wire.

In order to observe the above effects of localization on resistance we must satisfy two conditions (i) $L>L_o$ i.e. the length of the wire must be greater than the localization length. (ii) $\ell_{inel}>L_o$ i.e. the inelastic scattering length must be greater than the localization length. As an example consider No.50 Cu wire which has a resistance of 34 ohms/meter and a diameter of 20 μ. The localization length is that length with a resistance of $2\pi\hbar/e^2$ = 25000 ohms i.e. $L_o \sim 25000/34 = 1000$ meters. This is out of the question and we need a material of high resistivity and small cross section. We expect $L_o \sim A\ell/a^2$ where A = cross sectional area, ℓ = mean free path and a is a microscopic length ($\sim$ the interatomic spacing). Recently such thin wires of W-Re alloys have been fabricated by Chaudhari and Habermaier (1980) and Giordano et al. (1979) with resistivities of 10^6 ohms/cm. This corresponds to a localization length $L_o \sim 10^{-2} - 10^{-3}$ cm.

The second requirement that the inelastic scattering length $\ell_{inel} > L_o$ is more difficult to satisfy. Suppose that the time between inelastic collisions is τ_i. This could be due to phonons or electron-electron interactions and is the average time in which the energy of an electron changes by more than the energy difference between localized states. The electron diffuses a distance $\ell_{inel} = \sqrt{D\tau_i}$ in the time τ_i. The diffusion constant is related to the conductivity by the Einstein relation $\sigma = \frac{e^2}{2}\frac{dn}{dE} D$. Experimentally it is found that $\tau_i \sim \hbar/kT$. It is believed that this comes from scattering from localized two level impurity states. Thus for the W-Re wires $\ell_{inel} \sim 5\,10^{-6}/\sqrt{T}$ cm and $\ell_{inel} \sim L_o$ only at very low temperatures $T \sim 10^{-5}$°K. It is difficult to attain and make measurements at these temperatures because of heating effects.

We can estimate the effects of localization at higher temperatures using the following simple model, Thouless (1980). Divide the wire with sections of length ℓ_{inel}.

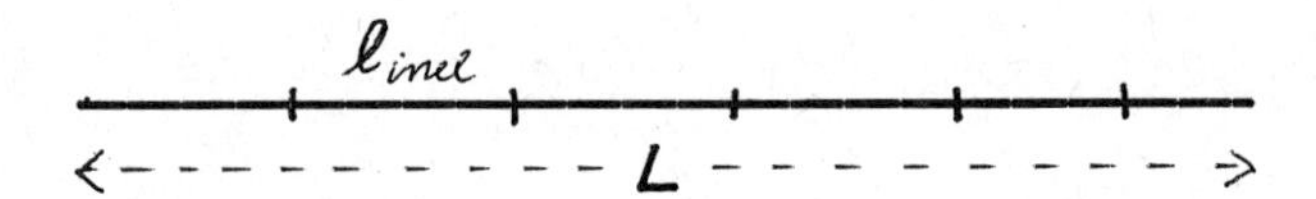

Each section has a resistance proportional to $e^{\ell_{inel}/L_o} - 1$ and the total resistance is

$$\rho_L \sim \frac{L}{\ell_{inel}} \left(e^{\ell_{inel}/L_o} - 1\right) \tag{23.7}$$

For $\ell_{inel} < L_o$ we expand the exponential and the first temperature correction to the resistance is

$$\frac{\Delta\rho}{\rho} = \frac{\ell_{inel}}{2L_o} \sim T^{-\frac{1}{2}} \tag{23.8}$$

Thus the resistance should increase at low temperatures. A small increase in ρ (of order 1-2%) has been observed by Chaudhari and Habermaier (1980) and Giordano et al. (1979) with the temperature dependence of Eq. (23.8) (this was how it was obtained). In addition L_o is predicted to be inversely proportional to the cross sectional area of the wire and this is also observed.

A final comment on the resistivity at T=0°K where it increases exponentially with length is that it is statistically not a well behaved quantity. This is almost obvious from the fact that it is a product of independent random quantities (the central limit theorem applies to a sum of random quantities). It can also be shown explicitly that the fluctuations in the resistance grow exponentially more rapidly than the average resistance (Abrahams and Stephen 1980). It has been shown Anderson et al. (1980) that $\log(1+\rho_L)$ is a statistically well behaved quantity obeying a central limit theorem. Further it is additive for large L

$$\log \rho_{L_1+L_2} = \log \rho_{L_1} + \log \rho_{L_2} \quad . \tag{23.9}$$

24. Scaling Theory

A successful scaling theory of the conductance has been developed by Abrahams, Anderson, Licciardello and Ramakrishnan (1979) and in this section we will give an outline of their work. We will use the Anderson model as an illustration. In this model, as discussed above, there are two important parameters: W which measures the degree of disorder and V which measures the coupling between nearest neighbor sites. For a given Fermi energy E_F for large W/V we expect localized states and small W/V extended states and somewhere in between is the mobility edge. The localized states are characterized by an exponential envelope $\sim e^{-r/\xi}$ where ξ is the localization length. At the mobility edge $\xi \to \infty$ and the states become extended. The extended states are characterized by a non-zero conductivity or diffusion constant. The distinction between localized and extended states is only clear on a scale $L > \xi$ i.e. it would be necessary to make measurements on a sample of size $L > \xi$ in order to determine whether we have localized or extended states. Thus the size of the sample is important.

We now consider what happens to the parameters W and V as we look at the system on larger and larger scales, Thouless (1978). Consider a block of material of side L and volume L^d in d dimensions. The average spacing between the energy levels determines the scaled parameter W' and is

$$W' = \left(\frac{dE}{dn}\right) \frac{1}{L^d} \tag{24.1}$$

where $\left(\frac{dn}{dE}\right)$ is the density of states per unit volume.

The new coupling V' of the states in the block to states in a neighboring block is determined by the hopping rate from one block to the next. It can be written, using Fermi's golden rule

$$V' = \frac{\hbar}{\tau} = 2\pi |M|^2 \frac{dn}{dE} \tag{24.2}$$

where M is the matrix element between states on the left and states on the right, $M = \langle \psi_L | T | \psi_R \rangle$ and T is the coupling between the blocks. This is a reasonable model for a granular metal.

We can now relate the ratio V'/W' to the conductance. Apply a potential ϕ to the left hand block. The excess number of electrons on the left is $\left(\frac{dn}{dE}\right) L^d e\phi$ and the current is

$$I = e^2 \phi \left(\frac{dn}{dE}\right) L^d \frac{1}{\tau} \tag{24.3}$$

where $\frac{1}{\tau}$ is the hopping rate. The conductance on the scale L

$$\begin{aligned} G(L) &= I/\phi = \left(\frac{e^2}{\hbar}\right)\left(\frac{\hbar}{\tau}\right)\left(\frac{dn}{dE} L^d\right) \\ &= \left(\frac{e^2}{\hbar}\right)(V'/W') \end{aligned} \tag{24.4}$$

Thus the dimensionless conductance $g(L) = \frac{\hbar}{e^2} G(L)$ determines the parameter of interest, V'/W'. These kind of arguments led to the proposal by Abrahams et al. that the only important parameter in the theory is the dimensionless conductance g(L). This parameter determines whether we have localized or extended states. For localized states the matrix element M in Eq. (24.2), which depends on the overlap of states on the left and right, will be exponentially small, i.e.

$$g(L) \sim e^{-L/\xi} \qquad \text{(localized states)} \tag{24.5}$$

For extended states $M \sim L^{d-1}/L^d = L^{-1}$, i.e. the area divided by the volume and $\frac{h}{\tau} \sim L^{-2}$. This gives

$$g(L) \sim L^{d-2} \qquad \text{(extended states)} \tag{24.6}$$

This is the result which we expect because the conductance of a metal is $g(L) \sim \frac{\hbar}{e^2}\sigma L^{d-2}$.

Now consider the conductance on the scale bL i.e. we join b^d blocks of size L together. The assumption is made that the new conductance only depends on the old conductance and b (if $L>\xi$)

$$g(bL) = f_b(g(L)) \tag{24.7}$$

By letting $b = 1+\varepsilon$ we can write this in differential form

$$\frac{d\ln g}{d\ln L} = \beta(g) \tag{24.8}$$

We have only considered systems of finite size L so we expect the beta function $\beta(g)$ is a smooth function with no singularities. β describes how g changes with scale.

The asymptotic forms of $\beta(g)$ follow from (24.5) and (24.6). Thus in the localized region from $\beta(g) \sim \ln g$ and in the extended region $\beta(g) = d-2$. If we smoothly extrapolate between these limits we obtain Fig. 24-1.

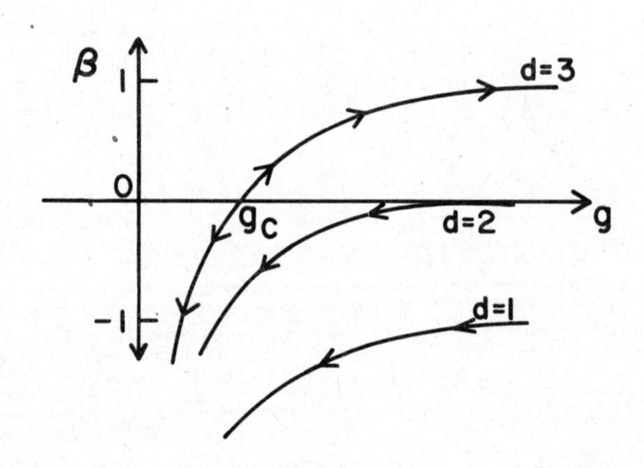

Fig. 24.1. Sketch of the expected form of $\beta = d\ln g/d\ln L$

We now consider some consequences of this picture in different dimensions.
(1) d=3. Suppose we begin at some length scale L_o with conductance g_o. If $g_o>g_c$ as we increase the size we eventually find $g\sim L$ i.e. metallic behavior. If $g_o<g_c$ as we increase the size we eventually find $g\sim e^{-L/\xi}$ i.e. insulating behavior. The mobility edge is g_c where $\beta(g_c) = 0$. The correlation length exponent ν is related to the slope of the beta function at g_c. Thus if we linearize the relation (24.8) about g_c we get

$$\frac{d\ln g}{d\ln L} \simeq (g-g_c)\beta'(g_c) \tag{24.9}$$

This is then integrated from initial values g_o, L_o to g, L giving

$$g(L) = g_c[1+(\frac{L}{L_o})^{\beta' g_c} \frac{g_o-g_c}{g_c} \ldots] \tag{24.10}$$

A scaling hypothesis now enables us to relate the correlation length exponent to the conductivity exponent. The scaling hypothesis is that

$$g(L) = f(L/\xi) \tag{24.11}$$

where ξ is the correlation length and diverges as $\xi \sim (\frac{E_F-E_c}{E_c})^{-\nu}$. Matching (24.10) and (24.11) (with $g_o-g_c \sim E_F-E_c$) we find $\nu = \frac{1}{\beta' g_c}$. Above the mobility edge for $L>>\xi$ we have metallic behavior and $f(L/\xi) \sim (L/\xi)^{d-2}$. The d.c. conductivity is given by

$$\sigma = \frac{e^2}{\hbar} L^{2-d} g \sim \xi^{2-d} \tag{24.12}$$

The conductivity exponent is thus $t = \nu\ (d-2)$ (Wegner 1976).

(2) d=2. This is the marginal case. Whatever conductance we begin at because $\beta<0$ we eventually end up with insulating behavior. Thus no true metallic behavior should occur in d=2.

It is possible to do perturbation theory for the conductivity in the weak scattering limit. The usual Drude formula for the conductivity comes from summing the ladder graphs. The next corrections to this result have been discussed by Langer and Neal (1966) and come from crossed graphs. These latter graphs have been summed and found to diverge logarithmically. The result for the conductivity is (Abrahams and Ramakrishnan (1980))

$$\sigma = \sigma_o - \frac{e^2}{\hbar\pi^2} \ln (L/\ell) \tag{24.13}$$

where ℓ is the mean free path. The 2d dimensionless conductance $g = g_o - \frac{1}{\pi^2} \ln L/\ell$.

This size dependence of the conductivity has not been observed but recent experiments on 2d films have observed a log dependence of the conductivity on temperature and voltage (Bishop et al. (1981)). These experiments can be interpreted as follows.

(i) At finite temperatures the length L should be replaced by the inelastic scattering length $\ell_{inel} \sim T^{-1/2}$ (see Eq. 23.8)). Thus the conductance should depend on temperature as

$$g = g_o + \frac{1}{2\pi^2} \ln T \tag{24.14}$$

The universal constant $\frac{1}{\pi^2}$ is also in good agreement with experiment (when the temperature dependence of $\ell_{inel} \sim T^{-1/2}$ is assumed). This, however, is complicated by the effects of electron-electron interactions which give a similar log temperature dependence to the conductance in 2d.

(ii) At finite voltages the experimental results have been interpreted in terms of a heating model (Abrahams et al. 1980). The Joule heating is related to the temperature by $\sigma E^2 = CT/\tau_{inel}$ where C is the specific heat ($C \sim T$). Thus $E^2 \sim T^3$ (using $\tau_{inel} \sim T^{-1}$) and substituting

$$g = g_o + \frac{1}{3\pi^2} \ln E \tag{24.15}$$

which again agrees with the observations.

25. The ε-expansion

The conductivity and correlation length exponents have been calculated by Wegner (1980) by an ε-expansion in $\varepsilon = d-2$. The leading term in this expansion can also be obtained from (24.1). The dimensionless conductance in d dimensions (d close to 2)

$$g = \frac{\hbar}{e^2} \sigma L^{d-2}$$

$$= \frac{\hbar}{e^2} \sigma_o L^{d-2} [1 - \frac{e^2}{\hbar \pi^2 \sigma_o} \ln L/\ell] \; . \tag{25.1}$$

The beta function calculated from this is

$$\beta = d-2 - \frac{1}{\pi^2 g} \tag{25.2}$$

Thus $\beta=0$ gives $g_c = \frac{1}{\pi^2(d-2)}$ and $\beta'(g_c) = \frac{1}{\pi^2 g_c^2} = \pi^2(d-2)^2$. The correlation length exponent is

$$\nu = \frac{1}{\beta' g_c} = \frac{1}{d-2} \tag{25.3}$$

and the conductivity exponent is

$$t = (d-2)\nu = 1 \quad . \tag{25.4}$$

References

Abrahams E. and Stephen M.J. 1980 J. Phys. C13, L377
Abrahams E., Anderson P.W., Licciardello D.C. and Ramakrishnan T.V. 1979 Phys. Rev. Lett. 42, 673
Abrahams E. and Ramakrishnan T.V. 1980 J. Non-Crystalline Solids 35, 15
Abrahams E., Anderson P.W. and Ramakrishnan T.V. 1980 Phil. Mag. 42, 827
Almeida J.R.L. and Thouless D. 1978 J. Phys. A11, 983
Anderson P.W., Thouless D.J., Abrahams E. and Fisher D.S. 1980 Phys. Rev. B22, 3519
Anderson P.W. 1958 Phys. Rev. 109, 1492
Birgeneau R.J., Cowley R.A., Shirane G. and Guggenheim H.J. 1976 Phys. Rev. Lett. 37, 940. See also Phys. Rev. B21 317 (1980)
Bishop P.J., Tsui D.C. and Dynes R.C. 1981 Phys. Rev. Lett. 46, 360
Borland R.E. 1963 Proc. Roy. Soc. A274, 529
Bray A.J. and Moore M.A. 1978 Phys. Rev. Lett. 41, 1068
Broadbent S.R. and Hammersley J.M. 1957 Proc. Camb. Phil. Soc. 53, 639
Bruggeman D.A.G. 1935 Ann. Physik 24, 636
Cannella V. and Mydosh J.A. 1972 Phys. Rev. B6, 4220
Chaudhari P. and Habermeier H.U. 1980 Solid State Commun. 34, 687
Coniglio A. 1981 Phys. Rev. Lett. to be published
De Dominicis C. and Garel T. 1979 J. Physique Lett. 40, L575
De Dominicis C. 1978 Phys. Rev. B18, 4913
De Dominicis C., Gabay M. and Duplantier B. 1981 to be published
De Gennes P.G. 1976 J. Phys. (Paris) Lett. 37, L1
Domany E. 1978 J. Phys. C 11, L337
Domb C. 1976 J. Phys. A9, 283
Edwards S.F. and Anderson P.W. 1975 J. Phys. F5, 965
Essam J.W. 1980 Rep. Prog. Phys. 43 833
Essam J.W. 1972 Phase Transitions and Critical Phenomena vol. 2, C. Domb and M.S. Green, Eds. (Academic Press)
Fischer K.H. 1975 Phys. Rev. Lett. 34, 1438
Fisher M.E. and Essam J.W. 1961 J. Math. Phys. 2, 609
Ford P.J. 1982 Contemporary Physics to be published
Frisch H.L. and Hammersley J.M. 1963 J. Soc. Indust. Appl. Math. 11, 894
Giordano N., Gibson W. and Prober D.E. 1979 Phys. Rev. Lett. 43, 725
Guy C.N. 1975 J. Phys. F5, L242 and F7, 1505
Harris A.B., Lubensky T.C., Holcomb W.K. and Dasgupta C. 1975 Phys. Rev. Lett. 35, 327
Harris A.B. 1974 J. Phys. C7, 1671
Harris A.B. and Fisch R. 1977 Phys. Rev. Lett. 38, 796
Ishii K. 1973 Prog. Theor. Phys. Suppl. 53, 77
Kasteleyn P.W. and Fortuin C. 1969 J. Phys. Soc. Japan (Suppl.) 26, 11
Kirkpatrick S. 1977 Phys. Rev. B15, 1533
Kirkpatrick S. 1973 Rev. Mod. Phys. 45, 574
Kogut P.M. and Straley J. 1978 in Electrical Transport and Optical Properties of Inhomogeneous Media, Ed. Garland and Tanner (Am. Inst. of Physics)
Kunz H. and Souillard B. 1978 J. Stat. Phys. 1, 77
Landauer R. 1978 in Electrical Transport and Optical Properties of Inhomogeneous Media, Eds. Garland and Tanner (Am. Inst. of Physics)
Landauer R. 1970 Phil. Mag. 21, 863
Langer J. and Neal T. 1966 Phys. Rev. 169, 508
Leath P.L. and Reich G.R. 1978 J. Phys. C. Solid State Phys. 11, 4017
Lubensky T.C. 1975 Phys. Rev. B11, 3573
Lubensky T.C. 1977 Phys. Rev. B15, 311
Luther A. and Grinstein G. 1976 Phys. Rev. B13, 1329
Mandelbrot R.B. 1977 Fractals: Form Chance and Dimension (W.H. Freeman, San Francisco)
Mott N.F. 1974 Metal-Insulator Transitions (Taylor and Francis, London)
Nagata S., Keesom P.H. and Harrison H.R. 1979 Phys. Rev. B19, 1633
Parisi G. 1979 Phys. Rev. Lett. 23, 1754 and J. Phys. A13, L115, L1887 (1980)
Reatto L. and Rastelli E. 1972 J. Phys. C5, 2785
Shante V.K.S. and Kirkpatrick S. 1971 Adv. in Physics 20, 325

Sherrington D. and Kirkpatrick S. 1975 Phys. Rev. Lett. 35, 1972. See also Phys. Rev. B17, 4385 (1978)
Skal A.S. and Shklovskii B.I. 1975 Sov. Phys. Semicond. 8, 1029
Stanley H.E., Birgeneau R.J., Reynolds P.J., and Nicoll J.F., J. Phys. (Paris) Colloq. 9, L553
Stauffer D. 1979 Physics Reports 54, No. 1
Stauffer D. 1976 Z. Physik B25, 391
Stauffer D. 1975 Z. Physik B22, 161
Stephen M.J. 1977 Phys. Rev. B15, 5674
Stephen M.J. and Grest G.S. 1977 Phys. Rev. Lett. 38, 567
Stephen M.J. 1978 Phys. Rev. 17, 4444
Stephen M.J. 1981 J. Stat. Phys. 25, 663
Sommers H.J. 1978 Zeit.f. Physik B31, 301
Sompolinsky H. and Zippelius A. 1981 Phys. Rev. Lett. 47, 359
Sompolinsky H. 1981 Phys. Rev. Lett. 47, 935
Stinchcombe R.B. 1974 J. Phys. C7, 179
Straley J.P. 1977a Phys. Rev. B15, 5733
Straley J.P. 1977b J.Phys. C10, 3009
Thouless D.J., Anderson P.W. and Palmer R.J., 1977 Phil. Mag. 35, 593
Thouless D.J. 1980 Solid State Commun. 34, 683
Thouless D.J. 1978 in "Ill Condensed Matter" Proc. of Les Houches Summer School Eds. Balian, Maynard and Toulous (North Holland)
Toulouse G. 1977 Comm. on Physics 2, 115
Wallace D.J. and Young P. 1978 Phys. Rev. B17, 2384
Wegner F. 1981 to be published
Wegner F. 1976 Z. Phys. 25, 327
Wegner F. 1980 Phys. Rep. 67, 15.

Lectures on Correlation Functions†

Presented at Advanced Course in Theoretical Physics: Critical Phenomena, January 1982, Stellenbosch, South Africa

Alexander L. Fetter
Institute of Theoretical Physics, Department of Physics,
Stanford University, Stanford, CA 94305

ITP-701 1/82

†Supported in part by the National Science Foundation through Grant No. DMR 78-25258.

LECTURES ON CORRELATION FUNCTIONS

Alexander L. Fetter
Institute of Theoretical Physics, Stanford University
Stanford, CA 94305 USA

I. X-RAY SCATTERING

Correlation functions play a central role in analyzing and understanding the properties of condensed matter. One familiar example is the use of x-rays in crystallography, but similar techniques also apply to scattering of visible light and neutrons, both of which have provided much basic information on various types of phase transitions. Although the relation between measured properties (for example, intensity of scattered radiation) and the correlation functions is relatively direct it is not always spelled out in detail, and it seems useful first to consider a simple concrete model of x-ray scattering from a collection of atoms. We shall work in a classical picture here. The physical quantity of interest is the electric field of the incident electromagnetic wave, which acts on the electrons, setting them into motion and thereby inducing electric dipole radiation (Fig. 1.1).

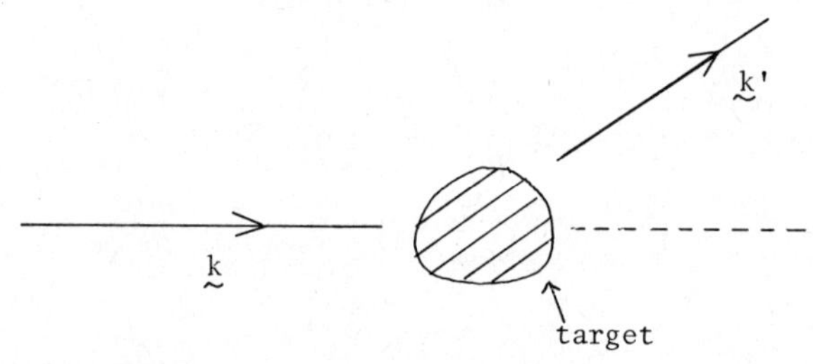

Fig. 1.1 Geometry of x-ray scattering

Suppose there is an incident plane wave (we mean the real part here, as is common in E. & M.)

$$\underset{\sim}{E}_i(\underset{\sim}{r},t) = E_0 \underset{\sim}{\varepsilon}\, e^{i(\underset{\sim}{k}\cdot\underset{\sim}{r}-\omega t)} . \tag{1.1}$$

The induced current $\underset{\sim}{j}(\underset{\sim}{r},t) = \underset{\sim}{j}(\underset{\sim}{r})e^{-i\omega t}$ oscillates at frequency ω with a harmonic factor $e^{-i\omega t}$. The corresponding radiated vector potential has the same factor $e^{-i\omega t}$ and an amplitude

$$\underset{\sim}{A}(\underset{\sim}{r}) = \frac{1}{c}\int d^3r' \,\frac{\underset{\sim}{j}(\underset{\sim}{r}')}{|\underset{\sim}{r}-\underset{\sim}{r}'|}\, e^{ik|\underset{\sim}{r}-\underset{\sim}{r}'|} \tag{1.2}$$

where the integral runs over the volume of the target. Now $\underset{\sim}{j}(\underset{\sim}{r})$ is a sum of contributions from all electrons of the form

$$\underset{\sim}{j}(\underset{\sim}{r}) = \Sigma_i \,(-e)\, \underset{\sim}{v}_i \,\delta(\underset{\sim}{r}-\underset{\sim}{r}_i) \tag{1.3}$$

where $\underset{\sim}{v}_i$ and $\underset{\sim}{r}_i$ are the velocity and position of the $\underline{ith}$ electron, and we have approximated $\underset{\sim}{r}_i$ by the average position ignoring the amplitude of the motion (which is permissible if the incident E field is weak). The velocity $\underset{\sim}{v}_i$ has a phase factor $e^{i\underset{\sim}{k}\cdot\underset{\sim}{r}_i}$ since it arises from the electric field at that point. Furthermore, the distance $|\underset{\sim}{r}-\underset{\sim}{r}'| \approx r-\hat{r}\cdot\underset{\sim}{r}' + \ldots$ for a distant observer at r, and the resulting vector potential takes the form

$$\underset{\sim}{A}(\underset{\sim}{r}) \sim \frac{e^{ikr}}{rc}\int d^3r'\, e^{-i\underset{\sim}{k}'\cdot\underset{\sim}{r}'}\,\underset{\sim}{j}(\underset{\sim}{r}')$$

$$= \frac{e^{ikr}}{rc} E_0\underset{\sim}{\varepsilon}\, f(\underset{\sim}{k},\underset{\sim}{k}')\sum_i e^{i\underset{\sim}{q}\cdot\underset{\sim}{r}_i} \tag{1.4}$$

where (Fig. 1.2)

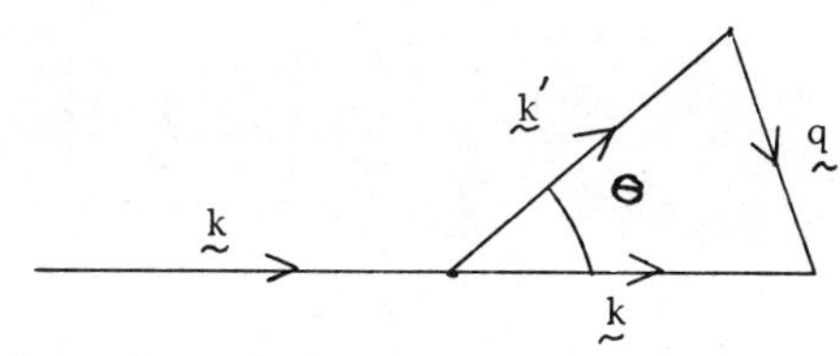

Fig. 1.2 Wave vector $\underset{\sim}{q}$ transferred to the target.

$$\underset{\sim}{q} = \underset{\sim}{k} - \underset{\sim}{k}'. \tag{1.5}$$

In quantum-mechanical language, $\hbar\underset{\sim}{q}$ is momentum transferred to the target. Note that

$$|\underset{\sim}{q}| = 2k\sin(\tfrac{1}{2}\theta) \tag{1.6}$$

where θ is the angle between $\underset{\sim}{k}$ and $\underset{\sim}{k}'$ (assuming that $|\underset{\sim}{k}| = |\underset{\sim}{k}'|$, which is generally accurate for x-rays). Thus $q \to 0$ for $\Theta \to 0$ (forward scattering) and $q \to 2k$ for $\Theta \to \pi$ (backward scattering). The scattered vector potential has several factors: e^{ikr}/rc is the outgoing spherical wave, $E_0\underset{\sim}{\varepsilon}$ is the incident E field, $f(\underset{\sim}{k},\underset{\sim}{k}')$ is the "scattering amplitude" for a single electron, and finally, the sum over i contains all the properties of the multielectronic system. This last factor is the quantity of interest here.

It is a straightforward exercise in electrodynamics to find the energy flux radiated by the system and, dividing by the incident energy, one can get the differential cross section for scattering by the target; it takes the form

$$\frac{d\sigma}{d\Omega} = \left.\frac{d\sigma}{d\Omega}\right|_T \left|\sum_i e^{i\underset{\sim}{q}\cdot\underset{\sim}{r}_i}\right|^2 \tag{1.7}$$

where $\left.\frac{d\sigma}{d\Omega}\right|_T$ is the Thomson cross section for scattering by a single electron[1]. For unpolarized light with frequency high enough that $\hbar\omega$ exceeds the binding energy of the electrons, the Thomson cross section[1] for 1 electron is $\frac{1}{2}r_0^2[1+(\hat{k}\cdot\hat{k}')^2]$, but the details do not matter here (note: $r_0 = e^2/mc^2 \simeq 2.8 \times 10^{-13}$ cm is the classical electron radius).

The basic point here is that the cross section for scattering from an assembly of electrons differs from that for one electron by the factor

$$\left| \sum_i e^{i\mathbf{q}\cdot\mathbf{r}_i} \right|^2 . \tag{1.8}$$

If we introduce the electron number density

$$n_{e\ell}(\mathbf{r}) = \sum_i \delta(\mathbf{r}-\mathbf{r}_i) \tag{1.9}$$

then the Fourier transform of $n_{e\ell}$ is given by

$$n_{e\ell}(\mathbf{k}) = \int d^3r\, e^{-i\mathbf{k}\cdot\mathbf{r}}\, n_{e\ell}(\mathbf{r}) = \sum_i e^{-i\mathbf{k}\cdot\mathbf{r}_i} \tag{1.10}$$

and it is clear that $n_{e\ell}(-\mathbf{k}) = n_{e\ell}(\mathbf{k})^*$. Thus the basic factor of interest is the squared modulus of $n_{e\ell}(\mathbf{q})^*$, allowing a direct study of the electron density in the sample merely by studying the angular dependence of the differential cross section.

The preceding picture is fine for a plasma, but in most cases of interest to condensed matter, the electrons are collected into atoms, whose charge density can be described by an atomic distribution $-en_0(\mathbf{r})$ for an atom at the origin. Consequently, the total electronic density can be written as a sum of the densities from individual atoms:

$$n_{e\ell}(\mathbf{r}) = \sum_j n_0(\mathbf{r}-\mathbf{R}_j) \tag{1.11}$$

where $\mathbf{R}_j$ is the position of the jth atom. Here we assume that each atom is identical. The Fourier transform of $n_{e\ell}(\mathbf{r})$ now becomes

$$\begin{aligned} n_{e\ell}(\mathbf{q}) &= \sum_j \int d^3r\, e^{-i\mathbf{q}\cdot\mathbf{r}}\, n_0(\mathbf{r}-\mathbf{R}_j) \\ &= \sum_j \exp(-i\mathbf{q}\cdot\mathbf{R}_j)\, n_0(\mathbf{q}) \end{aligned} \tag{1.12}$$

where $n_0(\mathbf{q}) \equiv \int d^3r e^{-i\mathbf{q}\cdot\mathbf{r}}\, n_0(\mathbf{r})$ is the Fourier transform of the electron density in the single atom. This is assumed to be known or calculable from atomic theory.

The quantity $n_0(\mathbf{q})$ has some simple properties:

1. $$n_0(\mathbf{q}=0) = \int d^3r\, n_0(\mathbf{r}) = Z \tag{1.13}$$

where Z is the number of electrons in the atom.

2. Define the "atomic form factor" $f_{\mathbf{q}}$ by the relation

$$f_{\mathbf{q}} \equiv n_0(\mathbf{q})/n_0(\mathbf{q}=0) = \frac{1}{Z}\, n_0(\mathbf{q}) \tag{1.14}$$

By definition:

$$f_{\underset{\sim}{q}} = \frac{\int d^3r\, e^{-i\underset{\sim}{q}\cdot\underset{\sim}{r}}\, n_0(\underset{\sim}{q})}{\int d^3r\, n_0(\underset{\sim}{q})} \tag{1.15}$$

and it is expected that $|f_q| \leqslant 1$ owing to extra interference in the numerator.

As a result, our basic quantity of interest may be written:

$$\begin{aligned} \left| \Sigma_i e^{i\underset{\sim}{q}\cdot\underset{\sim}{r}_i} \right|^2 &= \left| n_{el}(\underset{\sim}{q})^* \right|^2 \\ &= Z^2 |f_{\underset{\sim}{q}}|^2 \left| \Sigma_j e^{i\underset{\sim}{q}\cdot\underset{\sim}{R}_j} \right|^2 \end{aligned} \tag{1.16}$$

where the sum over j runs over all separate atoms. Here we act as if the atoms were fixed in space. In fact, they move about owing to thermal motion, and we really need to take an ensemble average, denoted $\langle \cdots \rangle$; in a quantum mechanical problem, the angular brackets will also include an appropriate quantum expectation value. The differential cross section for x-ray scattering thus becomes

$$\frac{d\sigma}{d\Omega} = \left.\frac{d\sigma}{d\Omega}\right|_T Z^2 |f_{\underset{\sim}{q}}|^2 \left\langle \left| \Sigma_j e^{i\underset{\sim}{q}\cdot\underset{\sim}{R}_j} \right|^2 \right\rangle \tag{1.17}$$

and the last factor contains all the information on the position of the atoms in the medium.

It is convenient to introduce an atomic density $n(\underset{\sim}{r})$ defined by

$$n(\underset{\sim}{r}) = \Sigma_j\, \delta(\underset{\sim}{r} - \underset{\sim}{R}_j) \tag{1.18}$$

Evidently, its Fourier transform is $n(\underset{\sim}{q}) = \Sigma_j e^{-i\underset{\sim}{q}\cdot\underset{\sim}{R}_j}$, and the desired factor in Eq. (1.17) is just the quantity

$$\langle n(\underset{\sim}{q})\, n(-\underset{\sim}{q}) \rangle = \langle n(\underset{\sim}{q})\, n(\underset{\sim}{q})^* \rangle \tag{1.19}$$

Alternatively, it can be written as the double Fourier transform of the density-density "correlation function" $\langle n(\underset{\sim}{r})\, n(\underset{\sim}{r}') \rangle$:

$$\langle n(\underset{\sim}{q})\, n(-\underset{\sim}{q}) \rangle = \int d^3r\, d^3r'\, e^{i\underset{\sim}{q}\cdot(\underset{\sim}{r}'-\underset{\sim}{r})} \langle n(\underset{\sim}{r})\, n(\underset{\sim}{r}') \rangle \tag{1.20}$$

Thus all one can ever get from x-ray scattering is a determination of the density-density correlation function.

For an extended target, the number of atoms N and volume V both become large, with their ratio N/V fixed. Hence $\langle n(\underset{\sim}{r}) \rangle = N/V = n$, and it is more interesting to consider the density deviations, defined as

$$\tilde{n}(\underset{\sim}{r}) = n(\underset{\sim}{r}) - \langle n(\underset{\sim}{r}) \rangle = n(\underset{\sim}{r}) - n . \tag{1.21}$$

It follows immediately that

$$\langle \tilde{n}(\underset{\sim}{r})\, \tilde{n}(\underset{\sim}{r}') \rangle = \langle n(\underset{\sim}{r})\, n(\underset{\sim}{r}') \rangle - n^2 \tag{1.22}$$

and this latter function is more properly called the density-density correlation function, for it vanishes as $|\underset{\sim}{r} - \underset{\sim}{r}'| \to \infty$ and, more generally becomes small for all $|\underset{\sim}{r} - \underset{\sim}{r}'|$ if the atoms are uniformly distributed with no correlations. Furthermore, for an extended system with translational invariance, this function depends only on the relative separation $\underset{\sim}{r}-\underset{\sim}{r}'$, except near the boundaries. It is conventional to introduce the "static structure function" $S(\underset{\sim}{q})$ by the definition

$$S(\underset{\sim}{q}) = n^{-1} \int d^3r \, e^{-i\underset{\sim}{q}\cdot\underset{\sim}{r}} \langle \tilde{n}(\underset{\sim}{r}) \, \tilde{n}(0) \rangle \tag{1.23}$$

where we make explicit that $\langle \tilde{n}(\underset{\sim}{r}) \, \tilde{n}(\underset{\sim}{r}') \rangle$ depends only on $\underset{\sim}{r}-\underset{\sim}{r}'$. We will see that all the x-ray scattering information can be expressed in this one function $S(\underset{\sim}{q})$.

II. TWO-BODY CORRELATIONS

To understand some of the properties of the static structure function $S(\underset{\sim}{q})$, it is helpful to introduce the "two-body correlation function" for N atoms in a medium with volume V

$$p(\underset{\sim}{r},\underset{\sim}{r}') = \langle \sum_{i=1}^{N} \delta(\underset{\sim}{r}-\underset{\sim}{R}_i) \sum_{j\neq i=1}^{N} \delta(\underset{\sim}{r}'-\underset{\sim}{R}_j) \rangle \tag{2.1}$$

This function is proportional to the joint probability of finding one atom at $\underset{\sim}{r}$ and another, distinct atom at $\underset{\sim}{r}'$ (note the restriction $j \neq i$). It can be written in terms of the density $n(\underset{\sim}{r})$ by adding and subtracting the terms with $i = j$.

$$\begin{aligned} p(\underset{\sim}{r},\underset{\sim}{r}') &= \langle n(\underset{\sim}{r})\, n(\underset{\sim}{r}')\rangle - \sum_{i=1}^{N} \langle \delta(\underset{\sim}{r}-\underset{\sim}{R}_i)\,\delta(\underset{\sim}{r}'-\underset{\sim}{R}_i)\rangle \\ &= \langle n(\underset{\sim}{r})n(\underset{\sim}{r}')\rangle - \delta(\underset{\sim}{r}-\underset{\sim}{r}')\,\langle n(\underset{\sim}{r})\rangle . \end{aligned} \tag{2.2}$$

For a uniform system, this function depends only on $\underset{\sim}{r}-\underset{\sim}{r}'$, and it is usually expressed in terms of the "pair distribution function" $g(\underset{\sim}{r}-\underset{\sim}{r}')$ according to the definition:

$$g(\underset{\sim}{r}) = n^{-2}\, p(\underset{\sim}{r},0) \tag{2.3}$$

In a real liquid, it is reasonable to assume that $g(\underset{\sim}{r})$ approaches 0 as $r \to 0$, since two atoms cannot occupy the same spatial region, and that $g(r) \to$ const for $r \to \infty$, because there is no long-range order in a fluid. [The situation is different in a truly infinite crystal, since the differential x-ray cross section has Bragg peaks, reflecting the long-range order in $g(\underset{\sim}{r},\underset{\sim}{r}')$.]

To study the behavior of $g(\underset{\sim}{r})$, consider the integral

$$\begin{aligned} \int d^3r\,[g(\underset{\sim}{r})-1] &= \int d^3r\; n^{-2}[\langle n(\underset{\sim}{r})n(0)\rangle - n\delta(\underset{\sim}{r}) - n^2] \\ &= n^{-2}[N\langle n(0)\rangle - n - Vn^2] \\ &= \frac{N}{n} - \frac{1}{n} - V \;=\; -\frac{1}{n} \end{aligned} \tag{2.3}$$

Note that $g(r)-1$ has a well-defined spatial integral even in the thermodynamic limit ($N \to \infty$, $V \to \infty$, but N/V fixed). Thus we conclude that $g(r)$ must approach 1 as $|\underset{\sim}{r}| \longrightarrow \infty$. Furthermore, $g(r)-1$ has a well-defined Fourier transform. Indeed, a simple calculation gives

$$\begin{aligned} &\int d^3r\; e^{-i\underset{\sim}{q}\cdot\underset{\sim}{r}}[g(\underset{\sim}{r})-1] \\ &= \frac{1}{n^2}\int d^3r\; e^{-i\underset{\sim}{q}\cdot\underset{\sim}{r}} \left[\langle n(\underset{\sim}{r})n(0)\rangle - n^2 - n\delta(\underset{\sim}{r})\right] \\ &= \frac{1}{n^2}\int d^3r\; e^{-i\underset{\sim}{q}\cdot\underset{\sim}{r}} \langle \tilde{n}(\underset{\sim}{r})\tilde{n}(0)\rangle - \frac{1}{n} . \end{aligned} \tag{2.4}$$

But comparison with Eq. (1.23) now shows that the integral is just $nS(\underset{\sim}{q})$, leading to the basic relation

$$S(\underset{\sim}{q}) - 1 = n \int d^3r\, e^{-i\underset{\sim}{q}\cdot\underset{\sim}{r}} [g(\underset{\sim}{r}) - 1] \tag{2.5}$$

Thus any measurement of $S(\underset{\sim}{q})$ is also a measurement of the pair distribution function (in a liquid, $g(\underset{\sim}{r})$ is isotropic and often called the radial distribution function). Equation (2.5) shows one important property of $S(q)$, namely that it approaches 1 as $|\underset{\sim}{q}| \longrightarrow \infty$. This arises from the rapid oscillations of the integrand on the right-hand side; more formally, it follows from the Riemann-Lebesgue lemma because $S(0)-1$ is well defined and $S(\underset{\sim}{q})-1$ must then tend to zero as $q \rightarrow \infty$. For an isotropic fluid, the angular integrals can be performed explicitly to give

$$S(q) - 1 = \frac{4\pi n}{q} \int_0^\infty r\, dr\, \sin qr\, [g(r) - 1] \tag{2.6}$$

but this form offers few advantages over (2.5).

One other property of $S(\underset{\sim}{q})$ follows from Eq. (1.23), by writing the integral as

$$S(\underset{\sim}{q}) = n^{-1} \int d^3(\underset{\sim}{r} - \underset{\sim}{r}')\, e^{-i\underset{\sim}{q}\cdot(\underset{\sim}{r}-\underset{\sim}{r}')} \langle \tilde{n}(\underset{\sim}{r})\, \tilde{n}(\underset{\sim}{r}') \rangle \tag{2.7}$$

Since $\langle n(\underset{\sim}{r}) n(\underset{\sim}{r}') \rangle$ depends on $\underset{\sim}{r}-\underset{\sim}{r}'$ only, there is no dependence on $(\underset{\sim}{r}+\underset{\sim}{r}')/2$ and we can integrate over this dummy variable, dividing by the volume V. A change of variables with unit Jacobian then yields

$$\begin{aligned} S(\underset{\sim}{q}) &= (nV)^{-1} \int d^3r\, d^3r'\, e^{-i\underset{\sim}{q}\cdot\underset{\sim}{r}}\, e^{i\underset{\sim}{q}\cdot\underset{\sim}{r}'} \langle \tilde{n}(\underset{\sim}{r})\, \tilde{n}(\underset{\sim}{r}') \rangle \\ &= (nV)^{-1} \int d^3r\, d^3r'\, e^{-i\underset{\sim}{q}\cdot\underset{\sim}{r}}\, e^{i\underset{\sim}{q}\cdot\underset{\sim}{r}'} [\langle n(\underset{\sim}{r})\, n(\underset{\sim}{r}') \rangle - n^2] \\ &= N^{-1} \langle n(\underset{\sim}{q})\, n(-\underset{\sim}{q}) \rangle - N\, \delta_{\underset{\sim}{q},0} \end{aligned} \tag{2.8}$$

where we use box normalization to introduce the Kronecker delta in the last line. Note that for $q \neq 0$, the quantity $N\, S(\underset{\sim}{q})$ is precisely that determined in x-ray scattering [see Eq. (1.19)]. Furthermore, q is guaranteed to be nonzero since otherwise the photons would not be scattered at all. Thus any x-ray experiment of the sort considered always measures $S(\underset{\sim}{q})$ directly. Working backwards, it also measures $g(\underset{\sim}{r})-1$ through Eq. (2.5). These relations are widely used.

To this point, all our correlation functions have been expressed in terms of a first-quantized description, but it is often preferable to introduce the standard second-quantized operators. It will be convenient to apply periodic boundary conditions and to use either $\hat{\psi}_\alpha(\underset{\sim}{r})$ that destroys a particle at $\underset{\sim}{r}$ or $a_{\underset{\sim}{k}\lambda}$ that destroys a particle with momentum $\hbar\underset{\sim}{k}$ and spin λ. These operators are related by the equation

$$\hat{\psi}_\alpha(\underset{\sim}{r}) = V^{-1/2} \sum_{\underset{\sim}{k}\lambda} e^{i\underset{\sim}{k}\cdot\underset{\sim}{r}} (\eta_\lambda)_\alpha\, a_{\underset{\sim}{k}\lambda} \tag{2.9}$$

where η_λ is a spinor for spin state λ and α denotes the particular component.

The particle density $n(\underset{\sim}{r})$ then becomes the second-quantized operator

$$\hat{n}(\underset{\sim}{r}) = \sum_\alpha \hat{\psi}_\alpha^\dagger(\underset{\sim}{r})\,\hat{\psi}_\alpha(\underset{\sim}{r}) \tag{2.10}$$

and its Fourier transform has the form

$$\hat{n}(\underset{\sim}{k}) = \int d^3r\, e^{-i\underset{\sim}{k}\cdot\underset{\sim}{r}}\,\hat{n}(\underset{\sim}{r}) = \sum_{\underset{\sim}{q}\lambda} a_{\underset{\sim}{q}-\underset{\sim}{k},\lambda}^\dagger\, a_{\underset{\sim}{q}\lambda} \tag{2.11}$$

In this language, it is easy to see that Eqs. (1.20)-(1.23) remain correct with n reinterpreted as a second-quantized operator, and that the two-body correlation function in Eq. (2.1) becomes

$$p(\underset{\sim}{r},\underset{\sim}{r}') = \sum_{\alpha\beta} \langle \hat{\psi}_\alpha^\dagger(\underset{\sim}{r})\,\hat{\psi}_\beta^\dagger(\underset{\sim}{r}')\,\hat{\psi}_\beta(\underset{\sim}{r}')\,\hat{\psi}_\alpha(\underset{\sim}{r})\rangle \tag{2.12}$$

It is an interesting exercise to evaluate $p(\underset{\sim}{r},\underset{\sim}{r}')$ for an ideal gas at temperature $T = (k_B\beta)^{-1}$ and chemical potential μ in the grand canonical ensemble. Use of Eq. (2.9) and the orthogonality of the spinors readily yields

$$p(\underset{\sim}{r},\underset{\sim}{r}') = V^{-2}\sum_{\underset{\sim}{k}\underset{\sim}{k}'\underset{\sim}{p}\underset{\sim}{p}'}\sum_{\lambda\lambda'} e^{i(\underset{\sim}{p}-\underset{\sim}{k})\cdot\underset{\sim}{r}}\, e^{i(\underset{\sim}{p}'-\underset{\sim}{k}')\cdot\underset{\sim}{r}'}\langle a_{\underset{\sim}{k}\lambda}^\dagger a_{\underset{\sim}{k}'\lambda'}^\dagger a_{\underset{\sim}{p}'\lambda'} a_{\underset{\sim}{p}\lambda}\rangle \tag{2.13}$$

Furthermore, it is not difficult to prove that grand-canonical ensemble averages in an ideal gas obey a form of Wick's theorem[2]. As a result, the operator factor may be rewritten

$$\begin{aligned}\langle a_{\underset{\sim}{k}\lambda}^\dagger a_{\underset{\sim}{k}'\lambda'}^\dagger a_{\underset{\sim}{p}'\lambda'} a_{\underset{\sim}{p}\lambda}\rangle &= \langle a_{\underset{\sim}{k}\lambda}^\dagger a_{\underset{\sim}{p}\lambda}\rangle\langle a_{\underset{\sim}{k}'\lambda'}^\dagger a_{\underset{\sim}{p}'\lambda'}\rangle \\ &\quad \pm \langle a_{\underset{\sim}{k}\lambda}^\dagger a_{\underset{\sim}{p}'\lambda'}\rangle\langle a_{\underset{\sim}{k}'\lambda'}^\dagger a_{\underset{\sim}{p}\lambda}\rangle \\ &= n_{\underset{\sim}{k}\lambda}^0\, n_{\underset{\sim}{k}'\lambda'}^0(\delta_{\underset{\sim}{k}\underset{\sim}{p}}\,\delta_{\underset{\sim}{k}'\underset{\sim}{p}'} \pm \delta_{\underset{\sim}{k}\underset{\sim}{p}'}\,\delta_{\underset{\sim}{p}\underset{\sim}{k}'}\,\delta_{\lambda\lambda'})\end{aligned} \tag{2.14}$$

Here $n_{\underset{\sim}{k}\lambda}^0 = \langle a_{\underset{\sim}{k}\lambda}^\dagger a_{\underset{\sim}{k}\lambda}\rangle$ is the momentum distribution in an ideal gas

$$n_{\underset{\sim}{k}\lambda}^0 = \left(e^{\beta(\epsilon_k^0-\mu)} \mp 1\right)^{-1} \tag{2.15}$$

and the upper and lower signs refer to bosons and fermions, respectively[3]. A combination of these results gives

$$\begin{aligned}p(\underset{\sim}{r},\underset{\sim}{r}') &= V^{-2}\sum_{\underset{\sim}{k}\underset{\sim}{k}'}\sum_{\lambda\lambda'} n_{k\lambda}^0\, n_{k'\lambda'}^0\left(1 \pm \delta_{\lambda\lambda'}\, e^{i(\underset{\sim}{k}-\underset{\sim}{k}')\cdot(\underset{\sim}{r}-\underset{\sim}{r}')}\right) \\ &= N^2V^{-2} \pm V^{-2}\sum_{\underset{\sim}{k}\underset{\sim}{k}'\lambda} n_{k\lambda}^0\, n_{k'\lambda}^0\, e^{i(\underset{\sim}{k}-\underset{\sim}{k}')\cdot(\underset{\sim}{r}-\underset{\sim}{r}')} \\ &= n^2 \pm (2s+1)\left|(2\pi)^{-3}\int d^3k\, e^{i\underset{\sim}{k}\cdot(\underset{\sim}{r}-\underset{\sim}{r}')}\, n_k^0\right|^2\end{aligned} \tag{2.16}$$

where 2s+1 is the spin degeneracy.

Two special cases are easily evaluated.

(1) For an ideal classical gas, Eq. (2.15) becomes

$$n^0_{k\lambda} = n^0_k = e^{\beta\mu} e^{-\beta\varepsilon^0_k} \tag{2.17}$$

where

$$e^{\beta\mu} = n\lambda^3 (2s+1)^{-1} \tag{2.18}$$

and

$$\lambda = (2\pi\beta\hbar^2/m)^{1/2} = (2\pi\hbar^2/mk_B T)^{1/2} \tag{2.19}$$

is the thermal wavelength. The resulting integral can be evaluated by completing the square in the exponent and gives

$$\begin{aligned} p(\underset{\sim}{r},\underset{\sim}{r}') &= p(\underset{\sim}{r}-\underset{\sim}{r}') = n^2 \left[1 \pm (2s+1)^{-1} \exp\left(-|\underset{\sim}{r}-\underset{\sim}{r}'|^2/\lambda^2\right)\right] \\ &\approx n^2 \qquad \text{as } \hbar \longrightarrow 0 \quad \text{for } |\underset{\sim}{r}-\underset{\sim}{r}'| \neq 0 \end{aligned} \tag{2.20}$$

Thus the pair distribution in an ideal classical gas is just $g(r) = 1$ [see Eq. (2.3)]. Correspondingly, $S(k)$ also $= 1$ for an ideal classical gas.

(2) The other simple case is an ideal Fermi gas at zero temperature, when the distribution functions again become simple

$$n^0_{k\lambda} = \theta(k_F - k) \tag{2.21}$$

with $\theta(x)$ a unit step function and $\varepsilon^0_F = \hbar^2 k_F^2/2m = \mu$. The corresponding integrals are easily evaluated in spherical polar coordinates to give

$$\begin{aligned} g(r) &= 1 - \frac{2s+1}{n}\left[\int_0^{k_F} \frac{k^2\,dk}{2\pi^2}\,\frac{\sin kr}{kr}\right]^2 \\ &= 1 - (2s+1)^{-1}\left[3 j_1(k_F r)/k_F r\right]^2 \end{aligned} \tag{2.22}$$

where $j_1(x) = x^{-2}(\sin x - x\cos x)$ is a spherical Bessel function and we have used the relation $n = k_F^3(2s+1)/6\pi^2$. Note that $g(r) \approx 2s(2s+1)^{-1}$ as $r \longrightarrow 0$ and approaches 1 for $k_F r \gg 1$. This reduction below the classical value of 1 reflects the Pauli exclusion principle which keeps particles with the same spin projection apart. Since $k_F^{-1} \approx n^{-1/3}$ is comparable with the interparticle spacing, each particle is surrounded by an "exchange hole" of radius roughly the nearest-neighbor spacing. For spin 1/2, the corresponding $g(r)$ has the approximate behavior shown in Fig. 2.1 with the first zero of $j_1(x)$ occurring at ≈ 4.49.

In principle, the Fourier transform of Eq. (2.22) gives the static structure function, but it is simpler to work directly with Eq. (2.8), interpreted as a second quantized expression. For $q \neq 0$, use of Eq. (2.11) and the generalized Wick's theorem gives

$$\begin{aligned} S(\underset{\sim}{q}) &= N^{-1}\langle \hat{n}(\underset{\sim}{q})\,\hat{n}(-\underset{\sim}{q})\rangle \\ &= N^{-1}\sum_{\underset{\sim}{p}\underset{\sim}{p}'}\sum_{\lambda\lambda'} \langle a^\dagger_{\underset{\sim}{p}'-\underset{\sim}{q},\lambda'}\, a_{\underset{\sim}{p}'\lambda'}\, a^\dagger_{\underset{\sim}{p}+\underset{\sim}{q},\lambda}\, a_{\underset{\sim}{p}\lambda}\rangle \end{aligned}$$

$$= N^{-1}(2s+1)\sum_{\underset{\sim}{p}} n^{\circ}_{\underset{\sim}{p}+\underset{\sim}{q}}(1-n^{\circ}_{p})$$

$$= 1 - N^{-1}(2s+1)\sum_{\underset{\sim}{p}} n^{\circ}_{\underset{\sim}{p}+\underset{\sim}{q}}\, n^{\circ}_{p} \tag{2.23}$$

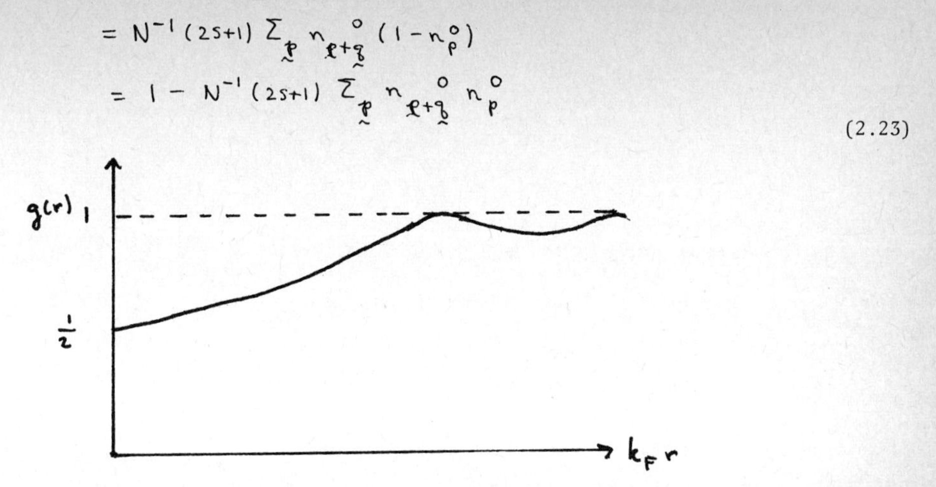

Fig. 2.1. Pair distribution function for ideal spin 1/2 Fermi gas at zero temperature.

The last sum is just that needed in evaluating the first correction to the ground-state energy of an electron gas [see Ref. 2, Eq. (3.35)] and gives

$$S(k) = \begin{cases} (3x - x^3)/2 & x = k/2k_F < 1 \\ 1 & x = k/2k_F > 1 \end{cases} \tag{2.24}$$

for the static structure function of a degenerate Fermi gas. Note that S(k) has a

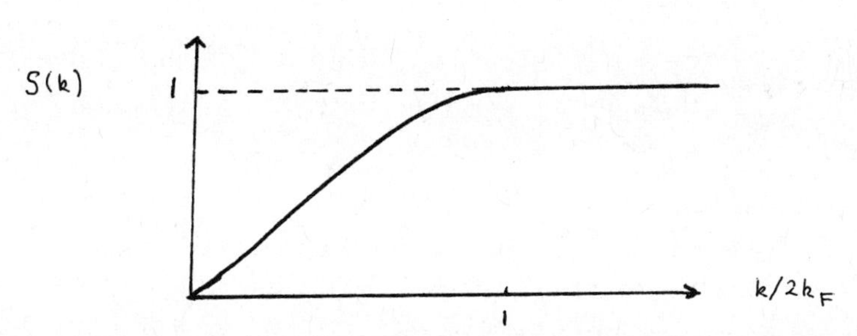

Fig. 2.2. Static structure function for ideal Fermi gas at zero temperature.

discontinuous second derivative at $2k_F$, and that it falls below the classical value (1) for $k \leqslant 2k_F$, again reflecting the Pauli exclusion principle.

It is interesting to compare these simple models with the behavior found in real systems. Figure 2.3 shows the form of S(k) for liquid ^{4}He at T ≈ 0K and the corresponding g(r) obtained by numerical integration. Here the hole in g(r) at small distances arises from the hard repulsive cores in the interparticle potential, and the peaks reflect the predominance of nearest neighbors and next-nearest neighbors. Typically, S(k) has peaks associated with the characteristic Fourier components of the density correlations. Similar features are seen in classical liquids such as Argon.

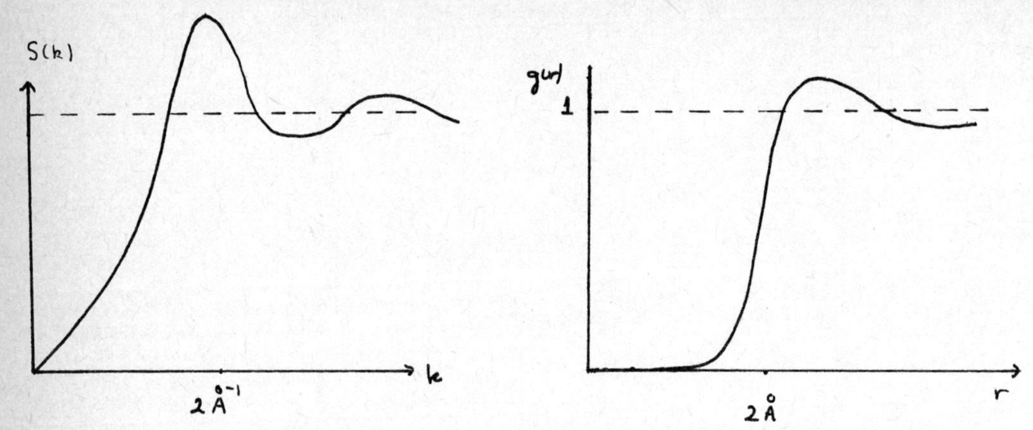

Fig. 2.3. Schematic static structure function and pair distribution function for liquid ^{4}He at T = 0 K.

III. Neutron Scattering

In addition to x-ray scattering, which is one of the oldest techniques for studying condensed matter, many other probes exist. Here we shall consider one particularly valuable and versatile one — neutron scattering. The basic observation is that the interaction between a neutron and a nucleus is strong but short range ($\approx 10^{-13}$cm). Thus the scattering amplitude $f(\underset{\sim}{k}',\underset{\sim}{k})$ for a neutron to scatter from $\underset{\sim}{k}$ to $\underset{\sim}{k}'$ has the characteristic wavenumber scale 10^{13} cm^{-1}, and can be approximated by its long-wavelength (low-energy) value $-a$ whenever the neutron wavelengths are large compared to 10^{-13} cm. Here a ($\approx 10^{-13}$ cm) is the s-wave scattering length for scattering off a bound nucleus. For example, a neutron with 1eV has $k \approx 2\text{x}10^{9}$ cm^{-1}, so that the product ka is indeed small. To study the scattering, it is convenient to follow Fermi and introduce a "pseudopotential"

$$V(\underset{\sim}{r}-\underset{\sim}{r}_j) = \frac{4\pi a\hbar^2}{2m_n}\,\delta(\underset{\sim}{r}-\underset{\sim}{r}_j) \tag{3.1}$$

for interaction between the neutron at $\underset{\sim}{r}$ and the fixed nucleus at $\underset{\sim}{r}_j$. In Born approximation, the scattering length a_B for Eq. (3.1) is precisely the exact value a. Thus this potential, <u>used in Born approximation only</u>, gives a correct low-energy description. For an assembly of identical spinless atoms, the total interaction potential with the neutron is

$$H^{ex}(\underset{\sim}{r}) = \frac{4\pi a\hbar^2}{2m_n}\sum_j \delta(\underset{\sim}{r}-\underset{\sim}{r}_j) \tag{3.2}$$

and the sum over j may be recognized as the atomic number density $n(\underset{\sim}{r})$. Hence H^{ex} has the second-quantized representation

$$\hat{H}^{ex}(\underset{\sim}{r}) = \frac{4\pi a\hbar^2}{2m_n}\,\hat{n}(\underset{\sim}{r}) \tag{3.3}$$

To study the scattering by a target with volume V, we introduce incident and final neutron wave functions $\Omega^{-1/2}e^{i\underset{\sim}{k}\cdot\underset{\sim}{r}}\eta_\lambda$ and $\Omega^{-1/2}e^{i\underset{\sim}{k}'\cdot\underset{\sim}{r}}\eta_{\lambda'}$ where Ω is the quantization volume for the neutron and presumably exceeds V. The corresponding matrix element of Eq. (3.3) for a transition with the target states $|i\rangle \rightarrow |f\rangle$ is

$$\begin{aligned}
&\Omega^{-1}\int d^3r\, e^{-i\underset{\sim}{k}'\cdot\underset{\sim}{r}}\,\eta_{\lambda'}^{\dagger}\langle f|\hat{H}^{ex}(\underset{\sim}{r})|i\rangle\, e^{i\underset{\sim}{k}\cdot\underset{\sim}{r}}\,\eta_\lambda \\
&= \frac{4\pi a\hbar^2}{2m_n\Omega}\,\delta_{\lambda\lambda'}\int d^3r\, e^{i(\underset{\sim}{k}-\underset{\sim}{k}')\cdot\underset{\sim}{r}}\langle f|\hat{n}(\underset{\sim}{r})|i\rangle \\
&= \frac{4\pi a\hbar^2}{2m_n\Omega}\,\delta_{\lambda\lambda'}\langle f|\hat{n}^{\dagger}(\underset{\sim}{q})|i\rangle
\end{aligned} \tag{3.4}$$

where $\underset{\sim}{q} = \underset{\sim}{k}-\underset{\sim}{k}'$ is the wavevector transferred to the target. The generalization to spin-dependent scattering is straightforward, as is the inclusion of different nuclear isotopes[4]. The transition rate from an initial state $|i,\underset{\sim}{k}\lambda\rangle$ to a final state $|f,\underset{\sim}{k}'\lambda'\rangle$ is given by the golden rule as

$$2\pi\hbar^{-1}\left|\langle f,\underset{\sim}{k}'\lambda'|\hat{H}^{ex}|i,\underset{\sim}{k}\lambda\rangle\right|^2\delta(\hbar\omega-E_f+E_i) \tag{3.5}$$

where E_f and E_i are the target energies and $\hbar\omega=(\hbar^2/2m_n)(k^2-k'^2)$ is the energy lost by the neutron. A typical experiment does not measure the final target state, so we sum over f. Moreover, the initial target states are distributed according to the probability $e^{-\beta E_i}/Z$, where Z is the partition function. The cross section $d^2\sigma$ for a transition to a final neutron state in an interval $\Delta\underset{\sim}{k}'\equiv\Omega d^3k'(2\pi)^{-3}$ is obtained by summing over these states, averaging over the initial neutron states (assumed unpolarized), and dividing by the incident neutron flux $v_{rel}/\Omega=\hbar k/m_n\Omega$

$$d^2\sigma=\frac{2\pi}{\hbar\Omega^2}\left(\frac{4\pi a\hbar^2}{2m_n}\right)^2\frac{1}{2}\sum_{\lambda\lambda'}\delta_{\lambda\lambda'}\sum_{fi}\left|\langle f|\hat{n}(\underset{\sim}{q})^\dagger|i\rangle\right|^2 \times\delta(\hbar\omega-E_f+E_i)\,Z^{-1}e^{-\beta E_i}\,\frac{\Omega\,d^3k'}{(2\pi)^3}\,\frac{m_n\Omega}{\hbar k}\;; \tag{3.6}$$

we see that Ω cancels out. Simple manipulationsthen lead to the double differential cross section

$$\frac{d^2\sigma}{d\Omega' d\omega}=\frac{Na^2k'}{k}S(\underset{\sim}{q},\omega) \tag{3.7}$$

for scattering into a final solid angle at Ω', with energy loss $\hbar\omega$. Here $S(\underset{\sim}{q},\omega)$ is the "dynamic structure function"

$$S(\underset{\sim}{q},\omega)=(NZ)^{-1}\sum_{fi}e^{-\beta E_i}\left|\langle f|\hat{n}^\dagger(\underset{\sim}{q})|i\rangle\right|^2\delta\left(\omega-\frac{E_f-E_i}{\hbar}\right) \tag{3.8}$$

Note that Eq. (3.7) factors into a projectile part and a target part, with all the properties of the target contained in $S(\underset{\sim}{q},\omega)$, which is effectively the relative probability that if the neutron transfers momentum $\hbar\underset{\sim}{q}$ to the target, it will also transfer energy $\hbar\omega$.

Evidently, $S(\underset{\sim}{q},\omega)$ is very complicated, for it involves the exact target eigenstates. Nevertheless, it has some very simple properties that can help guide one's intuition.

1. If $\underset{\sim}{q}=0$, the neutron does not scatter; thus we may take $\underset{\sim}{q}\neq 0$ in general. In that case, the Fourier component $\hat{n}(\underset{\sim}{q})$ is that for the density fluctuation operator $\tilde{n}(\underset{\sim}{q})$ because the constant density $\langle n\rangle$ has only the $\underset{\sim}{q}=0$ component in a extended uniform medium. The frequency integral of S then becomes

$$\begin{aligned}\int_{-\infty}^{\infty}d\omega\,S(\underset{\sim}{q},\omega)&=(NZ)^{-1}\sum_{fi}e^{-\beta E_i}\left|\langle f|\tilde{n}(\underset{\sim}{q})^\dagger|i\rangle\right|^2\\&=(NZ)^{-1}\sum_{fi}e^{-\beta E_i}\langle i|\tilde{n}(\underset{\sim}{q})|f\rangle\langle f|\tilde{n}(\underset{\sim}{q})^\dagger|i\rangle\\&=(NZ)^{-1}\sum_i e^{-\beta E_i}\langle i|\tilde{n}(\underset{\sim}{q})\,\tilde{n}(\underset{\sim}{q})^\dagger|i\rangle\\&=N^{-1}\langle\tilde{n}(\underset{\sim}{q})\,\tilde{n}(\underset{\sim}{q})^\dagger\rangle\end{aligned} \tag{3.9}$$

where the third line follows from completeness of the target states and the last from the definition of a thermal ensemble. Comparison with Eq. (2.8) shows that this quantity is just the static structure function

$$\int_{-\infty}^{\infty} d\omega \, S(\underset{\sim}{q},\omega) = S(\underset{\sim}{q}) \tag{3.10}$$

Hence neutron scattering also can measure $S(\underset{\sim}{q})$, which is effectively the relative probability of scattering with momentum loss $\hbar\underset{\sim}{q}$, irrespective of energy loss.

2. The delta function in Eq. (3.8) can be written as a Fourier integral over a dummy variable t:

$$S(\underset{\sim}{q},\omega) = (2\pi N Z)^{-1} \int_{-\infty}^{\infty} dt \, e^{i\omega t} \sum_{fi} e^{-\beta E_i} \langle i|\tilde{n}(\underset{\sim}{q})|f\rangle\langle f|\tilde{n}(\underset{\sim}{q})^{\dagger}|i\rangle \times \exp[i(E_i - E_f)t/\hbar] \tag{3.11}$$

Furthermore the last factor can be rewritten in terms of the exact hamiltonian $\hat{H}$ of the target, including all the interactions:

$$\langle i|\tilde{n}(\underset{\sim}{q})|f\rangle \exp[i(E_i - E_f)t/\hbar] = \langle i| e^{i\hat{H}t/\hbar} \tilde{n}(\underset{\sim}{q}) e^{-i\hat{H}t/\hbar}|f\rangle$$

This last expression is just the matrix element of the time-dependent Heisenberg operator $\langle i|\tilde{n}_H(\underset{\sim}{q},t)|f\rangle$. In this way, Eq. (3.11) becomes

$$\begin{aligned} S(\underset{\sim}{q},\omega) &= (2\pi N Z)^{-1} \int_{-\infty}^{\infty} dt \, e^{i\omega t} \sum_i e^{-\beta E_i} \langle i|\tilde{n}_H(\underset{\sim}{q},t)\, \tilde{n}_H^{\dagger}(\underset{\sim}{q},0)|i\rangle \\ &= (2\pi N)^{-1} \int_{-\infty}^{\infty} dt \, e^{i\omega t} \langle \tilde{n}_H(\underset{\sim}{q},t)\, \tilde{n}_H^{\dagger}(\underset{\sim}{q},0)\rangle \end{aligned} \tag{3.12}$$

again using completeness. Thus $S(\underset{\sim}{q},\omega)$ is the time Fourier transform of a time-dependent density-density correlation function. Introducing the space-time operators $\tilde{n}_H(\underset{\sim}{r},t)$, we obtain

$$S(\underset{\sim}{q},\omega) = (2\pi N)^{-1} \int dt \, e^{i\omega t} \int d^3r \, d^3r' \, e^{i\underset{\sim}{q}\cdot(\underset{\sim}{r}'-\underset{\sim}{r})} \langle \tilde{n}_H(\underset{\sim}{r},t)\, \tilde{n}_H(\underset{\sim}{r}',0)\rangle \tag{3.13}$$

expressed in terms of the space-time correlation between density fluctuations at $(\underset{\sim}{r},t)$ and $(\underset{\sim}{r}',0)$. Thus neutron scattering provides a direct measure of time-dependent correlations, in contrast to the static correlations seen in x-ray scattering.

The dynamic structure function has an important symmetry under the transformation $(\underset{\sim}{q},\omega) \to (-\underset{\sim}{q},-\omega)$. To derive the relation, consider

$$S(-\underset{\sim}{q},-\omega) = (NZ)^{-1} \sum_{if} e^{-\beta E_i} \langle i|\hat{n}(-\underset{\sim}{q})|f\rangle\langle f|\hat{n}^{\dagger}(-\underset{\sim}{q})|i\rangle \, \delta\left(\omega + \frac{E_f - E_i}{\hbar}\right)$$

Since $\hat{n}(\underset{\sim}{q}) = \hat{n}^{\dagger}(-\underset{\sim}{q})$, this double sum can be rewritten as follows by interchanging the dummy variables i and f:

$$S(-\underset{\sim}{q},-\omega) = (NZ)^{-1} \sum_{if} e^{-\beta E_f} \langle f|\hat{n}^{\dagger}(\underset{\sim}{q})|i\rangle\langle i|\hat{n}(\underset{\sim}{q})|f\rangle \, \delta\left(\omega - \frac{E_f - E_i}{\hbar}\right)$$

The delta function requires $E_f = \hbar\omega + E_i$, so that this last equation may be recast in the desired form

$$S(-\mathbf{q},-\omega) = e^{-\beta\hbar\omega} S(\mathbf{q},\omega) \tag{3.14}$$

In an isotropic medium, S depends only on $|\mathbf{q}|$, so that Eq. (3.14) then takes the simpler form $S(q,-\omega)=e^{-\beta\hbar\omega}S(q,\omega)$. At low temperature, the factor $e^{-\beta\hbar\omega}$ vanishes for $\omega > 0$, indicating that $S(q,\omega)$ is exponentially small for negative frequencies. This merely reflects the simple observation that the neutron can gain energy only if the target is initially in an excited state, which becomes very improbable as $T \to 0K$.

One final general property of $S(q,\omega)$ is an important sum rule for the first frequency moment. Consider the integral

$$\int_{-\infty}^{\infty} d\omega \; \omega \; S(\mathbf{q},\omega) = (NZ\hbar)^{-1} \sum_{fi} e^{-\beta E_i} \left| \langle f | \hat{n}^\dagger(\mathbf{q}) | i \rangle \right|^2 (E_f - E_i)$$

Assuming that the system has inversion symmetry under the transformation $\mathbf{q} \to -\mathbf{q}$, we can rewrite the right-hand side as

$$\begin{aligned}(2NZ\hbar)^{-1} \sum_{fi} e^{-\beta E_i} \{ &\langle i | \hat{n}(\mathbf{q}) | f \rangle (E_f - E_i) \langle f | \hat{n}^\dagger(\mathbf{q}) | i \rangle \\ &+ \langle i | \hat{n}^\dagger(\mathbf{q}) | f \rangle (E_f - E_i) \langle f | \hat{n}(\mathbf{q}) | i \rangle \} \\ = (2N\hbar)^{-1} &\langle [\hat{n}(\mathbf{q}), [\hat{H}, \hat{n}^\dagger(\mathbf{q})]] \rangle \end{aligned} \tag{3.15}$$

where $\hat{H}$ is again the full hamiltonian of the interacting target. Since the density operator $\hat{n}(\mathbf{r})$ commutes with the potential energy operator $\hat{V}$ for a velocity independent interparticle potential, the only contribution to Eq. (3.15) is from the kinetic energy $\hat{T} = \sum_{\mathbf{p}\lambda} (\hbar^2 p^2 / 2m_n) a^\dagger_{\mathbf{p}\lambda} a_{\mathbf{p}\lambda}$. A straightforward calculation using Eq. (2.11) eventually yields the remarkably simple answer

$$\int_{-\infty}^{\infty} d\omega \; \omega \; S(\mathbf{q},\omega) = \hbar q^2 / 2m_n \tag{3.16}$$

which must hold for every value of $\mathbf{q}$. It places quite severe limits on the allowed form of $S(\mathbf{q}, \omega)$.

It is instructive to consider the form of $S^0(\mathbf{q},\omega)$ for an ideal gas with particle mass m. The relevant matrix elements of the number operator (2.11) can be written

$$\langle i | \hat{n}(\mathbf{q}) | f \rangle \langle f | \hat{n}(-\mathbf{q}) | i \rangle = \sum_{\mathbf{k}\lambda} \sum_{\mathbf{p}\lambda'} \langle i | a^\dagger_{\mathbf{k}\lambda} a_{\mathbf{k}+\mathbf{q},\lambda} | f \rangle \langle f | a^\dagger_{\mathbf{p}+\mathbf{q},\lambda'} a_{\mathbf{p}\lambda'} | i \rangle$$

Furthermore, the exact eigenstates of an ideal gas are direct products of normal-mode states, so that $|f\rangle$ must have one less particle with quantum numbers $\mathbf{p}\lambda'$ and one more with $\mathbf{p} + \mathbf{q},\lambda'$, and the operators with $\mathbf{k}\lambda$ must then restore the initial conditions. Hence we require $\mathbf{k} = \mathbf{p}$ and $\lambda = \lambda'$. In addition, $E_f - E_i = (\hbar^2/2m)[(\mathbf{k}+\mathbf{q})^2 - k^2] = (\hbar^2/2m)(\mathbf{k}\cdot\mathbf{q} + \frac{1}{2}q^2)$ independent of any details of $|f\rangle$. We may now use completeness of the states $|f\rangle$ to evaluate $S^0(q,\omega)$ as

$$S^0(\underset{\sim}{q},\omega) = N^{-1}\sum_{\underset{\sim}{k}\lambda} \langle a^\dagger_{\underset{\sim}{k}\lambda} a_{\underset{\sim}{k}+\underset{\sim}{q},\lambda} a^\dagger_{\underset{\sim}{k}+\underset{\sim}{q},\lambda} a_{\underset{\sim}{k}\lambda}\rangle \,\delta(\omega - \hbar \underset{\sim}{q}\cdot\underset{\sim}{k}/m - \hbar q^2/2m)$$

$$= (2s+1)N^{-1}\sum_{\underset{\sim}{k}} \delta(\omega - \hbar\underset{\sim}{q}\cdot\underset{\sim}{k}/m - \hbar q^2/2m)\, n^0_{\underset{\sim}{k}}(1 \pm n^0_{\underset{\sim}{k}+\underset{\sim}{q}}) \tag{3.17}$$

where n_k^0 is the ideal gas distribution function (2.15) and Wick's theorem has been used as in Eq. (2.14). Although Eq. (3.17) can be evaluated explicitly for any T and μ, the expressions are cumbersome and we shall here treat only the classical limit in detail. In that case, the occupation numbers are much less than unity, so that $S^0(\underset{\sim}{q},\omega)$ has the classical limit [see Eq. (2.17)]

$$S^0(\underset{\sim}{q},\omega) \approx \frac{2s+1}{n}\int\frac{d^3k}{(2\pi)^3}\,\delta(\omega - \hbar\underset{\sim}{k}\cdot\underset{\sim}{q}/m - \hbar q^2/2m)\, e^{\beta\mu} e^{-\beta\varepsilon^0_k} \tag{3.18}$$

A straightforward calculation gives

$$S^0(\underset{\sim}{q},\omega) = \left(\frac{m}{2\pi q^2 k_B T}\right)^{3/2} \exp\left[-\frac{m}{2q^2 k_B T}\left(\omega - \frac{\hbar q^2}{2m}\right)^2\right] \tag{3.19}$$

where Eq. (2.18) has been used. It is easy to verify that this expression satisfies the symmetry relation (3.14) and the two sum rules (3.10) and (3.16). Considered as a function of ω at fixed q, Eq. (3.19) has a peak at the quasielastic value $\omega_0(q) = \hbar q^2/2m$, which is just the kinematic relation between the energy and momentum of a single free recoiling target particle initially at rest. The width of this peak is of order $q(k_B T/m)^{1/2} \approx q\langle v^2\rangle^{1/2}$, where $\langle v^2\rangle^{1/2}$ is the rms velocity in the gas. Hence the "quasielastic peak" in $S^0(q,\omega)$ measures the internal motions in the target.

A similar picture of an ideal gas holds at all temperatures, with the quasielastic peak at $\omega_0(q)$ broadeded by the mean velocity of the constituents. In particular, an ideal Fermi gas at zero temperature has a quasielastic peak with a width of order qv_F, where v_F is the Fermi velocity. For $q > 2k_F$, the corresponding $S^0(q,\omega)$ is parabolic, and vanishes identically outside certain kinematic regions. This behavior arises from the sharp Fermi surface and differs considerably from the Gaussian tail in the classical limit. The detailed calculation of $S^0(q,\omega)$ for this case can be found in Ref. 2, pp. 159-163, and the comparison with data from electron scattering off nuclei is illustrated on p. 194.

IV. LINEAR - RESPONSE THEORY

X-ray scattering and neutron scattering have one important feature in common, for each can be treated as a weak external perturbation on the target. It is helpful to take a more general view of this situation, by considering an arbitrary weak perturbation $\hat{H}^{ex}(t)$ that is turned on at some time t_0 (hence $\hat{H}^{ex} = 0$ for $t < t_0$)[5].

It is convenient to work in the Schrödinger picture, where the state vector $|\Psi_S(t)\rangle$ is explicitly time dependent. This is just the usual description of elementary quantum mechanics (see Ref. 2, pp. 53-59 for a discussion of various pictures). If $\hat{H}$ is the full hamiltonian of the interacting system in the absence of the perturbation $\hat{H}^{ex}$, then the exact state vector obeys the Schrödinger equation

$$i\hbar\frac{\partial}{\partial t}|\Psi_S(t)\rangle = \hat{H}|\Psi_S(t)\rangle \tag{4.1}$$

before $\hat{H}^{ex}$ is turned on. Thus for $t < t_0$, $|\Psi_S(t)\rangle$ is given explicitly by

$$|\Psi_S(t)\rangle = e^{-i\hat{H}t/\hbar}|\Psi_S(0)\rangle \tag{4.2}$$

For $t > t_0$, however, the total hamiltonian is $\hat{H} + \hat{H}^{ex}(t)$, where $\hat{H}^{ex}$ may have explicit time dependence associated with finite frequencies or other behavior. As a result, Eq. (4.1) must be altered to

$$i\hbar\frac{\partial}{\partial t}|\overline{\Psi_S(t)}\rangle = \left[\hat{H} + \hat{H}^{ex}(t)\right]|\overline{\Psi_S(t)}\rangle \qquad t > t_0 \tag{4.3}$$

where $|\overline{\Psi_S(t)}\rangle$ is the new state vector that satisfies the modified Schrödinger equation.

Since $\hat{H}^{ex}$ is to be considered weak, it is natural to try to express $|\overline{\Psi_S}\rangle$ in terms of $|\Psi_S\rangle$, given in Eqs. (4.1) and (4.2). In particular, we shall seek a solution in the form

$$|\overline{\Psi_S(t)}\rangle = e^{-i\hat{H}t/\hbar}\,\hat{A}(t)\,|\Psi_S(0)\rangle \tag{4.4}$$

where $\hat{A}(t)$ is an operator to be determined. Evidently, $\hat{A}(t) = 1$ for $t < t_0$. The equation for $\hat{A}$ follows by differentiating with respect to t:

$$i\hbar\frac{\partial}{\partial t}|\overline{\Psi_S(t)}\rangle = i\hbar\frac{\partial}{\partial t}\left[e^{-i\hat{H}t/\hbar}\hat{A}(t)\right]|\Psi_S(0)\rangle$$

$$= e^{-i\hat{H}t/\hbar}\left[\hat{H}\hat{A}(t) + i\hbar\frac{\partial}{\partial t}\hat{A}(t)\right]|\Psi_S(0)\rangle$$

Since the Schrödinger equation (4.3) implies that the left-hand side is also equal to

$$\left[\hat{H} + \hat{H}^{ex}(t)\right]e^{-i\hat{H}t/\hbar}\hat{A}(t)|\Psi_S(0)\rangle$$

a little manipulation yields

$$i\hbar\frac{\partial}{\partial t}\hat{A}(t) = \hat{H}^{ex}_H(t)\,\hat{A}(t) \tag{4.5}$$

where

$$\hat{H}_H^{ex}(t) = e^{i\hat{H}t/\hbar}\,\hat{H}^{ex}(t)\,e^{-i\hat{H}t/\hbar} \tag{4.6}$$

expresses $\hat{H}^{ex}$ in the Heisenberg picture with respect to the fully interacting but unperturbed $\hat{H}$. Equation (4.5) has the same form as that for the operator $\hat{U}$ in the interaction picture, and it may be iterated in exactly the same way. To lowest order, we find

$$\hat{A}(t) = 1 - \frac{i}{\hbar}\int_{t_0}^{t} dt'\,\hat{H}_H^{ex}(t') + \cdots \tag{4.7}$$

where the integral vanishes if $t < t_0$ because the integrand does so. It follows from Eq. (4.4) that the corresponding exact state vector becomes

$$|\overline{\Psi_S(t)}\rangle \approx e^{-i\hat{H}t/\hbar}\left[1 - \frac{i}{\hbar}\int_{t_0}^{t} dt'\,\hat{H}_H^{ex}(t')\right]|\Psi_S(0)\rangle \tag{4.8}$$

The physically interesting quantity is the matrix element of some operator $\hat{O}(t)$ in the presence of the perturbation $\hat{H}^{ex}$. This may be expressed in the Schrödinger picture as

$$\begin{aligned}
\langle\hat{O}(t)\rangle_{ex} &= \langle\overline{\Psi_S(t)}|\,\hat{O}_S(t)\,|\overline{\Psi_S(t)}\rangle \\
&\approx \langle\Psi_S(0)|\left[1 + \frac{i}{\hbar}\int_{t_0}^{t} dt'\,\hat{H}_H^{ex}(t')\right] e^{i\hat{H}t/\hbar}\,\hat{O}_S(t) \\
&\qquad \times e^{-i\hat{H}t/\hbar}\left[1 - \frac{i}{\hbar}\int_{t_0}^{t} dt'\,\hat{H}_H^{ex}(t')\right]|\Psi_S(0)\rangle \\
&\approx \langle\Psi_H(0)|\,\hat{O}_H(t)\,|\Psi_H(0)\rangle \\
&\qquad - i\hbar^{-1}\langle\Psi_H(0)|\int_{t_0}^{t} dt'\,[\hat{O}_H(t), \hat{H}_H^{ex}(t')]\,|\Psi_H(0)\rangle
\end{aligned} \tag{4.9}$$

where $\hat{O}_H(t) = e^{i\hat{H}t/\hbar}\,\hat{O}_S(t)\,e^{-i\hat{H}t/\hbar}$ is the exact Heisenberg operator of the unperturbed system, and we have noted that $|\Psi_S(0)\rangle = |\Psi_H(0)\rangle$. In particular, the <u>change</u> in the expectation value of the operator $\hat{O}(t)$ is given by

$$\begin{aligned}
\delta\langle\hat{O}(t)\rangle &= \langle\hat{O}(t)\rangle_{ex} - \langle\hat{O}(t)\rangle \\
&= -i\hbar^{-1}\int_{t_0}^{t} dt'\,\langle[\hat{O}_H(t), \hat{H}_H^{ex}(t')]\rangle
\end{aligned} \tag{4.10}$$

and this formula remains correct for finite temperatures if the angular brackets are interpreted as an ensemble average. In this way, the linear response of the operator $\hat{O}(t)$ is directly expressible in terms of a retarded commutator (since $t' < t$).

To make this abstract expression more concrete, it is helpful to consider the special case of a one-component charged system (electrons in a metal, for example) with a charge q per particle and subject to an external scalar potential $\phi^{ex}(\underset{\sim}{r},t)$. that is turned on at t_0. It follows from elementary electrostatics that the interaction energy is given by

$$\hat{H}^{ex}_H(t) = q \int d^3r' \, \hat{n}_H(\underset{\sim}{r}',t) \, \phi^{ex}(\underset{\sim}{r}',t) \tag{4.11}$$

To characterize the linear response, we may consider the change in the density itself

$$\delta\langle\hat{n}(\underset{\sim}{r},t)\rangle = -i\hbar^{-1}\int_{-\infty}^{t} dt' \int d^3r' \langle[\tilde{n}_H(\underset{\sim}{r},t), \tilde{n}_H(\underset{\sim}{r}',t')]\rangle \, q\phi^{ex}(\underset{\sim}{r}',t') \tag{4.12}$$

where the time integral can run from $-\infty$ since ϕ^{ex} vanishes for $t < t_0$, and the change from $\hat{n}$ to the fluctuation in density does not affect the commutator. We now introduce the retarded density correlation function

$$iD^R(\underset{\sim}{r}t,\underset{\sim}{r}'t') = \theta(t-t')\langle[\tilde{n}_H(\underset{\sim}{r},t), \tilde{n}_H(\underset{\sim}{r}',t')]\rangle \tag{4.13}$$

where θ is the unit positive step function. Equation (4.12) then becomes

$$\delta\langle\hat{n}(\underset{\sim}{r},t)\rangle = \hbar^{-1}\int_{-\infty}^{\infty} dt' \int d^3r' \, D^R(\underset{\sim}{r}t,\underset{\sim}{r}'t') \, q\phi^{ex}(\underset{\sim}{r}',t') \tag{4.14}$$

where the step function in D^R allows the t' integral to run to $+\infty$.

In a uniform homogeneous bulk system, D^R will depend only on $\underset{\sim}{r}-\underset{\sim}{r}'$ and $t-t'$, so that it has the Fourier expansion

$$D^R(\underset{\sim}{r}-\underset{\sim}{r}',t-t') = \int\frac{d^3k}{(2\pi)^3}\int\frac{d\omega}{2\pi} \, e^{i\underset{\sim}{k}\cdot(\underset{\sim}{r}-\underset{\sim}{r}')} \, e^{-i\omega(t-t')} \, D^R(\underset{\sim}{k},\omega) \tag{4.15}$$

Expressing $\phi^{ex}(\underset{\sim}{r},t)$ as a Fourier integral, we readily find the corresponding response of the system at wavevector $\underset{\sim}{k}$ and frequency ω:

$$\delta\langle\hat{n}(\underset{\sim}{k},\omega)\rangle = \hbar^{-1} D^R(\underset{\sim}{k},\omega) \, q\phi^{ex}(\underset{\sim}{k},\omega) \tag{4.16}$$

Note that D^R is the exact retarded correlation function for the fully interacting system at some temperature T. Thus it is not exactly calculable except in special simple models. Nevertheless, the important property of causality [the step function in Eq. (4.13)] implies that $D^R(\underset{\sim}{k},\omega)$ is an analytic function throughout the upper half ω plane. This follows by inverting Eq. (4.15)

$$D^R(\underset{\sim}{k},\omega) = \int_0^\infty dt \, e^{i\omega t} D^R(\underset{\sim}{k},t) \tag{4.17}$$

If $\omega = \omega_1 + i\omega_2$, then $D^R(\underset{\sim}{k},\omega_1+i\omega_2)$ contains the exponential factor $\exp(-\omega_2 t)$, which vanishes rapidly for $t \to \infty$ and $\omega_2 > 0$. Furthermore, the nth derivative

$$\left(\frac{d}{d\omega}\right)^n D^R(\underset{\sim}{k},\omega) = i^n\int_0^\infty dt \, t^n e^{i\omega_1 t} e^{-\omega_2 t} D^R(\underset{\sim}{k},t) \tag{4.18}$$

is well defined for $\omega_2 > 0$, so that one may construct a Taylor series at any point in the upper half ω plane, indicating that $D^R(\underset{\sim}{k},\omega)$ is indeed analytic for Im $\omega > 0$.

Although $D^R(\underset{\sim}{k},\omega)$ can be quite complicated, one's intuition can often help in inferring its detailed form. For example, a charged plasma is known to exhibit a long-wavelength resonant response at the plasma frequency $\Omega_p = (4\pi nq^2/m)^{1/2}$ with weak damping constant γ. As a result, $D^R(\underset{\sim}{k} \to 0, \omega)$ has a pole near $\omega \approx \pm\Omega_p - i\gamma$

properly located in the lower-half plane. One corollary of this behavior is that the cross section for inelastic scattering of an electron by a metallic film with the emission of a plasma oscillation has a Lorentzian peak near Ω_p with width γ (see Ref. 2, Secs. 15, 17, and 34).

V. DYNAMIC COMPRESSIBILITY

In contrast to the finite-frequency plasma oscillations found in a charged system, a neutral system is expected to exhibit compressional density waves with a long-wavelength dispersion relation $\omega \simeq uk$, where u is the speed of sound. To study such phenomena with linear-response theory, it is necessary to formulate the appropriate $\hat{H}^{ex}$ that incorporates the response to an external pressure variation. Consider a small volume element V with N particles. The usual expression for the work done in a process that changes the pressure and volume by small amounts is

$$\delta W = -p\,\delta V \tag{5.1}$$

where δV is the change in the volume V of the element. Since the number density is just $n = N/V$, the change in volume is related to a change in density by $\delta V = -N\delta n/n^2$. Thus the work done on the small volume element becomes

$$\delta W = V p\,\delta n/n \tag{5.2}$$

For an extended region, the total work is just the spatial integral

$$\delta W = \int d^3r\,(p/n_0)\,\delta\tilde{n} \tag{5.3}$$

where we think of keeping the mean density n_0 fixed and hence have introduced the density fluctuation $\tilde{n}$ measured from the mean background.

When Eq. (5.3) is combined with the first and second laws of thermodynamics, we may write the change of internal energy E as

$$\delta E = T\,\delta S + \int d^3r\,(p/n_0)\,\delta\tilde{n} \tag{5.4}$$

Correspondingly, the change in Helmholtz free energy $F = E - TS$ becomes

$$\delta F = -S\,\delta T + \int d^3r\,(p/n_0)\,\delta\tilde{n} \tag{5.5}$$

These relations have the important consequence that the internal energy is to be considered a function of the entropy S and the number density $\tilde{n}$ (or the Helmholtz free energy a function of T and $\tilde{n}$). Unfortunately, $\tilde{n}$ is not a very convenient variable to control externally, and it is therefore preferable to make a Legendre transformation from $\tilde{n}$ to its conjugate variable p/n_0. Thus we consider the "enthalpy"

$$\tilde{E} = E - \int d^3r\,p\tilde{n}/n_0 \tag{5.6}$$

whose change is immediately given by

$$\delta\tilde{E} = T\,\delta S - \int d^3r\,(\tilde{n}/n_0)\,\delta p \tag{5.7}$$

This last term is to be interpreted as the appropriate perturbation on the system when the pressure is the variable under our direct control, and we infer that the correct perturbation hamiltonian for an applied pressure variation $\delta p(\underset{\sim}{r},t)$ is

$$\hat{H}^{ex}(t) = -\int d^3r\, \tilde{n}(\underset{\sim}{r},t)\, n_0^{-1}\, \delta p(\underset{\sim}{r},t) \tag{5.8}$$

As expected, the system lowers its energy by increasing $\tilde{n}$ in regions where δp is positive. This formula (and its derivation) is wholly analogous to the familiar magnetic expression $-\underset{\sim}{\mu}\cdot\underset{\sim}{B}$ for the perturbation energy of a magnetic dipole $\underset{\sim}{\mu}$ placed in an external magnetic field $\underset{\sim}{B}$.[6]

Equation (5.8) is now in a form analogous to Eq. (4.11), and we may immediately conclude that the induced density response is given by

$$\delta\tilde{n}(\underset{\sim}{k},\omega) \equiv \delta\langle\tilde{n}(\underset{\sim}{k},\omega)\rangle = -(\hbar n_0)^{-1} D^R(\underset{\sim}{k},\omega)\,\delta p(\underset{\sim}{k},\omega) \tag{5.9}$$

Here $\delta p(\underset{\sim}{k},\omega)$ is the Fourier transform of the perturbation in pressure, characterizing the component at wave vector $\underset{\sim}{k}$ and frequency ω. In the long-wavelength, low-frequency limit, the ratio between $\delta\tilde{n}$ and δp is related to the compressibility $\varkappa = -V^{-1}\,\partial V/\partial p$ as follows

$$\frac{\delta\tilde{n}}{\delta p} = \frac{\partial n}{\partial p} = \frac{\partial}{\partial p}\frac{N}{V} = -\frac{n_0}{V}\frac{\partial V}{\partial p} = n_0\varkappa \tag{5.10}$$

Alternatively, we may define a generalized compressibility $\varkappa(\underset{\sim}{k},\omega)$ for particular Fourier components $(\underset{\sim}{k},\omega)$ by the relation

$$\varkappa(\underset{\sim}{k},\omega) = n_0^{-1}\frac{\delta\tilde{n}(\underset{\sim}{k},\omega)}{\delta p(\underset{\sim}{k},\omega)} = -\frac{1}{\hbar n_0^2} D^R(\underset{\sim}{k},\omega) \tag{5.11}$$

This equation is very important, for it shows how a generalized susceptibility (here the linear density response to an external pressure perturbation) is expressible in terms of an exact correlation function for the unperturbed but interacting many-particle system. We shall study the corresponding magnetic example subsequently, relating the generalized magnetic susceptibility to the retarded spin-density correlation function. Since these correlation functions undergo profound changes in the vicinity of phase transitions, one anticipates that the corresponding generalized susceptibilities will exhibit similar alterations, typically characterized by various familiar critical exponents. In this way, calculations of critical exponentsare related to calculations of the corresponding correlation functions.

One simple case of Eq. (5.11) is to consider the low-frequency or static limit, when the particle motion is expected to be isothermal rather than adiabatic or isentropic. Thus we obtain an expression for the wavenumber dependent isothermal compressibility

$$\varkappa_T(\underset{\sim}{k},0) = -\frac{1}{\hbar n_0^2} D^R(\underset{\sim}{k},0) \tag{5.12}$$

which has the long-wavelength limit

$$\varkappa_T = -\frac{1}{\hbar n_0^2}\lim_{k\to 0} D^R(\underset{\sim}{k},0) \tag{5.13}$$

expressing the thermodynamic derivative $-V^{-1}(\partial V/\partial p)_T$ in terms of the correlation function D^R. A similar relation will be shown later for the static magnetic susceptibility χ. Other linear-response quantities are describable in terms of transport coefficients; for example, the heat current induced by a temperature gradient defines the thermal conductivity, which may therefore be expressed in terms of a correlation function of two heat currents, and similarly for the viscosity and diffusion constant in a mixture. In this way, the singular features of phase transitions may be expected to appear in the transport properties as well as in the thermodynamic derivatives.

So far, we have related $\chi(\underset{\sim}{k},\omega)$ to the Fourier transform of the retarded density-density correlation function $D^R(\underset{\sim}{k},\omega)$. On the other hand, the neutron and x-ray scattering also measure the density correlations in the target, and it is natural to expect some connection between these various quantities. To make this precise, it is helpful to use the Lehmann representation for $D^R(\underset{\sim}{k},\omega)$, which can be derived just as for the single particle Green's function (see Ref. 2, pp. 72-79 for zero temperature, and Secs. 31 and 32 for finite temperature). In particular, the retarded correlation function can be proved to have the simple integral representation

$$D^R(\underset{\sim}{k},\omega) = \int_{-\infty}^{\infty} \frac{d\omega'}{2\pi} \frac{\Delta(\underset{\sim}{k},\omega')}{\omega - \omega' + i\eta} \tag{5.14}$$

which demonstrates that D^R is indeed an analytic function in the upper-half frequency plane. Furthermore, for real ω, the imaginary part of D^R is given by

$$\operatorname{Im} D^R(\underset{\sim}{k},\omega) = -\tfrac{1}{2}\Delta(\underset{\sim}{k},\omega) \tag{5.15}$$

Detailed analysis shows that $\Delta(\underset{\sim}{k},\omega)$ is positive for $\omega > 0$ and related to squares of matrix elements of $\tilde{n}$, and comparison of this expansion with the original definition in Eq. (3.8) yields the useful relation

$$\begin{aligned} S(\underset{\sim}{k},\omega) &= -(\pi n)^{-1}\left(1 - e^{-\beta\hbar\omega}\right)^{-1} \operatorname{Im} D^R(\underset{\sim}{k},\omega) \\ &= (2\pi n)^{-1}\left(1 - e^{-\beta\hbar\omega}\right)^{-1} \Delta(\underset{\sim}{k},\omega) \end{aligned} \tag{5.16}$$

Thus measurement of $S(\underset{\sim}{k},\omega)$ (in principle, at all frequency) fixes $\operatorname{Im} D^R(\underset{\sim}{k},\omega)$ and thus the whole function $D^R(\underset{\sim}{k},\omega)$ through the integral representation (5.14). In practice, of course, $S(\underset{\sim}{k},\omega)$ can be determined only for a limited range of $\underset{\sim}{k}$ and ω, so that this procedure is not feasible. In addition, $\Delta(\underset{\sim}{k},\omega)$ obeys the symmetry relation

$$\Delta(\underset{\sim}{k},\omega) = -\Delta(-\underset{\sim}{k},-\omega) \tag{5.17}$$

which is equivalent to the symmetry relation (3.14) derived directly from the definition of $S(\underset{\sim}{k},\omega)$.

Equation (5.12) can now be used to obtain an important result concerning the structure function $S(\underset{\sim}{k})$. A combination with Eqs. (5.14)-(5.17) shows that

$$\varkappa_T(\underset{\sim}{k},0) = (\hbar n_0^2)^{-1} \int_{-\infty}^{\infty} \frac{d\omega'}{2\pi} \frac{\Delta(\underset{\sim}{k},\omega')}{\omega'} \tag{5.18}$$

where a principal-value symbol is unnecessary since $\Delta(\underset{\sim}{k},\omega=0)$ must vanish for a system with inversion symmetry. Moreover, Eq. (5.16) implies that

$$\varkappa_T(\underset{\sim}{k},0) = (\hbar n_0)^{-1} \int_{-\infty}^{\infty} d\omega' \left(\frac{1 - e^{-\beta\hbar\omega'}}{\omega'} \right) S(\underset{\sim}{k},\omega') \tag{5.19}$$

In the limit $\underset{\sim}{k} \to 0$, the function $S(\underset{\sim}{k},\omega')$ is dominated by low-frequency excitations, so that the remaining factor in parenthesis may be approximated by its value at $\omega' = 0$. Thus we obtain the result

$$\varkappa_T \equiv \lim_{\underset{\sim}{k}\to 0} \varkappa_T(\underset{\sim}{k},0) = \frac{\beta}{n_0} \lim_{\underset{\sim}{k}\to 0} \int_{-\infty}^{\infty} d\omega'\, S(\underset{\sim}{k},\omega')$$

or, equivalently [see Eq. (3.10)]:

$$\lim_{\underset{\sim}{k}\to 0} S(\underset{\sim}{k}) = n_0 k_B T \varkappa_T \tag{5.20}$$

Such behavior is easily seen in the x-ray scattering from liquid helium, which has significant temperature dependence at long wavelengths (Fig. 5.1).

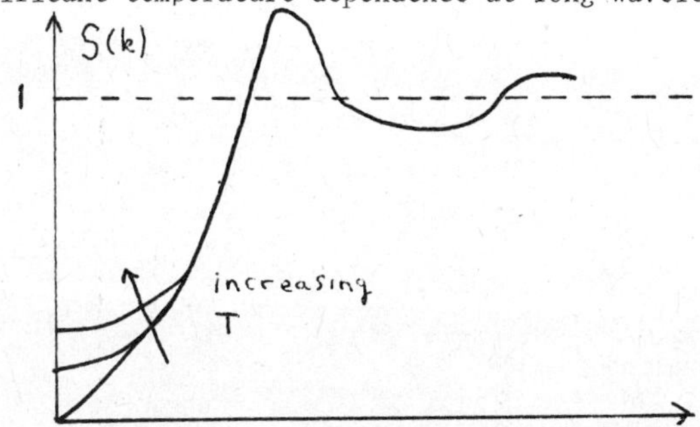

Fig. 5.1. Temperature dependence of S(k) for liquid ^{4}He.

This expression also provides insight into the behavior near a critical point in a fluid (Fig. 5.2), when the isotherms in the pV plane become flat as $T \to T_c$. Since the corresponding κ_T becomes singular, the scattered intensity of visible light becomes very large, accounting for the phenomenon of critical opalescence. An equivalent result follows from Eq. (1.23) rewritten as

$$\begin{aligned} S(q) &= N^{-1} \int d^3r\, d^3r'\, e^{i\underset{\sim}{q}\cdot(\underset{\sim}{r}-\underset{\sim}{r}')} \langle \tilde{n}(\underset{\sim}{r})\, \tilde{n}(\underset{\sim}{r}') \rangle \\ &= N^{-1} \int d^3r\, d^3r'\, e^{i\underset{\sim}{q}\cdot(\underset{\sim}{r}-\underset{\sim}{r}')} \left[\langle n(\underset{\sim}{r})\, n(\underset{\sim}{r}') \rangle - n^2 \right] \end{aligned} \tag{5.21}$$

In the limit $q \to 0$, it describes the fluctuations in the number of particles in some volume V probed by a beam of x-rays or visible light:

$$\lim_{q\to 0} S(q) = N^{-1} \left[\langle N^2 \rangle - \langle N \rangle^2 \right]$$

or equivalently:

$$\langle N^2 \rangle - \langle N \rangle^2 = V^{-1} N^2 k_B T \kappa_T \tag{5.22}$$

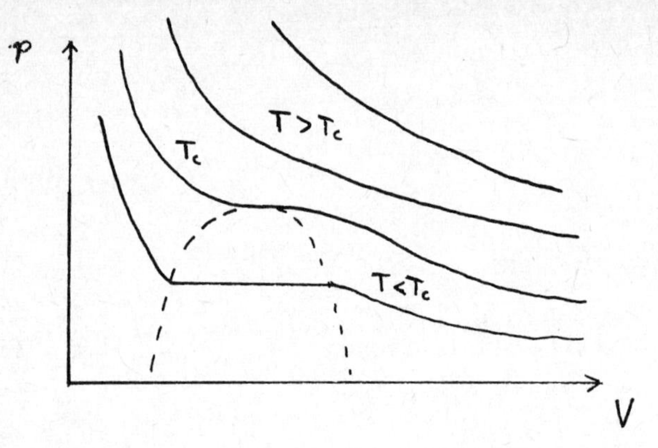

Fig. 5.2. Schematic phase diagram showing liquid-gas critical point.

Note that the fractional fluctuations in a volume V with density n become small as $V \to \infty$ but they increase with increasing κ_T .

VI. Model Calculations for Density Correlation Functions

To think about the exact $D^R(\underset{\sim}{r}t,\underset{\sim}{r}'t')$ in an interacting system, it is helpful to return to its definition in terms of second quantized operators

$$\tilde{n}_H(\underset{\sim}{r},t) = e^{i\hat{H}t/\hbar}\,\tilde{n}(\underset{\sim}{r})\,e^{-i\hat{H}t/\hbar} \tag{6.1}$$

where

$$\begin{aligned}\tilde{n}(\underset{\sim}{r}) &= \hat{n}(\underset{\sim}{r}) - \langle \hat{n}(\underset{\sim}{r})\rangle \\ &= \sum_\alpha \hat{\psi}^\dagger_\alpha(\underset{\sim}{r})\,\hat{\psi}_\alpha(\underset{\sim}{r}) - n\end{aligned} \tag{6.2}$$

in a uniform system. Thus $\tilde{n}(\underset{\sim}{r})$ is associated with the simultaneous creation and destruction of particles at $\underset{\sim}{r}$, or, alternatively, creation of a particle-hole pair. The density-density correlation functions involve the propagation of this particle-hole pair from $\underset{\sim}{r}t$ to $\underset{\sim}{r}'t'$, where they then annihilate. To develop a diagrammatic description, it is necessary to introduce imaginary times τ and a grand canonical hamiltonian $\hat{K} = \hat{H} - \mu\hat{N}$, which allows us to treat the "time" propagation on the same footing with the ensemble average over the operator $e^{-\beta\hat{K}}$. The details would take too long to go through here, but they are treated in detail in Chaps. 7 and 9 of Ref. 2. In essence, one studies a τ-ordered correlation function

$$\mathcal{D}(\underset{\sim}{r}\tau,\underset{\sim}{r}'\tau') = -\langle T_\tau[\tilde{n}(\underset{\sim}{r}\tau)\,\tilde{n}(\underset{\sim}{r}'\tau')]\rangle \tag{6.3}$$

which can be readily analyzed with perturbation theory, and is simply related to the desired quantity $D^R(\underset{\sim}{r}t,\underset{\sim}{r}'t')$. Indeed, the function $\mathcal{D}(\underset{\sim}{r}\tau,\underset{\sim}{r}'\tau')$ turns out to be periodic in the variable $\tau-\tau'$, and it therefore has a <u>Fourier series</u> representation, in contrast to the Fourier integral (4.15) appropriate for the continuous variable $t-t'$. Thus we may write

$$\mathcal{D}(\underset{\sim}{r}\tau,\underset{\sim}{r}'\tau') = \int \frac{d^3k}{(2\pi)^3}\, e^{i\underset{\sim}{k}\cdot(\underset{\sim}{r}-\underset{\sim}{r}')} \sum_{j=-\infty}^{\infty} e^{-i\nu_j(\tau-\tau')}\,\mathcal{D}(\underset{\sim}{k},\nu_j) \tag{6.4}$$

where

$$\nu_j = 2j\pi k_B T/\hbar = 2j\pi/\beta\hbar \tag{6.5}$$

The basic relation between $\mathcal{D}$ and D^R arises because $\mathcal{D}(\underset{\sim}{k},\nu_j)$ has a spectral representation analogous to Eq. (5.14)

$$\mathcal{D}(\underset{\sim}{k},\nu_j) = \int_{-\infty}^{\infty} \frac{d\omega'}{2\pi}\,\frac{\Delta(\underset{\sim}{k},\omega')}{i\nu_j - \omega'} \tag{6.6}$$

<u>with the same weight function</u> $\Delta(\underset{\sim}{k},\omega')$. Thus any approximation to $\mathcal{D}(\underset{\sim}{k},\nu_j)$ (which can be evaluated by resummation of Feynman diagrams) leads directly to a corresponding approximation to $D^R(\underset{\sim}{k},\omega)$ and hence to the various dynamical properties and scattering phenomena inherent in the various density-density correlation functions.

To be more precise, the function $\mathcal{D}(\underset{\sim}{k},\nu_j)$ is the sum of all connected Feynman diagrams of the form shown in Fig. 6.1.

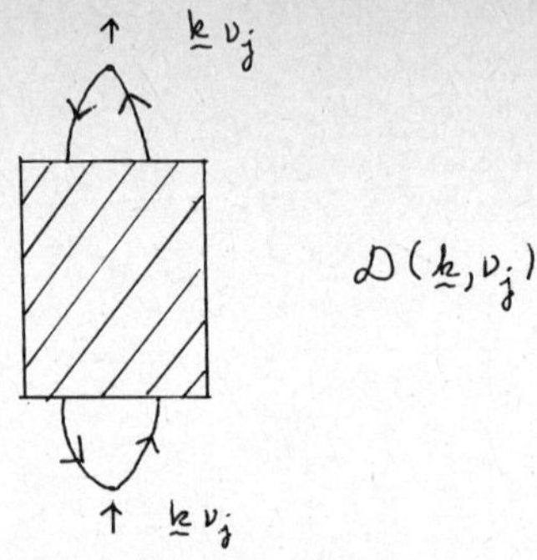

Fig. 6.1. Structure of density correlation function.

The continuous momentum $\underset{\sim}{k}$ and the discrete frequency ν_j are both conserved at any vertex. These diagrams have precisely the form of a polarization insertion (see p. 110 of Ref. 2) and may be separated into proper and improper contributions, with $\mathcal{D}^*$ the sum of all proper diagrams (Fig. 6.2). The full $\mathcal{D}(\underset{\sim}{k},\nu_j)$ then obeys Dyson's integral equation, which here takes the simple form

$$\mathcal{D}(\underset{\sim}{k},\nu_j) = \mathcal{D}^*(\underset{\sim}{k},\nu_j) + \hbar^{-1}\mathcal{D}^*(\underset{\sim}{k},\nu_j)\,\mathcal{V}(\underset{\sim}{k},\nu_j)\,\mathcal{D}(\underset{\sim}{k},\nu_j) \tag{6.7}$$

with $\mathcal{V}$ the Fourier transform of the interparticle potential. As a result, one can express $\mathcal{D}$ explicitly in terms of $\mathcal{D}^*$ as

$$\mathcal{D}(\underset{\sim}{k},\nu_j) = \frac{\mathcal{D}^*(\underset{\sim}{k},\nu_j)}{1 - \hbar^{-1}\mathcal{V}(\underset{\sim}{k},\nu_j)\,\mathcal{D}^*(\underset{\sim}{k},\nu_j)} \tag{6.8}$$

For some purposes, it is more convenient to introduce a different symbol π for the polarization, which is related to $\mathcal{D}$ by the factor $\hbar$ $(\mathcal{D} = \hbar\pi)$, but we shall not do so here.

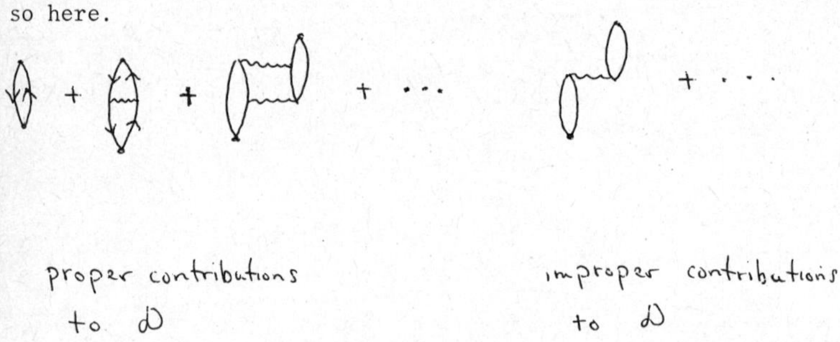

proper contributions to $\mathcal{D}$ improper contributions to $\mathcal{D}$

Fig. 6.2. Proper and improper contributions to density correlation function.

A very simple yet widely used approximation (often called random phase approximation or RPA) is to approximate $\mathcal{D}^*$ in Eq. (6.8) by the quantity for an ideal gas (Fig. 6.3) [see Eq. (30.9) of Ref. 2]

$$\mathcal{D}^{0}(\underset{\sim}{k}, \nu_j) = -(2s+1) \int \frac{d^3p}{(2\pi)^3} \frac{n^0_{p+k} - n^0_{p}}{i\nu_j - \hbar^{-1}(\varepsilon^0_{p+k} - \varepsilon^0_{p})} \tag{6.9}$$

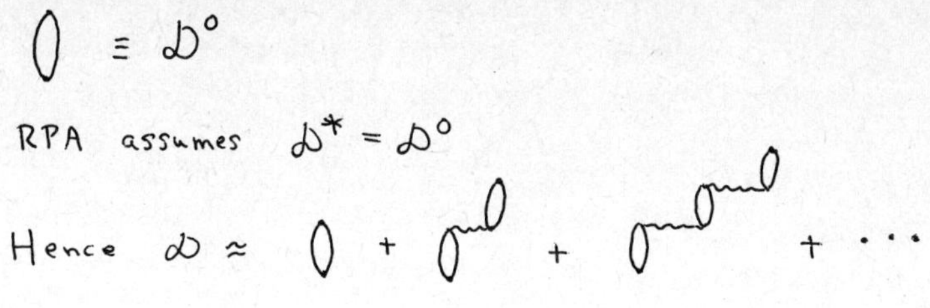

$0 \equiv \mathcal{D}^0$

RPA assumes $\mathcal{D}^* = \mathcal{D}^0$

Hence $\mathcal{D} \approx 0 + 0\sim0 + 0\sim0\sim0 + \cdots$

Fig. 6.3. Random Phase Approximation (RPA) to density correlation function.

In a charged gas, the resulting approximate form for $\mathcal{D}$ can be shown to contain the dynamical phenomena of plasma oscillations and Landau damping, as well as the static screening about a point change, not only in the classical limit but also at all temperatures, including the zero-temperature limit of a degenerate Fermi gas. For our purposes, we shall instead consider only the simpler case of a spin-1/2 Fermi gas interacting through short-range repulsive potentials with $\mathcal{V}(k,\nu_j) = V(k) \approx V(k=0)$. Thus, we approximate the Fourier transform by a single positive constant. The corresponding wave-number and frequency dependent compressibility can then be found by comparing Eqs. (5.11), (5.14), and (6.6). A fairly straightforward calculation gives the result

$$\kappa(\underset{\sim}{k},\omega) = -n^{-2} \frac{F(\underset{\sim}{k},\omega)}{1 - V(k)\, F(\underset{\sim}{k},\omega)} \tag{6.10}$$

where

$$F(\underset{\sim}{k},\omega) = -2 \int \frac{d^3p}{(2\pi)^3} \frac{n^0_{p+k} - n^0_{p}}{\hbar\omega + i\eta - (\varepsilon^0_{p+k} - \varepsilon^0_{p})} \tag{6.11}$$

In particular, the isothermal compressibility (5.13) becomes

$$\kappa_T = \lim_{k \to 0} (-n^{-2}) \frac{F(\underset{\sim}{k},0)}{1 - V(0)\, F(\underset{\sim}{k},0)} \tag{6.12}$$

In the classical limit, $F(\underset{\sim}{k},0)$ has the form

$$F(\underset{\sim}{k},0) = -n\beta\phi(k\lambda) = -n\phi(k\lambda)/k_B T \tag{6.13}$$

where $\phi(0) = 1$ and λ is the thermal wavelength (2.19) (see Ref. 2, pp. 277 and 306), so that

$$\varkappa_T \approx (nk_BT)^{-1}(1 + nV(0)/k_BT)^{-1}$$
$$= n^{-1}(k_BT + nV(0))^{-1} \tag{6.14}$$

This result is readily interpreted as the ideal classical gas expression $(nk_BT)^{-1}$, reduced by a temperature-dependent factor that vanishes in the high-temperature limit. At zero temperature, $F(\underset{\sim}{k},0)$ again has a simple form (see Ref. 2, p. 162)

$$F(\underset{\sim}{k},0) = -2N(0)\,\phi_F(k/k_F) \tag{6.15}$$

where $N(0) = mk_F/2\pi^2\hbar^2 = \frac{3}{4}\,n/\varepsilon_F^0$ is the density of states of one spin projection at the Fermi surface and

$$\phi_F(x) = \frac{1}{2} - \frac{1}{2x}\left(1 - \frac{1}{4}x^2\right)\ln\left|\frac{2-x}{2+x}\right| \tag{6.16}$$

At present, the most important feature is that $\phi_F(0) = 1$, so that the isothermal compressibility in the ground state becomes

$$\varkappa_T \simeq 2N(0)\,n^{-2}(1 + 2N(0)V(0))^{-1}$$
$$= \frac{3}{2}(n\varepsilon_F^0)^{-1}(1 + 2N(0)V(0))^{-1} \tag{6.17}$$

The first factor is the value for an ideal degenerate gas, and the repulsive potentials again act to reduce the compressibility. It is easy to sketch $\varkappa_T$ as a function of T for given density, for it increases with decreasing temperature until the

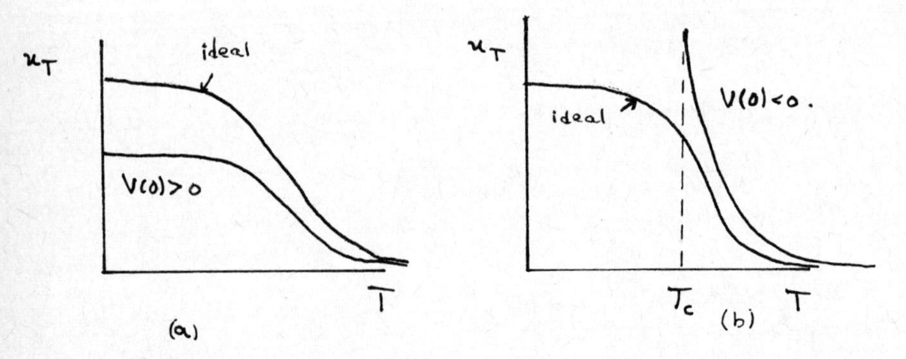

Fig. 6.4. Compressibility of weakly interacting Fermi gas for (a) repulsive and (b) attractive short-range potentials.

degeneracy starts to appear (at T_F, where $k_BT_F \approx \varepsilon_F^0$) (see Fig. 6.4a).

It is interesting to ask what would be the corresponding situation for an attractive potential, when V(0) can be negative. In that case, the compressibility would again grow with decreasing temperature, exceeding the ideal-gas value and diverging at a critical temperature T_c (see Fig. 6.4b), which is given approximately by

$$k_B T_c \approx n\,|V(0)| \tag{6.18}$$

(assuming $n|V(0)| > k_B T_F$). As noted in connection with Eq. (5.20), this behavior is roughly that found at the critical point of a fluid, where the basic condensation phenomenon arises from the attraction between molecules. In detail, however, simple models of the sort considered here are inadequate, for they always predict that κ_T diverges linearly at the transition temperature, whereas the observed behavior is of the form

$$\kappa_T \propto (T - T_c)^{-\gamma} \tag{6.19}$$

with $\gamma \approx 4/3$ for a 3-dimensional fluid.

In this context, it is valuable to recall Eqs. (5.20) and (5.21). It is clear from these results that any growth of κ_T near T_c necessarily reflects a growing scale length for the correlations in $\langle \hat{n}(\mathbf{r})\hat{n}(\mathbf{r}')\rangle$. To make this precise, we may parametrize this function with a Yukawa-like form in 3 dimensions

$$\langle \hat{n}(\mathbf{r})\,\hat{n}(0)\rangle \propto \frac{\exp(-r/\xi(T))}{r^{1+\eta}} \tag{6.20}$$

where η is another critical exponent and $\xi(T)$ is the correlation length. The corresponding compressibility thus has the form

$$\kappa_T \propto \int_0^\infty r^{1-\eta}\,dr\,e^{-r/\xi} \propto \xi^{2-\eta} \tag{6.21}$$

explicitly showing the relation between the critical exponent γ for κ_T and a corresponding singularity in the correlation length:

$$\xi(T) \propto |T - T_c|^{-\gamma/(2-\eta)} \tag{6.22}$$

It is conventional to introduce the additional exponent ν by writing

$$\xi(T) \propto |T - T_c|^{-\nu} \tag{6.23}$$

and these relations thus predict

$$\nu(2-\eta) = \gamma \tag{6.24}$$

The divergence of the correlation length $\xi(T)$ as $T \to T_c$ is a central feature of modern theories of phase transitions.

VII. Fluctuation-Dissipation Theorem

We have now seen a few specific examples of the density-density correlation function, evaluated from simple models. It is often preferable to attempt to build in as much information as possible at the beginning, and we therefore turn to a more general description that has been emphasized by Martin.[7,8] The basic start is a set of observables $\hat{A}_i(\underset{\sim}{r}t)$ where i runs over the various quantities; for example, a fluid can be described by its density, momentum density, and energy density, and these then form a convenient set. To study such quantities, there must be some coupling to external parameters at our control, such as the pressure in Eq. (5.8). In general, the external coupling hamiltonian will be written

$$\hat{H}^{ex}(t) = -\sum_i \int d^3r' \hat{A}_i(\underset{\sim}{r}',t)\, \delta a(\underset{\sim}{r}',t) \tag{7.1}$$

where δa is some small external applied field. As shown in Sec. IV, the change in the expectation value of $\hat{A}_i$ in the presence of these fields is given by

$$\delta\langle \hat{A}_i(\underset{\sim}{r},t)\rangle = i\hbar^{-1}\sum_j \int d^3r' \int_{-\infty}^{t} dt' \langle [\hat{A}_i(\underset{\sim}{r},t), \hat{A}_j(\underset{\sim}{r}',t')]\rangle\, \delta a_j(\underset{\sim}{r}',t') \tag{7.2}$$

This commutator of two operators determines the linear response. It appears frequently and is used to define the conventional quantity

$$\chi''_{ij}(\underset{\sim}{r},t;\underset{\sim}{r}',t') = (2\hbar)^{-1}\langle [\hat{A}_i(\underset{\sim}{r},t), \hat{A}_j(\underset{\sim}{r}',t')]\rangle \tag{7.3}$$

Note that χ'' is defined for all t-t', without a step function. In a translationally invariant system at thermal equilibrium, χ''_{ij} will depend only on $\underset{\sim}{r}-\underset{\sim}{r}'$ and t-t'. These equations can be combined to give the expression

$$\delta\langle \hat{A}_i(\underset{\sim}{r},t)\rangle = \sum_j \int d^3r' \int_{-\infty}^{t} dt'\, 2i\chi''_{ij}(\underset{\sim}{r},t;\underset{\sim}{r}',t')\, \delta a_j(\underset{\sim}{r}',t') \tag{7.4}$$

In many cases of importance, the applied fields have a well-defined wavenumber $\underset{\sim}{k}$ and frequency ω. In that case, it is natural to define the Fourier transform of χ''_{ij} (assuming an infinite translationally invariant medium)

$$\chi''_{ij}(\underset{\sim}{r}-\underset{\sim}{r}', t-t') = \int \frac{d^3k'}{(2\pi)^3} \int_{-\infty}^{\infty} \frac{d\omega'}{2\pi}\, \chi''_{ij}(\underset{\sim}{k}',\omega')\, e^{i\underset{\sim}{k}'\cdot(\underset{\sim}{r}-\underset{\sim}{r}')}\, e^{-i\omega'(t-t')} \tag{7.5}$$

Taking the inverse transform of Eq. (7.4), we find the induced response in $\hat{A}_i$ at wavevector $\underset{\sim}{k}$ and frequency ω

$$\delta\langle \hat{A}_i(\underset{\sim}{k},\omega)\rangle = \int d^3r\, e^{-i\underset{\sim}{k}\cdot\underset{\sim}{r}} \int_{-\infty}^{\infty} dt\, e^{i\omega t}\, \delta\langle \hat{A}_i(\underset{\sim}{r},t)\rangle$$

If we represent $\delta a(\underset{\sim}{r}',t')$ as a Fourier integral, the spatial parts all combine simply, but the finite upper limit on the time integral in Eq. (7.4) requires special treatment. Thus Eqs. (7.4) and (7.5) yield

$$\delta\langle \hat{A}_i(\underset{\sim}{k},\omega)\rangle = \sum_j \int_{-\infty}^{\infty} dt\, e^{i\omega t} \int_{-\infty}^{t} dt'\, (2\pi)^{-1} \int_{-\infty}^{\infty} d\omega'\, 2i\chi''_{ij}(\underset{\sim}{k},\omega')\, e^{-i\omega'(t-t')}$$
$$\times (2\pi)^{-1} \int_{-\infty}^{\infty} d\omega''\, e^{-i\omega'' t'}\, \delta a_j(\underset{\sim}{k},\omega'')$$

Define the new variable $\tau = t'-t$, so that this expression becomes

$$\delta\langle \hat{A}_i(\underset{\sim}{k},\omega)\rangle = \sum_j \int_{-\infty}^{\infty} dt \int_{-\infty}^{0} d\tau\, (2\pi)^{-2} \iint_{-\infty}^{\infty} d\omega'\, d\omega''\, e^{i(\omega'-\omega'')\tau}\, e^{i(\omega-\omega'')t}$$

$$\times\, 2i\chi''_{ij}(\underset{\sim}{k},\omega')\, \delta a_j(\underset{\sim}{k},\omega'')$$

The integral over t is just $2\pi\delta(\omega-\omega'')$, and the integral over τ is made convergent by inserting a convergence factor $e^{\eta\tau}$ with $\eta \to 0^+$. In this way we find

$$\delta\langle \hat{A}_i(\underset{\sim}{k},\omega)\rangle = \sum_j \int_{-\infty}^{\infty} \frac{d\omega'}{\pi} \frac{\chi''_{ij}(\underset{\sim}{k},\omega')}{\omega'-\omega-i\eta}\, \delta a_j(\underset{\sim}{k},\omega) \tag{7.6}$$

It is natural to introduce the following function of a complex variable z:

$$\chi_{ij}(\underset{\sim}{k},z) = \int_{-\infty}^{\infty} \frac{d\omega'}{\pi} \frac{\chi''_{ij}(\underset{\sim}{k},\omega')}{\omega'-z} \tag{7.7}$$

It has the following limiting form as z approaches the real axis from <u>above</u>

$$\chi^R_{ij}(\underset{\sim}{k},\omega) = \lim_{\eta\to 0} \chi_{ij}(\underset{\sim}{k},\omega+i\eta) \tag{7.8}$$

Evidently, χ^R_{ij} is the retarded or causal response function, since Eq. (7.6) shows that it determines the induced response

$$\delta\langle \hat{A}_i(\underset{\sim}{k},\omega)\rangle = \sum_j \chi^R_{ij}(\underset{\sim}{k},\omega)\, \delta a_j(\underset{\sim}{k},\omega) \tag{7.9}$$

It has the explicit form

$$\chi^R_{ij}(\underset{\sim}{k},\omega) = \int_{-\infty}^{\infty} \frac{d\omega'}{\pi} \frac{\chi''_{ij}(\underset{\sim}{k},\omega')}{\omega'-\omega-i\eta} \tag{7.10}$$

which shows that χ^R is an analytic function for $\text{Im}\,\omega > 0$. It may be separated into its real and imaginary parts

$$\chi^R_{ij}(\underset{\sim}{k},\omega) = \chi'_{ij}(\underset{\sim}{k},\omega) + i\,\chi''_{ij}(\underset{\sim}{k},\omega) \tag{7.11}$$

and use of the identity

$$(\omega'-\omega-i\eta)^{-1} = P(\omega'-\omega)^{-1} + i\pi\delta(\omega'-\omega) \tag{7.12}$$

shows that χ' and χ'' obey the Kramers-Kronig relation

$$\chi'_{ij}(\underset{\sim}{k},\omega) = P\int_{-\infty}^{\infty} \frac{d\omega'}{\pi} \frac{\chi''_{ij}(\underset{\sim}{k},\omega')}{\omega'-\omega} \tag{7.13}$$

Spectral functions of the form (7.7) play an important part in many of our discussions. Unfortunately, they are more complicated than the functions considered in introductory courses on complex variables. To understand their properties, it is helpful to consider briefly the function

$$f(z) = \int_{-\infty}^{\infty} \frac{dx'}{\pi} \frac{\rho(x')}{x'-z} \tag{7.14}$$

where x' is a real variable and $\rho(x') \geq 0$ is real and integrable. It is clear that f(z) is bounded and differentiable for $\text{Im}\, z \neq 0$, so that f(z) is analytic in both the upper- and lower-half plane. In general, however, there is a branch cut along the real z axis, which follows from the limit as z approaches the real axis from above or below [use Eq. (7.12)]

$$f(x \pm i\eta) = P\int_{-\infty}^{\infty} \frac{dx'}{\pi} \frac{\rho(x')}{x' - x} \pm i\rho(x) \tag{7.15}$$

Thus f(z) is discontinuous across the real axis at any point where $\rho \neq 0$, and $\rho(x)$ may be identified as the imaginary part of $f(x + i\eta)$. This analytic structure requires that f(z) consist of two distinct sheets that intersect along the real axis. For Im $z > 0$, f(z) coincides with a function $f_I(z)$ that is analytic for Im $z > 0$ but that, in general, has singularities in the lower-half plane. Similarly, for Im $z < 0$, f(z) coincides with a distinct function $f_{II}(z)$ that is analytic for Im $z < 0$ but that, in general, has singularities in the upper-half plane. If one starts in the upper-half plane (say), then $f(z) = f_I(z)$, and any subsequent motion of z (even into the lower-half plane) stays on the sheet f_I, if necessary analytically continued across the real axis; note, particularly, that such motion does not reach the other branch f_{II}.

It is helpful to exhibit a simple example of this behavior, taking $\rho(x') = \gamma(x'^2 + \gamma^2)^{-1}$ in Lorentzian form (with $\gamma > 0$). In this case, f(z) becomes

$$f(z) = \int_{-\infty}^{\infty} \frac{dx'}{\pi} \frac{\gamma}{(x'^2+\gamma^2)(x'-z)}$$

and the integrand has poles at $x' = \pm i\gamma$ and at $x' = z$. If z is in the upper-half plane, then the integral is readily evaluated by contour techniques closing the contour in the lower half plane, to yield

$$f(z) = f_I(z) \equiv -(z + i\gamma)^{-1} \qquad \text{Im}\, z > 0 \tag{7.16}$$

Note that f_I is analytic for Im $z > 0$ but has a pole in the lower half plane at $z = -i\gamma$. Similarly, if z lies in the lower half plane, we find

$$f(z) = f_{II}(z) \equiv -(z - i\gamma)^{-1} \qquad \text{Im}\, z < 0 \tag{7.17}$$

Again, f_{II} is analytic in the desired region but has a pole at $z = i\gamma$ in the upper half plane. Note that

$$\text{Im}\, f(x+i\eta) = \text{Im}\, f_I(x+i\eta) = \gamma(x^2+\gamma^2)^{-1} = \rho(x)$$

in accordance with the general property (7.15), and that $f_I \neq f_{II}$, displaying explicitly the two-sheeted structure. In some situations, $\rho(x')$ may vanish for a finite interval, in which case the branch cut does not extend along the whole real axis, and the functions f_I and f_{II} are then related through analytic continuation.

We now return to the more general description and show that $\omega\chi''_{ij}(\mathbf{k},\omega)$ is a positive semi-definite matrix in the indices ij for a dissipative system. Under a small change $d\delta a_j$ in the external fields, the change in the energy of the system is given by

$$dE = -\Sigma_j \int d^3r \, \langle \hat{A}_j(\mathbf{r},t)\rangle \, d\delta a_j(\mathbf{r},t) \tag{7.18}$$

where the bracket denotes the ensemble average in the full time-dependent density matrix. To lowest order, the only contribution can be shown to arise from the first-order change in $\langle \hat{A}_i \rangle$ (see Ref. 8, Sec. 3.3)

$$dE = -\sum_j \int d^3r\, \delta\langle \hat{A}_j(\underset{\sim}{r},t)\rangle\, d\,\delta a_j(\underset{\sim}{r},t) \tag{7.19}$$

The time integral of the time derivative thus becomes

$$\Delta E = \int_{-T}^{T} dt\, \frac{dE}{dt} = \sum_j \int_{-T}^{T} dt\, \delta a_j(\underset{\sim}{r},t) \frac{\partial}{\partial t} \delta\langle \hat{A}_j(\underset{\sim}{r},t)\rangle \tag{7.20}$$

on integrating by parts. Use of Eq. (7.9) then gives

$$\Delta E = \int \frac{d^3k}{(2\pi)^3} \int_{-\infty}^{\infty} \frac{d\omega}{2\pi} \sum_{ij} (-i)\, \delta a_i(\underset{\sim}{k},\omega)^* \omega \chi^R_{ij}(\underset{\sim}{k},\omega)\, \delta a_j(\underset{\sim}{k},\omega) \tag{7.21}$$

for purely harmonic perturbations $e^{i(\underset{\sim}{k}\cdot\underset{\sim}{r}-\omega t)}$. Here we note that $\delta a_i(\underset{\sim}{r},t)$ is real, so that $\delta a_i(\underset{\sim}{k},\omega) = \delta a_i(-\underset{\sim}{k},-\omega)^*$. It is easy to show that $\chi_{ij}'(\underset{\sim}{k},\omega)$ is even under the transformation of $\underset{\sim}{k},\omega,ij \to -\underset{\sim}{k},-\omega,ji$, whereas $\chi''_{ij}(\underset{\sim}{k},\omega)$ is odd. Hence only χ'' survives in Eq. (7.21), and we find

$$\Delta E = \int \frac{d^3k}{(2\pi)^3} \int_{-\infty}^{\infty} \frac{d\omega}{2\pi} \sum_{ij} \delta a_i(\underset{\sim}{k},\omega)^* \omega \chi''_{ij}(\underset{\sim}{k},\omega)\, \delta a_j(\underset{\sim}{k},\omega) \tag{7.22}$$

Since ΔE must be positive in a dissipative system, we infer that $\omega\chi''_{ij}$ is indeed a positive-definite matrix.

Notice that the dissipation in the system is directly related to χ'', which in turn is the <u>commutator</u> of the relevant variables [see (7.3)]. On the other hand, the fluctuations in the operators may be characterized by the quantity [see Eq. (3.12)]

$$S_{ij}(\underset{\sim}{r}t,\underset{\sim}{r}'t') = \langle \tilde{A}_i(\underset{\sim}{r},t)\, \hat{A}_j(\underset{\sim}{r}',t')\rangle \tag{7.23}$$

where $\tilde{A}_i = \hat{A}_i - \langle \hat{A}_i \rangle$ is the deviation operator [see Eq. (1.21)]. For a translationally invariant system, S_{ij} depends only on $\underset{\sim}{r}-\underset{\sim}{r}'$ and $t-t'$, and its Fourier transform is given by

$$S_{ij}(\underset{\sim}{k},\omega) = \int d^3r \int dt\, e^{-i\underset{\sim}{k}\cdot\underset{\sim}{r}}\, e^{i\omega t}\, S_{ij}(\underset{\sim}{r},t) \tag{7.24}$$

An analysis very similar to that of Eq. (3.14) yields the general result

$$S_{ji}(-\underset{\sim}{k},-\omega) = e^{-\beta\hbar\omega} S_{ij}(\underset{\sim}{k},\omega) \tag{7.25}$$

As a result, the Fourier transform of the commutator $\chi''_{ij}(\underset{\sim}{r},t)$ may be rewritten in terms of $S_{ij}(\underset{\sim}{k},\omega)$ to give the fluctuation-dissipation theorem

$$\begin{aligned} 2\hbar\chi''_{ij}(\underset{\sim}{k},\omega) &= S_{ij}(\underset{\sim}{k},\omega) - S_{ji}(-\underset{\sim}{k},-\omega) \\ &= (1-e^{-\beta\hbar\omega})\, S_{ij}(\underset{\sim}{k},\omega) \end{aligned} \tag{7.26}$$

Similarly, the anticommutator may also be expressed in terms of $(1 + e^{-\beta\hbar\omega})\, S_{ij}(\underset{\sim}{k},\omega)$. In the classical limit, Eq. (7.26) becomes

$$\chi''_{ij}(\underset{\sim}{k},\omega) = \tfrac{1}{2}\beta\omega\, S_{ij}(\underset{\sim}{k},\omega) \tag{7.27}$$

VIII. Magnetic Phenomena; Perturbation Calculations

The preceding discussions have dealt solely with the density correlations in condensed media, but there are many other types of long-range order, the most familiar being the spontaneous magnetization in a ferromagnet. Since a neutron has a magnetic moment, it can also interact with the magnetic moment of an atom, leading to inelastic scattering because of the internal magnetization in the target. The detailed analysis is quite similar to that for $S(\underset{\sim}{k},\omega)$ developed in lecture III.[9] The net result is that one can measure the space-time Fourier transform of magnetic correlation functions of the form

$$\langle M_i(\underset{\sim}{r},t)\, M_j(\underset{\sim}{r}',t')\rangle \tag{8.1}$$

where M is the magnetization (magnetic moment per unit volume) and i and j refer to spatial vector components. Such inelastic studies can be used to investigate spin-wave excitations in a magnetic material, just as the usual neutron scattering has enabled investigators to map out the phonon dispersion relations in crystals and liquids. In addition, elastic magnetic scattering of neutrons can provide a picture of the spatial distribution of magnetization, just as the more usual Bragg scattering leads to the spatial density distribution. Such studies have been especially important in understanding the behavior of antiferromagnets, where there is long-range magnetic order at finite wave number but no net macroscopic magnetization.

In addition to the measurability of the magnetic correlation functions, they also are important in the theory of linear response to an external magnetic field. This second aspect is wholly analogous to the appearance of density correlation functions in the generalized compressibility $\kappa(\underset{\sim}{k},\omega)$. To analyze the situation in detail, we shall consider a spin 1/2 Fermi gas with magnetic moment μ_0 per particle. The corresponding magnetization operator (magnetic moment per unit volume) is given by $\mu_0\hat{\underset{\sim}{\sigma}}(\underset{\sim}{r})$, where

$$\hat{\underset{\sim}{\sigma}}(\underset{\sim}{r}) = \sum_{\alpha\beta} \hat{\psi}_\alpha^\dagger(\underset{\sim}{r})\, (\underset{\sim}{\sigma})_{\alpha\beta}\, \hat{\psi}_\beta(\underset{\sim}{r}) \tag{8.2}$$

and $\underset{\sim}{\sigma}$ denotes the 3 Pauli matrices. Note that μ_0 is not simply the gyromagnetic ratio but differs by a factor $\frac{1}{2}\hbar$ needed to relate the true spin density to $\underset{\sim}{\sigma}$. In the presence of a weak magnetic field $\underset{\sim}{\mathcal{H}}(\underset{\sim}{r},t)$ along $\hat{z}$, there is an additional energy specified by the hamiltonian

$$\hat{H}^{ex}(t) = -\mu_0 \int d^3r\; \hat{\sigma}_z(\underset{\sim}{r},t)\, \mathcal{H}(\underset{\sim}{r},t) \tag{8.3}$$

This is usually written down as obvious, for it properly orients the magnetic moment along $\underset{\sim}{\mathcal{H}}$ to lower the energy, but it actually takes a bit of thought and thermodynamics to understand the - sign.[6]

The response to $\underset{\sim}{\mathcal{H}}$ may be characterized by the induced magnetization (assuming a paramagnetic material)

$$\langle \hat{M}_z(\underset{\sim}{r},t)\rangle = \mu_0 \langle \hat{\sigma}_z(\underset{\sim}{r},t)\rangle \tag{8.4}$$

and the general result (4.10) yields the basic expression

$$\langle \hat{M}_z(\underset{\sim}{r},t)\rangle = -\mu_0^2 \hbar^{-1} \int d^3r'\, dt'\, D_\sigma^R(\underset{\sim}{r}-\underset{\sim}{r}', t-t')\, \mathcal{H}_z(\underset{\sim}{r}',t') \tag{8.5}$$

where D_σ^R is a retarded spin-spin correlation function

$$i D_\sigma^R(\underset{\sim}{r}t, \underset{\sim}{r}'t') = \Theta(t-t') \langle [\hat{\sigma}_{zH}(\underset{\sim}{r},t), \hat{\sigma}_{zH}(\underset{\sim}{r}',t')]\rangle \tag{8.6}$$

The corresponding Fourier coefficients obey the simpler multiplicative equation

$$M_z(\underset{\sim}{k},\omega) = \langle \hat{M}_z(\underset{\sim}{k},\omega)\rangle = -\mu_0^2 \hbar^{-1} D_\sigma^R(\underset{\sim}{k},\omega)\, \mathcal{H}_z(\underset{\sim}{k},\omega) \tag{8.7}$$

The coefficient may now be taken to define the generalized susceptibility

$$\chi^R(\underset{\sim}{k},\omega) = -\mu_0^2 \hbar^{-1} D_\sigma^R(\underset{\sim}{k},\omega) \tag{8.8}$$

which is obviously similar to Eqs. (5.11) and (7.9).

Near a phase transition to an ordered magnetic state, the function $D_\sigma^R(\underset{\sim}{k},\omega)$ may be expected to develop important correlations, near $\underset{\sim}{k} \simeq 0$ for a ferromagnet and at finite $\underset{\sim}{k}$ for an antiferromagnet. Thus these features will appear both in neutron scattering with wavevector transfer $\underset{\sim}{k}$ and also in the measured susceptibility (unfortunately it is not simple to measure χ directly for finite $\underset{\sim}{k}$). Although we shall not deal with other systems, very similar ideas can be used to describe various phase transitions, such as the appearance of charge-density waves or the onset of displacements in regular crystals, often leading to the onset of ferroelectricity.

To develop a theoretical framework for understanding the many-body features of magnetic correlations, it is helpful to introduce a generalized "polarization" function[10]

$$i D^R_{\lambda\lambda',\mu'\mu}(\underset{\sim}{r}t, \underset{\sim}{r}'t') = \Theta(t-t') \langle [\hat{\psi}^\dagger_\lambda(\underset{\sim}{r}t)\hat{\psi}_{\lambda'}(\underset{\sim}{r}t), \hat{\psi}^\dagger_{\mu'}(\underset{\sim}{r}'t')\hat{\psi}_\mu(\underset{\sim}{r}'t')]\rangle \tag{8.9}$$

associated with the following diagram Fig. 8.1

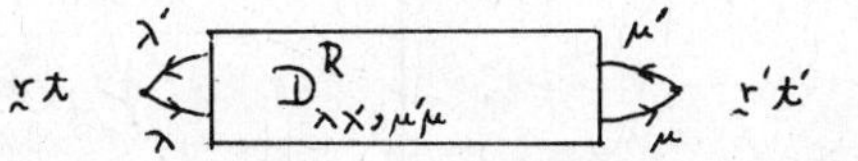

λ' μ'
$\underset{\sim}{r}t$ $D^R_{\lambda\lambda',\mu'\mu}$ $\underset{\sim}{r}'t'$
λ μ

Fig. 8.1. Structure of general polarization in position space.

If someone gave you such a function with 4 spin indices, it is easy to recover the density correlation function by setting $\mu' = \mu$ and $\lambda' = \lambda$ and summing over λ and μ. Similarly, the spin-density correlation function follows on multiplying by $(\sigma_z)_{\lambda\lambda'}(\sigma_z)_{\mu'\mu}$ and summing over all four indices. Thus it is sufficient to study the perturbation expansion for the general function.

As noted in Lecture VI, the Feynman diagrams must actually be analyzed in terms of slightly different functions, with Fourier transforms in the spatial variables and Fourier series in the times. Formally, we merely introduce the quantities $\mathcal{D}_{\lambda\lambda',\mu'\mu}(\underset{\sim}{k},\nu_j)$ as in Eq. (6.4), which represents the sum of all Feynman diagrams of the form in Fig. 8.2

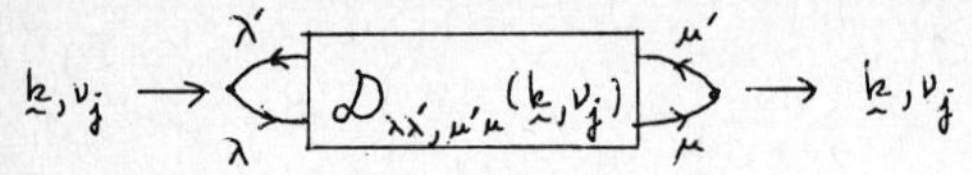

$$\underset{\sim}{k},\nu_j \longrightarrow \boxed{\mathcal{D}_{\lambda\lambda',\mu'\mu}(\underset{\sim}{k},\nu_j)} \longrightarrow \underset{\sim}{k},\nu_j$$

Fig. 8.2. Structure of general polarization in momentum space.

For spin-independent interactions, the spin component of the particle cannot change on undergoing an interaction. Thus any piece of a Feynman diagram involving a potential must have the spin structure shown in Fig. 8.3

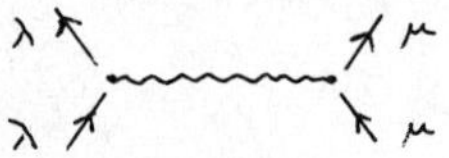

Fig. 8.3. Spin-independent interactions conserve spin projections.

As a result, all the proper diagrams of $\mathcal{D}$ for spin-independent interactions (those that cannot be separated into two parts, one at $\underset{\sim}{r}$ and one at $\underset{\sim}{r}'$, by cutting a single interaction line) necessarily have one of two spin structures

$$\mathcal{D}^*_{\lambda\lambda',\mu'\mu} = \mathcal{D}^*_A\,\delta_{\lambda\mu}\,\delta_{\lambda'\mu'} + \mathcal{D}^*_B\,\delta_{\lambda\lambda'}\,\delta_{\mu\mu'} \tag{8.10}$$

To see this result, we note that one class of proper diagrams has the structure of a single fermion loop with various decorations (Fig. 8.4)

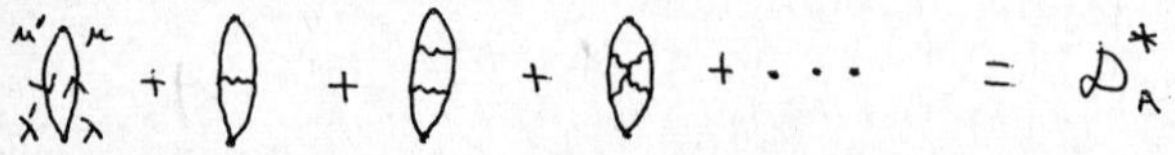

$$+ \; + \; + \; + \cdots = \mathcal{D}^*_A$$

Fig. 8.4. One-loop contributions to proper polarization.

Since the interactions cannot alter the spin of each Fermion line, these must all add up to give $\mathcal{D}^*_A\,\delta_{\lambda\mu}\,\delta_{\lambda'\mu'}$, so that each line retains its own spin projection. Initially, one might think that this exhausts the class of diagrams, but there is another type, shown in Fig. 8.5, in which two separate loops are joined by at least 2 interactions (hence they are still proper contributions).

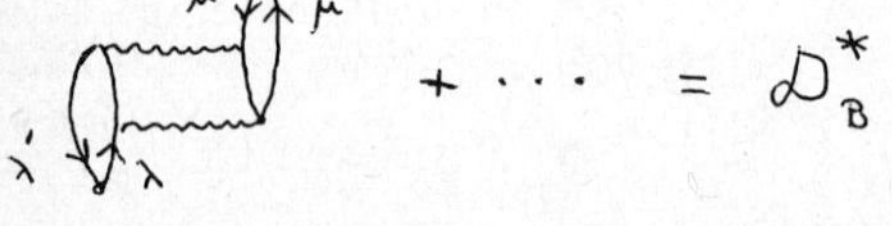

$$+ \cdots = \mathcal{D}^*_B$$

Fig. 8.5. Multi-loop contributions to proper polarization.

It is clear that each of these must have the spin structure $\delta_{\lambda\lambda'}\,\delta_{\mu\mu'}$ accounting for the structure $\mathcal{D}^*_B$ found in Eq. (8.10).

For spin-independent potentials, Dyson's equation in momentum space becomes

$$\mathcal{D}_{\lambda\lambda',\mu'\mu} = \mathcal{D}^*_{\lambda\lambda',\mu'\mu} + \hbar^{-1}\sum_{\alpha\beta}\mathcal{D}^*_{\lambda\lambda',\alpha\alpha}\,\mathcal{V}\,\mathcal{D}_{\beta\beta,\mu'\mu} \tag{8.11}$$

and this equation may be solved by contracting on the indices $\lambda\lambda'$. A straightforward calculation with Eq. (8.10) then gives the general structure of the full function

$$\mathcal{D}_{\alpha\beta,\lambda\mu} = \mathcal{D}^*_A\,\delta_{\alpha\mu}\delta_{\beta\lambda} + \left[\mathcal{D}^*_B + \frac{\mathcal{D}^{*2}}{4}\,\frac{\hbar^{-1}\mathcal{V}}{1-\hbar^{-1}\mathcal{V}\mathcal{D}^*}\right]\delta_{\alpha\beta}\delta_{\lambda\mu} \tag{8.12}$$

where

$$\mathcal{D}^* = \sum_{\alpha\lambda}\mathcal{D}^*_{\alpha\alpha,\lambda\lambda} = 2\mathcal{D}^*_A + 4\mathcal{D}^*_B \tag{8.13}$$

for spin -1/2 fermions. It is easy to construct $\mathcal{D}(\underset{\sim}{k},\nu_j) = \sum_{\alpha\lambda}\mathcal{D}_{\alpha\alpha,\lambda\lambda}(\underset{\sim}{k},\nu_j)$ and to see that it precisely agrees with Eq. (6.8). More interesting is the spin correlation function

$$\mathcal{D}_\sigma = \sum_{\alpha\beta\lambda\mu}\mathcal{D}_{\alpha\beta,\lambda\mu}(\sigma_z)_{\alpha\beta}(\sigma_z)_{\lambda\mu} = 2\mathcal{D}^*_A \tag{8.14}$$

so that none of the diagrams of the type B contributes to the spin correlations. In essence, this is because spin information cannot be propagated across a spin-independent potential line joining two separate fermion loops. In particular, note that the RPA approximation, which works reasonably for the number density function $\mathcal{D}$ [see discussion at Eq. (6.9) and Fig. 6.3] amounts to taking $\mathcal{D}^*_A = \mathcal{D}^0$ and $\mathcal{D}^*_B = 0$. Hence it fails entirely to give any many-body corrections, so that $\chi^R(\underset{\sim}{k},\omega)$ in RPA is that for an ideal gas. Thus one must study an improved approximation.

One widely used model is to replace $V(\underset{\sim}{q})$ by a constant $V(0)$ (approximating $V(\underset{\sim}{r})$ by a short range repulsive potential). In this case, one can actually evaluate $\mathcal{D}^*$ for the following set of diagrams

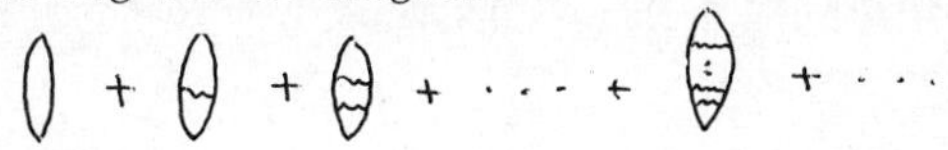

and finds

$$\mathcal{D}^*_A(\underset{\sim}{k},\nu_j) = \frac{\mathcal{D}^0(\underset{\sim}{k},\nu_j)}{2+\hbar^{-1}V(0)\,\mathcal{D}^0(\underset{\sim}{k},\nu_j)}\,; \tag{8.15}$$

evidently, $\mathcal{D}^*_B$ is still zero. In this case, the computation of $\mathcal{D}(\underset{\sim}{k},\nu_j)$ and $\mathcal{D}_\sigma(\underset{\sim}{k},\nu_j)$ is straightforward and gives

$$\mathcal{D}(\underset{\sim}{k},\nu_j) = \frac{2\mathcal{D}^0(\underset{\sim}{k},\nu_j)}{2-\hbar^{-1}V(0)\mathcal{D}^0(\underset{\sim}{k},\nu_j)} \qquad \mathcal{D}_\sigma(\underset{\sim}{k},\nu_j) = \frac{2\mathcal{D}^0(\underset{\sim}{k},\nu_j)}{2+\hbar^{-1}V(0)\mathcal{D}^0(\underset{\sim}{k},\nu_j)} \tag{8.16}$$

Note the crucial difference in signs in the denominator. As a result, this particular model gives the compressibility and susceptibility [see Eq. (6.11)]

$$\kappa_T = -n^2\lim_{k\to 0}\frac{2F(k,0)}{2-V(0)F(k,0)}$$

$$\chi = -\mu_0^2 \lim_{k\to 0} \frac{2F(k,0)}{2+V(0)F(k,0)} \tag{8.17}$$

In particular, the classical limit becomes

$$\varkappa_T = n^{-1}\left(k_B T + \tfrac{1}{2} n V(0)\right)^{-1}$$
$$\chi = \mu_0^2 n\left(k_B T - \tfrac{1}{2} n V(0)\right)^{-1} \tag{8.18}$$

and the zero temperature behavior is

$$\varkappa_T = 2N(0) n^{-2}\left(1 + N(0)V(0)\right)^{-1}$$
$$\chi = 2\mu_0^2 n\left(1 - N(0)V(0)\right)^{-1} \tag{8.19}$$

Note that repulsive interactions enhance the magnetic susceptibility yet reduce the compressibility. This behavior is qualitatively that found in liquid ^{3}He below 1K, for the magnetic susceptibility is considerably enhanced above its ideal-gas value, and the compressibility is correspondingly reduced. Detailed studies, however, indicate that a one-parameter model (here N(0)V(0) as $T \to 0$K) is inadequate in describing the data.

The enhanced magnetic behavior arises from the Fermi statics, which enforces the condition of overall antisymmetric states. The repulsive potential tends to keep the particles apart, with a preference for spatially antisymmetric states. The Fermi statistics then requires a preference for spin symmetric states, namely a tendency toward magnetic ordering. This also explains Hund's rules in atoms. Just the opposite effect occurs in nuclei, where the net potentials are attractive and low spin states are favored.

If the interaction is sufficiently strong that N(0)V(0) exceeds 1, then it is clear that the repulsive interaction can cause a phase transition to a magnetized state at some critical temperature T_c. As in the discussion of Eq. (6.18), the susceptibility would diverge linearly as $T \to T_c^+$. More generally, an expansion of $\chi(k,0)$ for small k and $T-T_c$ yields a denominator of the form $(T-T_c)/T_c + \xi^2 k^2$, whose Fourier transform reproduces the strict Yukawa form (6.20) with $\eta = 0$. Once again, comparison with the experimental observations indicates an inadequacy of the present description, which is essentially that of the mean-field picture.

IX. Magnetic Phenomena; Hydrodynamics Description

The preceding section treated magnetic phenomena in the context of many-body perturbation theory, and we now take a wider view of the same problem using the approach of Sec.VII, generalized slightly to include the conservation laws in the problem. As will be seen, the spin-spin correlation function must have a complicated structure with nonuniform limiting behavior near $k \simeq 0$ and $\omega \approx 0$. This structure is not readily reproduced, even in infinite-order perturbation theory, and it suggests alternative ways[7,8] to construct approximations that are frequently very useful. Such a technique will be used in Sec. X for light scattering by a fluid.

The basic problem of interest is the magnetization $\langle \hat{M}(\underset{\sim}{r},t)\rangle$ induced by an applied magnetic field $\mathcal{H}(\underset{\sim}{r},t)$. Equation (7.4) indicates that the relevant quantity is the Fourier transform $\chi''(\underset{\sim}{k},\omega)$ of the magnetization-magnetization commutator [see also Eqs. (8.4) - (8.8)]

$$\chi''(\underset{\sim}{r},t) = (2\hbar)^{-1}\langle[\hat{M}(\underset{\sim}{r}t),\hat{M}(\underset{\sim}{r}'t')]\rangle \tag{9.1}$$

As proved in Sec. VII, $\chi''(\underset{\sim}{k},\omega)$ is the imaginary part of the retarded (causal) susceptibility $\chi^R(\underset{\sim}{k},\omega)$. The full causal response to an applied field $\mathcal{H}(\underset{\sim}{r},t)$ is given by the Fourier transforms

$$\langle\hat{M}(\underset{\sim}{k},\omega)\rangle = \chi^R(\underset{\sim}{k},\omega)\,\mathcal{H}(\underset{\sim}{k},\omega) \tag{9.2}$$

where $\chi^R(\underset{\sim}{k},\omega)$ is the limiting value of an analytic function

$$\chi(\underset{\sim}{k},z) = \int_{-\infty}^{\infty}\frac{d\omega'}{\pi}\;\frac{\chi''(\underset{\sim}{k},\omega')}{\omega'-z} \tag{9.3}$$

as $z \to \omega + i\eta$ and $\eta = 0^+$. Thus

$$\chi^R(\underset{\sim}{k},\omega) = \lim_{\eta\to 0}\chi(\underset{\sim}{k},\omega+i\eta) \tag{9.4}$$

To proceed further, it is helpful to introduce an alternative way of thinking about $\chi^R(\underset{\sim}{k},\omega)$. As seen in Eq. (9.2), $\chi^R(\underset{\sim}{k},\omega)$ gives directly the response to an applied field $\mathcal{H}(\underset{\sim}{k},\omega)$. If the system has some magnetic collective mode with frequency $\omega_0(\underset{\sim}{k})$ and damping, then $\chi^R(\underset{\sim}{k},\omega)$ for fixed $\underset{\sim}{k}$ is peaked at $\omega \approx \omega_0(\underset{\sim}{k})$ with a width determined by the damping. This picture is analogous to that in electromagnetic cavities where Q is determined by the width of the resonant response. In the theory of cavities, however, one can also determine Q by setting up an initial disturbance, which will oscillate with the resonant frequency and decay in a time related to Q. For many purposes, this second viewpoint is preferable. We shall now see how the information contained in $\chi^R(\underset{\sim}{k},\omega)$ can be used to study a similar situation, in which the system is perturbed and then left to decay. This latter picture has the advantage that there is no applied field in the problem.

To formulate the problem, return to Eq. (7.4) expressed here in terms of the magnetization

$$\langle \hat{M}(\underset{\sim}{r}, t)\rangle = \int d^3r' \int_{-\infty}^{t} dt' \; 2i\chi''(\underset{\sim}{r}-\underset{\sim}{r}', t-t')\,\mathcal{H}(\underset{\sim}{r}', t') \tag{9.5}$$

In particular, suppose that $\mathcal{H}(\underset{\sim}{r},t)$ has a pure wavevector dependence $e^{i\underset{\sim}{k}\cdot\underset{\sim}{r}}$ and is turned on slowly, up to full strength $\mathcal{H}_0$ at time t = 0. At that time, $\mathcal{H}$ is turned off

$$\mathcal{H}(\underset{\sim}{r}, t) = \begin{cases} \mathcal{H}_0 e^{i\underset{\sim}{k}\cdot\underset{\sim}{r}} e^{\eta t} & t < 0 \\ 0 & t > 0 \end{cases} \tag{9.6}$$

For $t \geq 0$, Eq. (9.5) then gives

$$\langle \hat{M}(\underset{\sim}{r}, t>0)\rangle = \int d^3r' \int_{-\infty}^{0} dt' \; 2i\chi''(\underset{\sim}{r}-\underset{\sim}{r}', t-t')\,\mathcal{H}_0 e^{i\underset{\sim}{k}\cdot\underset{\sim}{r}'} e^{\eta t'} \tag{9.7}$$

Use of the representation (7.5) immediately leads to

$$\begin{aligned} \langle \hat{M}(\underset{\sim}{k}, t>0)\rangle &= \int_{-\infty}^{0} dt' \; 2i\chi''(\underset{\sim}{k}, t-t')\,\mathcal{H}_0 e^{\eta t'} \\ &= \int_{-\infty}^{\infty} \frac{d\omega'}{2\pi}\, 2i\chi''(\underset{\sim}{k},\omega')\,\mathcal{H}_0 \int_{-\infty}^{0} dt'\, e^{-i\omega'(t-t')} e^{\eta t'} \\ &= \int_{-\infty}^{\infty} \frac{d\omega'}{\pi}\, \frac{\chi''(\underset{\sim}{k},\omega')}{\omega'}\, \mathcal{H}_0\, e^{-i\omega' t} \end{aligned} \tag{9.8}$$

where the $i\eta$ in the denominator has been dropped since $\chi''(\underset{\sim}{k},\omega' = 0)$ vanishes. In particular, at t = 0, this equation becomes

$$\langle \hat{M}(\underset{\sim}{k}, t=0)\rangle = \int_{-\infty}^{\infty} \frac{d\omega'}{\pi}\, \frac{\chi''(\underset{\sim}{k},\omega')}{\omega'}\, \mathcal{H}_0 \tag{9.9}$$

and the integral on the right-hand side is just the full static susceptibility [see Eqs. (9.3) and (9.4)]

$$\chi(\underset{\sim}{k}) \equiv \chi^{R}(\underset{\sim}{k}, 0) = \int_{-\infty}^{\infty} \frac{d\omega'}{\pi}\, \frac{\chi''(\underset{\sim}{k},\omega')}{\omega'} \tag{9.10}$$

Thus these relations imply

$$\langle \hat{M}(\underset{\sim}{k}, t=0)\rangle = \chi(\underset{\sim}{k})\,\mathcal{H}_0 \;; \tag{9.11}$$

equivalently, we can eliminate $\mathcal{H}_0$ explicitly to express $\langle \hat{M}(\underset{\sim}{k}, t>0)\rangle$ in terms of the initial distribution of magnetization, using

$$\mathcal{H}_0 = \langle \hat{M}(\underset{\sim}{k}, t=0)\rangle / \chi(\underset{\sim}{k}) \tag{9.12}$$

Substitution into Eq. (9.8) gives

$$\langle \hat{M}(\underset{\sim}{k}, t>0)\rangle = \int_{-\infty}^{\infty} \frac{d\omega'}{\pi}\, \frac{\chi''(\underset{\sim}{k},\omega')\, e^{-i\omega' t}}{\omega'}\, \frac{\langle \hat{M}(\underset{\sim}{k}, t=0)\rangle}{\chi(\underset{\sim}{k})} \tag{9.13}$$

and causality requires that $t > 0$, so that there must be an implicit step function $\theta(t)$. If $\langle \hat{M}(\underset{\sim}{k}, t>0)\rangle$ is expressed as a Fourier integral, we therefore write

$$\langle \hat{M}(\underset{\sim}{k},\omega)\rangle \equiv \int_0^\infty dt\, e^{i\omega t} \langle \hat{M}(\underset{\sim}{k},t>0)\rangle$$
$$= \int_{-\infty}^{\infty} \frac{d\omega'}{\pi} \frac{\chi''(\underset{\sim}{k},\omega')}{\omega'} \frac{\langle \hat{M}(\underset{\sim}{k},t=0)\rangle}{\chi(\underset{\sim}{k})} \int_0^\infty dt\, e^{i(\omega-\omega')t}$$
$$= \int_{-\infty}^{\infty} \frac{d\omega'}{\pi} \frac{\chi''(\underset{\sim}{k},\omega')}{i\omega'(\omega'-\omega-i\eta)} \frac{\langle \hat{M}(\underset{\sim}{k},t=0)\rangle}{\chi(\underset{\sim}{k})} \tag{9.14}$$

Use of partial fractions on the ω' integral gives the simplification

$$\int_{-\infty}^{\infty} \frac{d\omega'}{\pi} \frac{\chi''(\underset{\sim}{k},\omega')}{i\omega'(\omega'-\omega-i\eta)} = \frac{\chi^R(\underset{\sim}{k},\omega) - \chi^R(\underset{\sim}{k},0)}{i(\omega+i\eta)}$$
$$= \frac{\chi^R(\underset{\sim}{k},\omega) - \chi(\underset{\sim}{k})}{i(\omega+i\eta)} \tag{9.15}$$

In this way, we find the final form for the Fourier transform of the response following an initial value $\langle \hat{M}(\underset{\sim}{k},t=0)\rangle$

$$\langle \hat{M}(\underset{\sim}{k},\omega)\rangle = \frac{1}{i(\omega+i\eta)} \left[\chi^R(\underset{\sim}{k},\omega)\,\chi(\underset{\sim}{k})^{-1} - 1\right] \langle \hat{M}(\underset{\sim}{k},t=0)\rangle \tag{9.16}$$

As anticipated, the function $\chi^R(\underset{\sim}{k},\omega)$ fully determines this behavior.

To understand the connection with hydrodynamics, we shall show that we may evaluate this response directly for a simple hydrodynamic model. Turning the argument around, one can then infer certain properties of $\chi^R(\underset{\sim}{k},\omega)$ that must occur if this model is correct. In the present example, we consider a magnetic fluid like ^{3}He, in which each atom carries a magnetic moment $\mu_0 \underset{\sim}{\sigma}$, where $\underset{\sim}{\sigma}$ is the Pauli matrix. The net magnetization is an ensemble average [see Eq. (8.2)]

$$M(\underset{\sim}{r},t) = \langle \Sigma_i \mu_0 (\sigma_z)_i \,\delta[\underset{\sim}{r} - \underset{\sim}{r}_i(t)]\rangle \tag{9.17}$$

In this model, each atom carries the magnetic moment, and M in some volume can change only by net influx of atoms. Introducing the magnetization current density

$$\underset{\sim}{j}_M(\underset{\sim}{r},t) = \langle \Sigma_i \mu_0 (\sigma_z)_i \underset{\sim}{v}_i \,\delta[\underset{\sim}{r} - \underset{\sim}{r}_i(t)]\rangle \tag{9.18}$$

we may write the local conservation law

$$\partial M(\underset{\sim}{r},t)/\partial t + \underset{\sim}{\nabla}\cdot \underset{\sim}{j}_M(\underset{\sim}{r},t) = 0 \tag{9.19}$$

This general equation must be augmented by a constitutive relation between $\underset{\sim}{j}_M$ and M. The simplest choice is a diffusive process, with

$$\underset{\sim}{j}_M(\underset{\sim}{r},t) = -D\,\underset{\sim}{\nabla} M(\underset{\sim}{r},t) \tag{9.20}$$

where D is a diffusion constant. Equations (9.19) and (9.20) together yield the diffusion equation

$$\partial M/\partial t = D\,\nabla^2 M \tag{9.21}$$

which describes slow space and time variations in M. Given an initial value $M(\underset{\sim}{r},t=0)$ the subsequent change is found by a spatial Fourier transform for $t > 0$

$$M(\underset{\sim}{k},t>0) = \int d^3r\, e^{-i\underset{\sim}{k}\cdot\underset{\sim}{r}}\, M(\underset{\sim}{r},t>0) \tag{9.22}$$

Simple manipulations give

$$\frac{\partial}{\partial t} M(\underset{\sim}{k},t>0) = -Dk^2 M(\underset{\sim}{k},t>0) \tag{9.23}$$

whose solution is

$$M(\underset{\sim}{k},t) = M(\underset{\sim}{k},t=0)\exp(-Dk^2t) \tag{9.24}$$

Note that $M(\underset{\sim}{k},t)$ decays with a characteristic time $(Dk^2)^{-1}$, showing that diffusion is very slow at long wavelengths.

Equation (9.24) solves the initial-value problem. Since $M = 0$ for $t < 0$, its Fourier transform becomes

$$\begin{aligned} M(\underset{\sim}{k},\omega) &= \int_0^\infty dt\, e^{i\omega t} M(\underset{\sim}{k},t) \\ &= M(\underset{\sim}{k},t=0)\int_0^\infty dt\, e^{i\omega t} e^{-Dk^2t} \\ &= \frac{iM(\underset{\sim}{k},t=0)}{\omega + iDk^2} \end{aligned} \tag{9.25}$$

which is analytic in the upper-half ω plane. Comparing with Eq. (9.16), we may identify

$$\frac{\chi^R(\underset{\sim}{k},\omega)}{\chi(\underset{\sim}{k})} = \frac{iDk^2}{\omega + iDk^2} \tag{9.26}$$

or

$$\chi^R(\underset{\sim}{k},\omega) = \frac{iDk^2\chi(\underset{\sim}{k})}{\omega + iDk^2} \tag{9.27}$$

Any model consistent with hydrodynamics must reproduce this structure at long wavelengths and low frequencies. In particular, the imaginary part is

$$\chi''(\underset{\sim}{k},\omega) = \frac{\omega Dk^2\chi(\underset{\sim}{k})}{\omega^2 + (Dk^2)^2}\,; \tag{9.28}$$

it is odd in ω, positive for $\omega > 0$, and has a very different limiting behavior depending on the ratio ω/k^2. More generally, the limit $\chi^R(\underset{\sim}{k},0)$ reproduces the static susceptibility $\chi(\underset{\sim}{k})$ for all k, and it is easy to verify Eq. (9.10) directly.

It is also interesting to find the two-particle correlation function for magnetic phenomena [see Eq. (7.26)]

$$S_{MM}(\underset{\sim}{k},\omega) = \frac{2\hbar\omega}{1 - e^{-\beta\hbar\omega}}\;\frac{Dk^2\chi(\underset{\sim}{k})}{\omega^2 + (Dk^2)^2} \tag{9.29}$$

This quantity characterizes the inelastic magnetic scattering by neutrons. For fixed $\underset{\sim}{k}$, it is peaked near $\omega = 0$, with width of order Dk^2. Thus this diffusive system has no well-defined collective modes with finite frequency; moreover, the damping depends on k, as noted below Eq. (9.24).

At high frequency, Eq. (9.28) suggests that $\chi''(\underset{\sim}{k},\omega) \sim \omega^{-1}$, which implies that only certain low moments are finite. This behavior is unphysical, since we anticipate that χ'' should vanish very rapidly for sufficiently high frequency because the medium cannot follow the oscillating fields. One interesting and physical way to avoid such difficulties is to retain the exact conservation law (9.19) but to generalize the constitutive relation (9.20) to a nonlocal one

$$\underset{\sim}{j}_M(\underset{\sim}{r},t) = -\int_{-\infty}^{t} dt'\, D(t-t')\, \underset{\sim}{\nabla} M(\underset{\sim}{r},t') \tag{9.30}$$

The function D(t-t') is called the memory function; it replaces the previous diffusion constant D. In the present case of an initial-value problem, Eq. (9.30) strictly runs only from t' = 0 to t' = t. A combination of Eqs. (9.19) and (9.30) with a spatial Fourier transform yields the integro-differential equation

$$\frac{\partial}{\partial t} M(\underset{\sim}{k},t) + k^2 \int_0^t dt'\, D(t-t')\, M(\underset{\sim}{k},t') = 0 \tag{9.31}$$

subject to the initial value $M(\underset{\sim}{k},t=0)$. It may be solved with a Fourier transform as in (9.25); a little manipulation gives the simple result

$$M(\underset{\sim}{k},\omega) = \frac{i\, M(\underset{\sim}{k},t=0)}{\omega + i k^2 D(\omega)} \tag{9.32}$$

where

$$D(\omega) = \int_0^\infty dt\, e^{i\omega t} D(t) \tag{9.33}$$

Comparison with (9.16) identifies the result

$$\chi^R(\underset{\sim}{k},\omega) = \frac{i k^2 D(\omega)\, \chi(\underset{\sim}{k})}{\omega + i k^2 D(\omega)} \tag{9.34}$$

If $D(\omega \to 0) = D$, then Eq. (9.34) reproduces the previous model (9.27) at low frequency, but it allows for more general high-frequency behavior. In addition since $D(\omega)$ itself is an analytic function of ω in the upper half plane, it must have a representation of the form

$$D(\omega) = \int_{-\infty}^{\infty} \frac{d\omega'}{\pi i} \frac{D'(\omega')}{\omega' - \omega - i\eta} = D'(\omega) + i D''(\omega) \tag{9.35}$$

where

$$D''(\omega) = -P \int_{-\infty}^{\infty} \frac{d\omega'}{\pi} \frac{D'(\omega')}{\omega' - \omega} \tag{9.36}$$

A combination with Eq. (9.34) provides the absorptive part

$$\chi''(\underset{\sim}{k},\omega) = \frac{\omega k^2 D'(\omega)\, \chi(\underset{\sim}{k})}{[\omega - k^2 D''(\omega)]^2 + [k^2 D'(\omega)]^2} \tag{9.37}$$

and we infer that $D'(\omega)$ must be positive. Various models for $D'(\omega)$ have been used, and this "memory-function" approach can also be generalized in several ways to include wavenumber dependence and to encompass other systems. Many applications are contained in Ref. 8.

X. Light Scattering in Fluids

The principal advantage of neutron scattering is that the energy and momentum transfers are comparable with those of collective modes in solids. For electromagnetic radiation, on the other hand, the kinematic relation $\omega = ck$ implies a very large frequency for k the order of inverse atomic sizes. This observation explains why x-ray scattering is unable to measure the very small energy shifts involved in inelastic processes. A similar relation applied to visible light, which has long wavelengths relative to atomic sizes, but the introduction of lasers altered this situation. Since a laser has a very narrow frequency profile, it became feasible to study inelastic scattering directly in the visible range, providing information on the dynamic structure factor $S(\underset{\sim}{k},\omega)$ [see Eq. (3.8)] in an entirely different kinematic regime from that probed by neutrons.

Because of the long wavelength, light scattering is insensitive to atomic variations. Thus a continuum or hydrodynamic description ought to suffice. Indeed, the exact hydrodynamic equations are well known, reflecting the conservation of mass, momentum, and energy. In this section, we shall use the treatment of Sec. IX to infer $S(\underset{\sim}{k},\omega)$ from the solution to an initial density disturbance $\tilde{n}(\underset{\sim}{k},t=0)$. The resulting spectrum of scattered radiation has provided crucial information on the behavior of fluids, especially near critical points. Good general references are Refs. 8, 11, and 14.

The basic strategy is precisely that in Eq. (9.16), generalized to a set of observables $\langle \hat{A}_i(\underset{\sim}{r},t)\rangle$ [see Eqs. (7.1)-(7.4)] coupled to external fields $\delta a_i(\underset{\sim}{r},t)$. A straightforward calculation analogous to that in Eqs. (9.6)-(9.16) yields the Fourier transform of the response for $t > 0$

$$\langle \hat{A}_i(\underset{\sim}{k},\omega)\rangle = \int d^3r\, e^{-i\underset{\sim}{k}\cdot\underset{\sim}{r}} \int_0^\infty dt\, e^{i\omega t} \langle \hat{A}_i(\underset{\sim}{r},t)\rangle \tag{10.1}$$

in terms of its initial value

$$\langle \hat{A}_i(\underset{\sim}{k},t=0)\rangle = \int d^3r\, e^{-i\underset{\sim}{k}\cdot\underset{\sim}{r}} \langle \hat{A}_i(\underset{\sim}{r},t=0)\rangle \tag{10.2}$$

through the equation

$$\langle \hat{A}_i(\underset{\sim}{k},\omega)\rangle = \frac{1}{i(\omega+i\eta)}\left\{\sum_{jl}\left[\chi^R_{ij}(\underset{\sim}{k},\omega)\,\chi(\underset{\sim}{k})^{-1}_{jl}\,\langle \hat{A}_l(\underset{\sim}{k},t=0)\rangle\right] - \langle \hat{A}_i(\underset{\sim}{k},t=0)\rangle\right\} \tag{10.3}$$

Here, as in Eq. (7.8), $\chi_{ij}{}^R(\underset{\sim}{k},\omega)$ is the ith response function for a perturbation δa_j, expressed in terms of the Fourier transform of the commutator $\chi''_{ij}(\underset{\sim}{k},\omega)$ [see Eqs. (7.4) and (7.5)]:

$$\chi^R_{ij}(\underset{\sim}{k},\omega) = \int_{-\infty}^{\infty} \frac{d\omega'}{\pi}\, \frac{\chi''_{ij}(\underset{\sim}{k},\omega')}{\omega'-\omega-i\eta} \tag{10.4}$$

Furthermore, $\chi_{ij}(\underset{\sim}{k})$ is the static susceptibility

$$\chi_{ij}(\underset{\sim}{k}) \equiv \chi^R_{ij}(\underset{\sim}{k},0) = \int_{-\infty}^{\infty} \frac{d\omega'}{\pi} \; \frac{\chi''_{ij}(\underset{\sim}{k},\omega')}{\omega'} \tag{10.5}$$

and $\chi^{-1}_{j\ell}$ is the $j\ell$ element of the inverse tensor. In particular, taking the density as the relevant operator, we use known hydrodynamic equations to infer $\chi^R_{nn}(\underset{\sim}{k},\omega)$ [which was called the generalized compressibility $\kappa(\underset{\sim}{k},\omega)$ in Eq. (5.11)] and, through its imaginary part, the dynamic structure factor [see Eqs. (5.16) and (7.26)].

The basic hydrodynamic equations[12,13] are well known. Conservation of mass implies the continuity equation for the mass density ρ and velocity $\underset{\sim}{v}$

$$\frac{\partial \rho}{\partial t} + \underset{\sim}{\nabla}\cdot(\rho \underset{\sim}{v}) = 0 \tag{10.6}$$

or, for a linearized small-amplitude motion

$$\frac{\partial \rho'}{\partial t} + \rho_0 \underset{\sim}{\nabla}\cdot \underset{\sim}{v}' = 0 \tag{10.7}$$

Here primed variables denote the small perturbations. The corresponding equation for momentum conservation may be written quite generally as

$$\frac{\partial}{\partial t}(\rho v_i) + \sum_j \frac{\partial}{\partial x_j} T_{ij} = 0 \tag{10.8}$$

where T_{ij} is the stress tensor, including convective and viscous terms. In the linearized form, it becomes

$$\frac{\partial \underset{\sim}{v}'}{\partial t} + \frac{1}{\rho_0}\underset{\sim}{\nabla} p' - \frac{\eta}{\rho_0}\nabla^2 \underset{\sim}{v}' - \frac{1}{\rho_0}\left(\zeta + \tfrac{1}{3}\eta\right)\underset{\sim}{\nabla}(\underset{\sim}{\nabla}\cdot\underset{\sim}{v}') = 0 \tag{10.9}$$

where p' is the pressure, and η and ζ are the shear and bulk viscosities. Finally, the equation ofconservation of energy can be rewritten (using thermodynamics) in terms of the entropy density s; its linearized form

$$T_0 \rho_0 \frac{\partial s'}{\partial t} - k_{th}\nabla^2 T' = 0 \tag{10.10}$$

merely states that the local increase in entropy must arise from an influx of heat through conduction, proportional to $-k_{th}\underset{\sim}{\nabla} T$, where k_{th} is the thermal conductivity.

Equations (10.7), (10.9), and (10.10) involve the velocity $\underset{\sim}{v}'$ and four thermodynamic variables ρ', S', T', p'. Since only two variables suffice to determine the state of a one-component system, these four quantities are connected by two thermodynamic relations, which we write as

$$p' = \left(\frac{\partial p}{\partial s}\right)_\rho s' + \left(\frac{\partial p}{\partial \rho}\right)_s \rho' \tag{10.11a}$$

$$T' = \left(\frac{\partial T}{\partial s}\right)_\rho s' + \left(\frac{\partial T}{\partial \rho}\right)_s \rho' \tag{10.11b}$$

Two of these coefficients have simple interpretations:

$$\left(\frac{\partial p}{\partial \rho}\right)_s = c^2 \tag{10.12a}$$

$$\left(\frac{\partial T}{\partial s}\right)_\rho = \frac{T_0}{c_V} \tag{10.12b}$$

where c is the speed of sound in a nonviscous isentropic fluid and c_V is the specific heat per unit mass at constant volume. Moreover, the remaining two are related through a Maxwell relation

$$\left(\frac{\partial T}{\partial \rho}\right)_s = \frac{1}{\rho_0^2}\left(\frac{\partial p}{\partial s}\right)_\rho \tag{10.13}$$

In this way, the linearized hydrodynamic equations can be written exactly in the form

$$\frac{\partial \rho'}{\partial t} + \rho_0 \nabla \cdot \underset{\sim}{v}' = 0 \tag{10.14}$$

$$\frac{\partial \underset{\sim}{v}'}{\partial t} + \frac{c^2}{\rho_0}\nabla\rho' + \rho_0\left(\frac{\partial T}{\partial \rho}\right)_s \nabla s' - \frac{\eta}{\rho_0}\nabla^2\underset{\sim}{v}' - \frac{1}{\rho_0}\left(\zeta + \tfrac{1}{3}\eta\right)\nabla(\nabla\cdot\underset{\sim}{v}') = 0 \tag{10.15}$$

$$\frac{\partial s'}{\partial t} - \gamma D_T \nabla^2 s' - \frac{c_p D_T}{T_0}\left(\frac{\partial T}{\partial \rho}\right)_s \nabla^2\rho' = 0 \tag{10.16}$$

where

$$D_T = \frac{k_{th}}{\rho_0 c_p} \tag{10.17}$$

is the thermal diffusivity and

$$\gamma \equiv c_p/c_V \geq 1 \tag{10.18}$$

is the ratio of the specific heats. Note the role of the thermodynamic derivative $(\partial T/\partial \rho)_s$, which can be shown to equal $Tc^2\beta/\rho c_p$ where $\beta = V^{-1}(\partial V/\partial T)_p$ is the thermal expansion coefficient. If $\beta = 0$, then these equations separate into two sets, the first involving the mechanical quantities ρ' and $\underset{\sim}{v}'$ and the second describing pure entropy variations. For $\beta \neq 0$, however, thermal expansion couples these various small amplitudes.

It is convenient to take a spatial Fourier transform, and Eq. (10.15), for example, becomes

$$\frac{\partial \underset{\sim}{v}'}{\partial t} + \frac{c^2}{\rho_0} i\underset{\sim}{k}\rho' + \rho_0\left(\frac{\partial T}{\partial \rho}\right)_s i\underset{\sim}{k} s' + \frac{\eta}{\rho_0}k^2\underset{\sim}{v}' + \frac{1}{\rho_0}\left(\zeta + \tfrac{1}{3}\eta\right)\underset{\sim}{k}(\underset{\sim}{k}\cdot\underset{\sim}{v}') = 0 \tag{10.19}$$

If $\underset{\sim}{v}'$ is separated into its transverse and longitudinal components as $\underset{\sim}{v}'(\underset{\sim}{k},t) = \underset{\sim}{v}'_\perp(\underset{\sim}{k},t) + \hat{k} v'_l(\underset{\sim}{k},t)$, then the transverse part of Eq. (10.19) is just

$$\partial \underset{\sim}{v}'_\perp/\partial t + \nu k^2 \underset{\sim}{v}'_\perp = 0 \tag{10.20}$$

where $\nu \equiv \eta/\rho_0$ is the kinematic viscosity. This equation has exactly the form (9.23), and we therefore infer that the transverse velocity correlation function is analogous to that for the magnetization in (9.26)-(9.29), with D replaced by ν. In principle, such behavior could be studied by exciting transverse shear waves in the fluid; they would be attenuated with a complex wavenumber $\delta^{-1}(1+i)$, where $\delta = (2\nu/\omega)^{1/2}$

is the viscous penetration depth.

Although the transverse velocity is uncoupled, the longitudinal part is connected to density and entropy fluctuations. Since these quantities vanish for $t < 0$, we may take the temporal Fourier transform as in Eq. (9.25). An integration by parts expresses the intial value problem in the form

$$\begin{bmatrix} \omega & -\rho_0 k & 0 \\ -kc^2/\rho_0 & \omega + i D_\ell k^2 & -k\rho_0 \left(\frac{\partial T}{\partial \rho}\right)_s \\ ikc_p T_0^{-1} D_T \left(\frac{\partial T}{\partial \rho}\right)_s & 0 & \omega + i D_T \gamma k^2 \end{bmatrix} \begin{bmatrix} \rho'(\underset{\sim}{k},\omega) \\ v_\ell'(\underset{\sim}{k},\omega) \\ s'(\underset{\sim}{k},\omega) \end{bmatrix} = i \begin{bmatrix} \rho'(\underset{\sim}{k},t=0) \\ v_\ell'(\underset{\sim}{k},t=0) \\ s'(\underset{\sim}{k},t=0) \end{bmatrix} \tag{10.20}$$

where $D_\ell = \rho_0^{-1}(\zeta + \frac{4}{3}\eta)$. The poles of these fluctuations are determined by setting the determinant of coefficients equal to zero. They lie in the lower-half ω plane, at values given approximately by $-i\Gamma_R k^2$ and $\pm ck - i\Gamma_B k^2$, where R and B denote Rayleigh and Brillouin. Direct evaluation to leading order in k (which is small) gives

$$\Gamma_R = D_T$$

$$2\Gamma_B = D_\ell + (\gamma - 1) D_T \tag{10.21}$$

where we have used the thermodynamic relation

$$\frac{c_p \rho_0^2}{T_0 c^2}\left(\frac{\partial T}{\partial \rho}\right)_s^2 = \gamma - 1 \tag{10.22}$$

Thus the various correlation functions $\chi^R_{ij}(\underset{\sim}{k},\omega)$ formed from the quantities ρ', v_ℓ', and s' all have terms proportional to $(\omega + iD_T k^2)^{-1}$ and $(\omega \pm ck + i\Gamma_B k^2)^{-1}$ with residues that depend on the specific choice of ij [compare Eq. (9.27)-(9.29)]. As a result, the imaginary parts χ''_{ij} and the corresponding $S_{ij}(\underset{\sim}{k},\omega)$ have denominators of the form $\omega^2 + D_T^2 k^4$ and $(\omega^2 - c^2k^2)^2 + (2\omega k^2 \Gamma_B)^2$.

In particular, the frequency spectrum of light scattered with fixed $\underset{\sim}{k}$ follows from $S_{nn}(\underset{\sim}{k},\omega)$. It contains a Lorentzian Rayleigh peak centered at $\omega = 0$ with width $D_T k^2$ and a pair of (not strictly Lorentzian) peaks centered at $\omega = \pm ck$ with widths of order $\Gamma_B k^2$ (Fig. 10.1). In the long-wavelength limit, these peaks are well separated and the

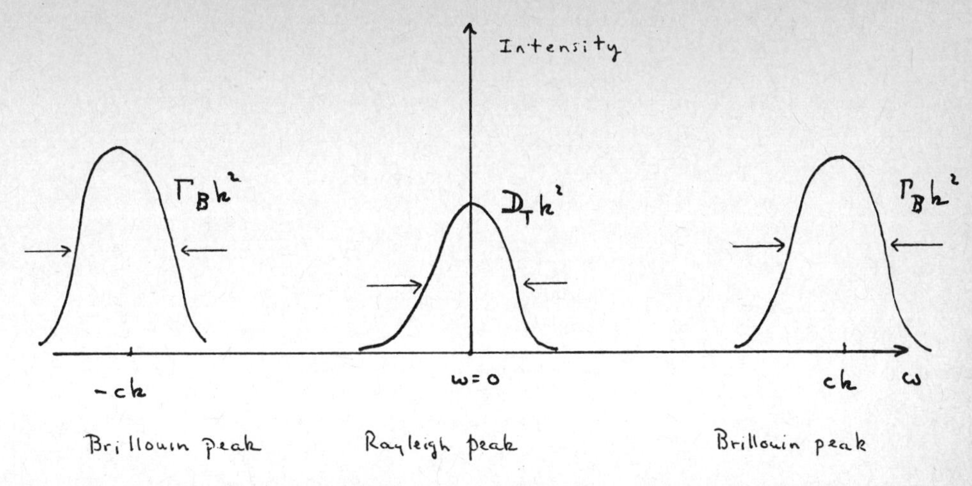

Fig. 10.1. Inelastic light-scattering intensity from a fluid for fixed k.

Brillouin peaks at $\omega = \pm ck$ are of equal height [see Eq. (7.27)]. The full expression for $S_{nn}(\underline{k}, \omega)$ is found in Ref. 8, p. 77, and we shall not reproduce the rather complicated structure. Physically, the Rayleigh peak is diffusive, as in Eq. (9.29), arising from entropy fluctuations that propagate by heat conduction (as seen by the width $D_T = k_{th}/\rho_0 c_p$). The Brillouin doublet is associated with density fluctuations that propagate with the adiabatic speed c from Eq. (10.12a) and decay with a damping constant $2\Gamma_B = D_\ell + (\gamma - 1) D_T$ containing both viscous contributions D_ℓ and thermal contributions D_T. Note the role of thermal expansion, which reduces the width of the Rayleigh peak from γD_T to D_T [see the 33 element of the matrix in Eq. (10.20) and Eq. (10.22)]. Detailed calculations show that the ratio of the areas under the Rayleigh peak and the Brillouin doublet is γ-1, an old result of Landau and Placzek [see Landau and Lifshitz, Electrodynamics of Continuous Media, Chap. XIV.]. In this way, light scattering can provide direct information on the thermodynamic parameters c and γ and the transport coefficients $\frac{4}{3}\eta + \zeta$ and k_{th} . Some of these quantities (especially the bulk viscosity ζ) are not readily accessible by other means.

Light scattering plays a particularly important role near the liquid-gas critical point, because the total intensity becomes large owing to critical opalescence (see the discussion at the end of Lecture V). In addition, the difference $c_p - c_V$ also becomes large, so that the Rayleigh peak dominates; this reflects the importance of entropy fluctuations near the phase transition. In contrast, the width of the Rayleigh peak becomes small, because $D_T = k_{th}/\rho_0 c_p$ varies inversely with c_p. As a result, entropy and temperature changes decay very slowly, leading to the behavior known as critical slowing down. Modern theories make various predictions for the power-law dependence on $T-T_c$ near the critical point, and light scattering has been central in verifying these critical exponents. We may also mention that light scattering has been important in studying the normal-superfluid transition in liquid ^4He. In that case, the Rayleigh peak below T_λ splits into a doublet associated with the excitation of second sound (see Ref. 8, Chap. 10).

REFERENCES

1. J. D. Jackson, Classical Electrodynamics (Wiley, New York, 1975), 2nd ed., Secs. 9.6 and 14.7.

2. A. L. Fetter and J. D. Walecka, Quantum Theory of Many-Particle Systems (McGraw-Hill, New York, 1971), Secs. 8 and 24.

3. Ref. 2, Sec. 5.

4. W. Marshall and S. W. Lovesey, Thermal Neutron Scattering (Oxford University Press, Oxford, 1971), Chaps. 1-3.

5. Ref. 2, Secs. 13 and 32.

6. L. D. Landau and E. M. Lifshitz, Electrodynamics of Continuous Media (Pergamon, Oxford, 1960), pp. 134-135.

7. P. C. Martin, in Many-Body Physics, edited by C. deWitt and R. Balian (Gordon and Breach, New York, 1968), pp. 41-136.

8. D. Forster, Hydrodynamic Fluctuations, Broken Symmetry, and Correlation Functions (Benjamin, New York, 1975), Chaps. 1-4.

9. Ref. 4, Chaps. 5-8.

10. Ref. 2, p. 110.

11. B. J. Berne and R. Pecora, Dynamic Light Scattering (Wiley, New York, 1976), Chap. 10.

12. L. D. Landau and E. M. Lifshitz, Fluid Mechanics (Pergamon, Oxford, 1959), Chaps. I, II, and V.

13. A. L. Fetter and J. D. Walecka, Theoretical Mechanics of Particles and Continua (McGraw-Hill, New York, 1980), Chaps. 9 and 12.

14. P. A. Fleury and J. P. Boon, Adv. Chem. Phys. XXIV, 1 (1973).

Texts and Monographs in Physics

J. Kessler: **Polarized Electrons.** 1976.

W. Rindler: **Essential Relativity: Special, General, and Cosmological.** Revised Second Edition. 1977.

K. Chadan, P. C. Sabatier: **Inverse Problems in Quantum Scattering Theory.** 1977.

C. Truesdell, S. Bharatha: **The Concepts and Logic of Classical Thermodynamics as a Theory of Heat Engines: Rigourously Constructed upon the Foundation Laid by S. Carnot and F. Reech.** 1977.

R. D. Richtmyer: **Principles of Advanced Mathematical Physics.** Volume I. 1978. Volume II. 1981.

R. M. Santilli: **Foundations of Theoretical Mechanics.** Volume I: The Inverse Problem in Newtonian Mechanics. 1978. Volume II: Birkhoffian Generalization of Hamiltonian Mechanics. 1983.

A. Böhm: **Quantum Mechanics.** 1979.

H. Pilkuhn: **Relativistic Particle Physics.** 1979.

M. D. Scadron: **Advanced Quantum Theory and Its Applications Through Feynman Diagrams.** 1979.

O. Bratteli, D. W. Robinson: **Operator Algebras and Quantum Statistical Mechanics.** Volume I: C*-and W*-Algebras. Symmetry Groups. Decomposition of States. 1979. Volume II: Equilibrium States. Models in Quantum Statistical Mechanics. 1981.

J. M. Jauch, F. Rohrlich: **The Theory of Photons and Electrons: The Relativistic Quantum Field Theory of Charged Particles with Spin One-half.** Second Expanded Edition. 1980.

P. Ring, P. Schuck: **The Nuclear Many-Body Problem.** 1980.

R. Bass: **Nuclear Reactions with Heavy Ions.** 1980.

R. G. Newton: **Scattering Theory of Waves and Particles.** Second Edition. 1982.

G. Ludwig: **Foundations of Quantum Mechanics I.** 1983.

G. Gallavotti: **The Elements of Mechanics.** 1983.

F. J. Yndurain: **Quantum Chromodynamics: An Introduction to the Theory of Quarks and Gluons.** 1983.

Springer-Verlag
Berlin
Heidelberg
New York
Tokyo